POLYMERIC MULTICOMPONENT MATERIALS

An Introduction

L. H. Sperling
Lehigh University, Bethlehem, Pennsylvania

A WILEY-INTERSCIENCE PUBLICATION
JOHN WILEY & SONS, INC.
New York / Chichester / Weinheim / Brisbane / Singapore / Toronto

Library of Congress Cataloging in Publication Data:

Sperling, L. H. (Leslie Howard), 1932–
    Polymeric multicomponent materials: an introduction / L. H. Sperling

        p.   cm.
    "A Wiley-Interscience publication."
    Includes bibliographical references and index.
    ISBN 0-471-04138-6 (cloth)
    1. Polymers.   2. Polymeric composites.   I. Title.
    QD381.S6348   1998
    620.1'92–DC21   97-6509
Printed in the United States of America

10 9 8 7 6 5 4 3 2

This book is dedicated to the memory of my teachers,

Dr. William R. Krigbaum

of Duke University, who was the most patient man I ever knew, and

Dr. Arthur V. Tobolsky

of Princeton University, who taught me how to organize a modern laboratory.

While the day-to-day setting in both laboratories was polymer science, both men had the overriding mission of teaching us the *Art of Research*. Their professional children, grandchildren, and even great-grandchildren now populate the ranks of scientists and engineers around the globe.

CONTENTS

II.   POLYMER SURFACES AND INTERFACES

4.   Basic Principles and Instruments   **119**

III.   SELECTED ENGINEERING POLYMER MATERIAL

PREFACE

Polymer blends and composites were born of sheer necessity: Although early polymers were light and inexpensive and promised to fill new application roles, the materials were weak or brittle and often unsatisfactory in performance. In about 1905, carbon black was discovered to toughen tire rubber. The earliest tires lasted about 5000 miles at speeds of 25 miles per hour. Now, they last 50,000 miles at 55 miles per hour. This has been due in significant measure to more finely dispersed carbon blacks, stronger tire cords, and improved adhesion of these materials to the rubber.

In 1914, rubber (then natural rubber, of course) was first added to phenol-formaldehyde phonograph records and other materials, providing a measure of toughness. In the mid 1940s, polybutadiene rubber was added to styrenic compositions, leading to high-impact polystyrene and ABS materials. From there, the list of commercial polymeric materials containing another phase grew rapidly.

Nearly all of this early work was empirical. People tried many things. If something worked better, they used it. There were few basic relationships, and fewer guiding principles.

In 1976, Plenum Press published *Polymer Blends and Composites*, by John Manson and the present author, in which we attempted to present the entire field as a unit for the first time. We organized what theory there was and sampled the best of the experimental data then available. Later, we considered a second edition of this book. Alas, John took sick and passed away. When I next taught out of the text, I realized it had now become hopelessly outdated. If the subject were to be covered again, it must be done from the beginning, fresh.

Here, I must confess one of the secrets of why I write books: It is a form of self-education. I laid out a map of current readings for me to do, conferences to participate in, people to speak with, and so on. This, indeed, was 80–90% of the work of preparing the book. The remaining 10–20% of the time was setting it down on paper so I, myself, could find this information later. The inverse was also true: I never would have systematically read all of this literature if I wasn't going to write a book about it.

Of course, there is another important reason why I wrote this book: No modern text covering the basics of the field seemed to exist. There was nothing to teach from except notes. It was obvious that professionals also lacked a modern book for an introduction to the field. Thus, this book is dedicated to the many workers in the field who have either provided information used in the book or who will want to read this book the better to advance themselves professionally. Who wants to write a book that isn't needed?

While *Polymer Blends and Composites* organized the field of that day, there have been many major areas of advance. Two of the most important subfields created were the thermodynamics of polymer blend miscibility, the introduction of phase diagrams, and so on, and the kinetics of phase separation in terms of spinodal decomposition versus nucleation and growth.

A third new subfield, now growing explosively, relates to polymer surfaces and interfaces. This area was born around 1989 with a series of articles appearing in *Macromolecules* and other journals. Suddenly, instruments and experiments were available to investigate the structure of interfaces, and the theorists were there to provide novel and exciting explanations of the results. When this book was started 3 years ago, I planned on having one chapter called "Polymer Surfaces and Interfaces." Now, the reader will find that there are four chapters in the center of this book, under the broader title, "Part II: Polymer Surfaces and Interfaces." The conformation of polymer chains at surfaces and interfaces is becoming clear, as are the important engineering areas of strengthening the interfaces mechanically via both chain entanglements and bonding schemes.

Now, the broader field demands a new title: *Polymeric Multicomponent Materials*. Since this book is intended as an introduction to this field, I trust that the title usage will be obvious. The book has 11 chapters, laid out in three parts. Part I is called "Fundamental Relationships," presenting a number of concepts, theories, nomenclature, and basic experimental facts common to both two-polymer and polymer–nonpolymer combinations. The effects of component mixing on the glass transitions, phase continuity and inversion, thermodynamics of mixing, and modeling of the modulus and other mechanical quantities are treated here. The mechanisms of fracture and fracture resistance are introduced, along with the concepts of how two-phased systems can impart toughness. Part II has already been discussed above. Part III, "Selected Engineering Polymer Materials," provides more advanced and/or engineering concepts and under-standing of a series of materials including rubber-toughened plastics, block copolymers, and interpenetrating polymer networks. The book concludes with a brief chapter titled, "Overview and Future." Overall, the coverage is about 80% relating to the fields of polymer blends, blocks, grafts, and IPNs, and about 20% relating to polymer composites, including carbon-black-reinforced rubber, glass fiber-reinforced systems, and the incorporation of particulates.

As always, there are many people to be thanked. First of all, the Lehigh University Fairchild–Martindale Library provided a carrel with a desk, a lock on the door, and no phone. This carrel is placed magnificently in the middle of the scientific and engineering stacks. Drafts of this book were read by my students, who repeatedly pointed out ways to improve the manuscript. Ms. Andrea Pressler provided many of the outstanding photographs in the book. Lastly, my secretaries, Ms. Virginia Newhard and Ms. Kathy Kennery, provided invaluable typing and administrative assistance.

L. H. Sperling

PART I

FUNDAMENTAL RELATIONSHIPS

1

BASIC CONCEPTS IN MULTICOMPONENT POLYMER MATERIALS

The field of multicomponent polymer materials includes all cases where polymers are either mixed with other polymers or with nonpolymeric solid materials. This book covers both scientific and engineering aspects. Professionals and students with interests in polymers, chemistry, chemical engineering, materials science and engineering, physics, and mechanical engineering will all recognize aspects of this book, yet all will find it truly interdisciplinary.

On the scientific side, the thermodynamics of mixing, kinetics of phase separation, morphology, modulus, and interfacial bonding are considered; on the engineering side, methods of producing higher modulus, tougher, lighter materials, improved adhesives and coatings, and many materials with synergistic properties. Prime among engineering studies are the characterization of fracture, failure, and fatigue. How do multicomponent polymer materials respond to imposed stress fields? How does a crack or craze grow, and how is the imposed energy consumed? Usually, if the energy can be partly absorbed by nondestructive mechanisms, the material will be tougher. Energy can be absorbed usefully in other ways as well, such as in sound and vibration damping. Today, one of the most important topics is the strengthening of interfaces, of interest to both scientists and engineers.

Thus, this book will present an integrated view of multicomponent polymer materials, emphasizing the interrelationships among synthesis, morphology, and mechanical behavior. The corresponding behavior of homopolymers will be briefly introduced as a basis for comparison.

1.1. DEFINITIONS AND NOMENCLATURE

The science of multicomponent polymer materials includes both polymer blends and composites, indeed all areas where polymers are combined with other materials, polymeric or not. The term *polymer blend* is used in two ways. First, it involves all multicomponent polymer materials composed primarily of two or more polymers. Second, it is used specifically to describe combinations of two or more polymers that are not bonded chemically to each other. Important combinations of two or more polymers bonded together include graft copolymers, block copolymers, AB-crosslinked polymers, and interpenetrating polymer networks (IPNs).

The term *polymer composite* was used originally to define combinations of polymers with nonpolymers, such as glass fibers, carbon black, and concrete (1). However, today there are many important examples of polymeric fibers that are used to reinforce plastics, multilayered polymer systems, and so forth, so the distinction has become a bit fuzzy. Of course, the term *composite* is used in materials science to mean any of a broad spectrum of material combinations: metals and ceramics, for example.

There are several approaches to polymer nomenclature. The two most important are the source-based and the structure-based schemes (2,3). The popular source-based nomenclature emphasizes the monomer name. Examples include: polystyrene, poly(methyl methacrylate), and poly(vinyl chloride). The reader should note that two-worded monomer names take parentheses. This book will use, where possible, the International Union of Pure and Applied Chemistry (IUPAC) (4), or proposed IUPAC nomenclature (5).

Nomenclature extends far beyond the formal names of compounds, however. There are lists of trade names, abbreviations, and terms relating to crystalline polymers as well as a growing nomenclature for copolymers.

1.1.1. Copolymer Nomenclature

In general, copolymers are defined as polymeric materials containing two or more kinds of mers (4–6). For the purposes of this work, it is important to distinguish between two kinds of copolymers, both of which have all of the mers together on the same chain. There are those copolymers with statistical distributions of mers, or at most short sequences of mers (Table 1.1), and those that contain long sequences of mers connected together in some way (Table 1.2). The term *mer* is defined as the individual unit derived from the monomer that makes up the polymer.

Table 1.1   Short sequence copolymer nomenclature

Type	Connective	Example
Homopolymer	None	PolyA
Unspecified	-*co*-	Poly(A-*co*-B)
Statistical	-*stat*-	Poly(A-*stat*-B)
Random	-*ran*-	Poly(A-*ran*-B)
Alternating	-*alt*-	Poly(A-*alt*-B)
Periodic	-*per*-	Poly(A-*per*-B-*per*-C)
Network	*net*-	*net*-PolyA

Table 1.2   Long sequence copolymer nomenclature

Type	Connective	Example
Polymer blend	*-blend-*	PolyX-*blend*-polyY
Block copolymer	*-block-*	PolyX-*block*-polyY
Graft copolymer	*-graft-*	PolyX-*graft*-polyY
Interpenetrating polymer network	*-ipn-*[a]	*net*-polyX-*ipn*-*net*-polyY
AB-crosslinked	*-net-*	PolyX-*net*-polyY
Starblock	*star-*	*star*-(polyX-*block*-polyY)
Segregated star	*star-*	*star*-(polyX; polyY)

[a] Some authors use *-inter-*.

In Tables 1.1 and 1.2, A, B, C, X, and Y represent the mers that make up the polymer. For example, polyA might represent polystyrene or polybutadiene. Referring first to Table 1.1, the connective *-co-* describes an unspecified sequence arrangement of different mers in a polymer. In the older literature, the term *-co-* was used to indicate a statistical copolymer, where the mers appeared in statistical order. The statistical arrangement is now indicated by *-stat-*, in which the sequential distribution of mers obeys known statistical laws. The connective *-stat-* embraces a large proportion of those copolymers that are prepared by simultaneous polymerization of two or more mers in admixture. A random copolymer is a statistical copolymer in which the probability of finding a given mer at a given site is independent of the nature of the neighboring units at the position. In an alternating copolymer, the opposite is true, with the mers distributed in an alternating sequence. An example of an alternating copolymer is

$$\text{poly[styrene-}alt\text{-(maleic anhydride)]} \tag{1.1}$$

which forms because maleic anhydride adds to the polystyrene free radical much faster than another styrene, but does not add to itself. An alternating copolymer is the simplest case of a periodic copolymer. Ordinarily, the term *-per-* is reserved for three or more mers that form a specific repeating sequence. The prefix *net-* is also included in Table 1.1 to indicate that the polymer is crosslinked to form a network. Other nomenclature available describes short and long branching, rings, and the like.

The various homopolymer and mer sequences delineated in Table 1.1 can be used to form the compositions of Table 1.2. Figure 1.1 provides a cartoon illustration of the compositions defined in Table 1.2. Polymer blends contain no chemical bonds between the various polymer species. Block copolymers are linked end on end. In complex sequences of block copolymers, the blocks are named from one end to another. For example,

$$\text{polystyrene-}block\text{-polybutadiene-}block\text{-polystyrene} \tag{1.2}$$

names the triblock copolymer used in a wide variety of commercial applications;

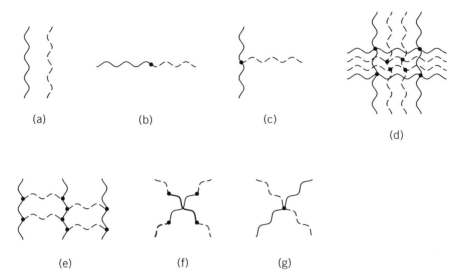

Figure 1.1. Long sequence copolymer structures. Polymer I /\/\/\/\, polymer II ⌒\/\⌒\/\⌒\, and crosslink or graft site •. (a) Polymer blend, (b) block copolymer, (c) graft copolymer, (d) interpenetrating polymer network, (e) AB-crosslinked copolymer, (f) starblock copolymer, and (g) segregated star copolymer.

see Chapter 9. In a graft copolymer, the backbone polymer is named first, and the side chain second. In an IPN, both polymers are crosslinked; if the order of polymerization is known, then the preferred name is given in the order of the polymerization sequence. If the two networks are polymerized simultaneously, then the designation simultaneous interpenetrating network (SIN) is sometimes given; see Chapter 10. Of course, there are two possible semi-IPNs where one or the other polymer is crosslinked. A semi-I IPN is sometimes used to describe a material where the first polymer polymerized is crosslinked and the second is linear, while a semi-II IPN has the reverse topology. The presence of the polymer in network form is indicated by the prefix *net*-. An alternate nomenclature is *cross*- for crosslinked.

For an AB-crosslinked copolymer, the two polymers are crosslinked to each other; that is, the two polymers form a single network. This has sometimes been called a conterminously crosslinked copolymer. In naming the star block copolymers, the inside block of an arm is written first. The segregated star copolymer contains multiple arms of different polymers.

Of course, the several long-sequence copolymers can be made of the various short-sequence copolymers as delineated in Table 1.1. While some important compositions have very irregular structures, the various possibilities may be distinguished through proper application of nomenclature. It must be noted that in the old days, a block copolymer was indicated by the connective -*b*-, as in poly(styrene-*b*-butadiene), which lacks clarity. Similarly, -*g*- was used to indicate a graft copolymer. In the above, no attempt to distinguish phase

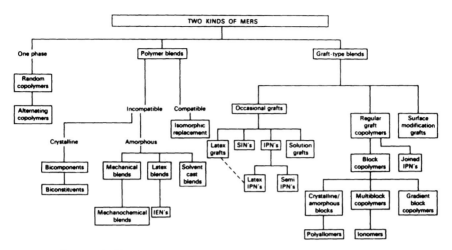

Figure 1.2. Classification scheme for multicomponent polymer materials.

morphology was given. Usually, those compositions of Table 1.1 are one-phased. The morphologies of Table 1.2 compositions are determined by thermodynamics, kinetics of phase separation, and processing conditions, among others, and it is these compositions that shall be emphasized in this book.

Figure 1.2 (7) classifies all of the major combinations of two or more kinds of mers. There are three main divisions: one-phase materials, polymers blends (with no bonds between the participating polymers), and bonded polymer combinations.

The nomenclature of polymers has been collected (4) and recently reviewed (8). ASTM has published a list of abbreviations and acronyms, many of which bear on multicomponent polymer materials; see Appendix 1.1.

At the end of this chapter, the reader will also find a list of recent monographs, edited works, and general reviews in the area of multicomponent polymer materials. Similar lists will appear at the ends of other chapters. A second list at the end of this chapter delineates handbooks and encyclopedias useful for calculations or broad reading on individual systems.

1.1.2. Kinds of Composite Materials

Most polymer-based composites do not have a formal nomenclature at this time, although it could well be argued that one is sorely needed. Figure 1.3 (9,10) provides a polymer composite classification scheme. There are three major types of polymer composites: those where the matrix is nonpolymeric (although wood, clearly polymeric, plays a matrix role here); those with the polymer as the matrix, usually being reinforced or filled with another material, such as fibers, platelets, or particulates; and those where some degree of dual-phase continuity prevails, as in the laminates and foams. Of course the continuous filament composites also exhibit dual-phase continuity. Figures 1.2 and 1.3 will

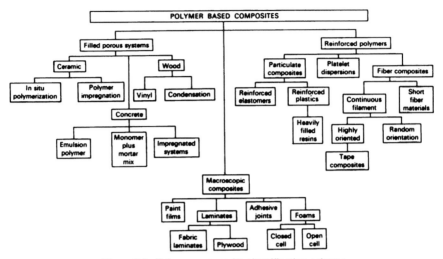

Figure 1.3. Polymer composite classification scheme.

be referred to throughout the book, directly and indirectly, as they provide maps interrelating the various materials of interest. More importantly, they may suggest to the reader yet new combinations of matter.

1.2. REASONS FOR MAKING MULTICOMPONENT POLYMER MATERIALS

The reasons for adding components or phases to homopolymers of commerce are many. Some deal with fire retardancy or with product color, for example. While these are surely important, they will not be emphasized in this book. Some of the reasons for making multicomponent polymer materials that will be discussed include:

1. *Higher Performance Through Synergism.* This includes impact resistance, toughness, high temperature performance, high modulus, lower expansion coefficients, dimensional stability, and high strength-to-weight ratio for plastics, and improved elasticity or damping for elastomers, better adhesion, blood compatibility, and/or a host of other properties. An example of a higher strength-to-weight ratio is the addition of unidirectionally oriented fibers to a plastic matrix.
2. *Achieve Desired Properties.* Frequently there is no single homopolymer with exactly the right properties for the job. Blending offers control over melt viscosity, softening point, processability, and solvent resistance.
3. *Recycling Industrial or Municipal Scrap Polymers (11,12).* This is one of the newest but most important reasons for dealing with multicomponent polymer materials: They come in a mixed state from the consumer.

Table 1.3   Principal properties claimed in polymer blend patents

Property	Frequency (%)
High-impact strength	38
Processability (including weld line)	18
Tensile strength	11
Rigidity/modulus	8
Heat deflection temperature	8
Flammability	4
Solvent resistance	4
Thermal stability	3
Dimensional stability	3
Elongation	2
Gloss	2
Others	4

4. *Adjust Composition to Customer Specifications.* Frequently, this involves blending of two batches of the same polymer, polymer blend, or composite, but, for example, of differing viscosities.

5. *Dilution of the Polymer for Reduced Cost.* Some polymers, especially the engineering polymers, are relatively expensive. If 90% of the properties can be retained while lowering the cost 30%, the economics clearly favor the dilution route. Some examples include adding of calcium carbonate filler to a plastic or low density polyethylene as a discontinuous phase to more expensive polymers and blends.

Utracki has summarized the polymer blend patent literature in Table 1.3 (13). It is interesting to note that high impact strength and processability dominate the field of thermoplastic blends, see Chapter 8.

1.3. CHEMISTRY OF BLENDING AND GRAFTING

Polymer blend preparation may be plain or fancy. Some of the simplest methods involve mechanical mixing in extruders or roll mills. The incorporation of a few percent of SBR (styrene–butadiene rubber) in polystyrene by this method will significantly improve its impact resistance. However, use of the solution graft method (see Section 8.2) results in much tougher materials. This section will delineate some of the basic chemistry of reactions between two polymers or between a polymer and a second monomer.

1.3.1. Grafting Onto and Grafting From

While early reactions on polymers included the acylation and nitration of cellulose, most synthetic polymers were historically treated as having fixed chemistry after polymerization. Kennedy (14) reviewed many of the possibilities of further reactions, most based on classical organic chemistry. He distinguished between grafting *onto*,

$$\left\{ +\cdot M-M-M\cdots \; \rightarrow \; \right\} -M-M-M\cdots \qquad (1.3)$$

and grafting *from*,

$$\left\{ \cdot + nM \; \rightarrow \; \right\} -M-M-M\cdots \qquad (1.4)$$

where M represents the monomer being added. Clearly, not only the chemistry, but the resultant morphology and properties will be different in these two cases. Kennedy (14) gives an example of grafting reactions involving polymers having α-unsaturation in the chain, which includes such polymers as polybutadiene, polyisoprene, gutta percha, and SBR compositions, distinguishing between an attack on the double bond by initiator radical $I^{\cdot}$ or polymer radical, $\sim P^{\cdot}$:

$$60 \pm 5\% \begin{cases} I^{\cdot} + -CH_2-CH=CH-CH_2- \rightarrow -CH_2-\overset{\cdot}{C}H-CH-CH_2- \\ \qquad\quad \text{1,4 enchainment} \qquad\qquad\qquad \overset{|}{I} \\ I^{\cdot} + -CH_2-\underset{\underset{CH=CH_2}{|}}{CH}- \qquad\quad \rightarrow -CH_2-\underset{\underset{\overset{\cdot}{C}H-CH_2I}{|}}{CH}- \\ \qquad\quad \text{1,2 enchainment} \end{cases} \qquad (1.5)$$

$$0\% \begin{cases} \sim P^{\cdot} + -CH_2-CH=CH-CH_2- \rightarrow \sim CH_2-CH-\overset{\cdot}{C}H-CH_2\sim \\ \qquad\qquad\qquad\qquad\qquad\qquad\qquad\qquad\quad \overset{|}{P}\sim \\ \sim P^{\cdot} + -CH_2-\underset{\underset{CH=CH_2}{|}}{CH} \qquad \rightarrow \sim CH_2-\underset{\underset{\overset{\cdot}{C}H-CH_2-P\sim}{|}}{CH}- \end{cases} \qquad (1.6)$$

and an attack on an allylic hydrogen:

$$40 \pm 5\% \begin{cases} I^{\cdot} + \sim CH_2-CH=CH-CH_2\sim \rightarrow IH + \sim \overset{\cdot}{C}H-CH=CH-CH_2\sim \\ I^{\cdot} + \sim CH_2-\underset{\underset{CH=CH_2}{|}}{CH}\sim \qquad\quad \rightarrow \sim CH_2-\underset{\underset{CH_2-CH_2-}{|}}{\overset{\cdot}{C}} \end{cases} \qquad (1.7)$$

$$0\% \begin{cases} \sim P^{\cdot} + \sim CH_2-CH=CH-CH_2\sim \rightarrow \sim PH + \sim \overset{\cdot}{C}H-CH=CH-CH_2- \\ \sim P^{\cdot} + \sim CH_2-\underset{\underset{CH=CH_2}{|}}{CH}\sim \qquad\quad \rightarrow \sim PH + \sim CH_2-\underset{\underset{CH=CH_2}{|}}{\overset{\cdot}{C}}\sim \end{cases} \qquad (1.8)$$

Odian (15) has also described such grafting reactions, pointing out that a chain transfer of a radical between the propagating radical and the polymer may take place. Taking the reaction between a growing polystyrene radical and 1,4-polybutadiene as an example,

$$
\text{wwCH}_2-\text{CH}=\text{CH}-\text{CH}_2\text{ww} + \text{wwCH}_2-\overset{\overset{\displaystyle H}{|}}{\underset{|}{C}}{}^{\cdot} \rightarrow
$$

(1.9)

$$
\text{ww}\overset{\cdot}{\text{C}}\text{H}-\text{CH}=\text{CH}-\text{CH}_2\text{ww} + \text{wwCH}_2-\overset{\overset{\displaystyle H}{|}}{\underset{|}{C}}-\text{H}
$$

The diene radical then initiates polymerization *from*, forming a graft co-polymer.

Odian (15) also points out that grafting *onto* predominates for polybuta-dienes containing high contents of vinyl groups, produced by 1,2-polymeriza-tion. In that case, the reaction proceeds through the double bond,

$$
\text{wwP}^{\cdot} + \text{wwCH}_2-\underset{\underset{\displaystyle \text{CH}=\text{CH}_2}{|}}{\text{CH}}\text{ww} \rightarrow \text{wwCH}_2-\underset{\underset{\underset{\underset{\displaystyle H}{|}}{\text{P}-\text{C}-\overset{\cdot}{\text{C}}}}{\overset{\displaystyle H}{|}}}{\text{CH}}\text{ww}
$$

(1.10)

contrary to reaction (1.6). This leads to *grafting from*, not *onto*.

It must be emphasized that the presence of free radicals in a solution of a polymer in a different monomer sometimes leads to a mixture of the two polymers, with only a slight extent of grafting. However, as will be developed below, even a low percentage of grafts in a polymer blend may profoundly influence the behavior of the system. This arises from the con-centration of such grafts at the interphase, as will be discussed in Part II of this book.

Another interesting reaction involves the peroxidation of polymers contain-ing active hydrogens. For example, polystyrene can be copolymerized with a small amount of isopropyl styrene. The isopropyl group readily incorporates oxygen to form a peroxide. In the presence of heat or ferrous ions, this decomposes as follows:

$$\text{(1.11)}$$

This leads to 50% of graft copolymer and 50% of homopolymer formation, on a molar basis. A host of similar reactions are known (14), some involving backbone reactions in addition to side chain reactions.

1.3.2. High-Energy Irradiation

Other methods of importance involve high-energy irradiation and mechano-chemical methods. In the former, alpha, beta, or gamma rays may be employed. Various groups such as hydrogen may be knocked off, leaving free radicals, anions, and/or cations. These activated sites then can initiate reactions. If two immiscible polymers are irradiated, the result is sometimes called a co-cross-linked polymer, since grafting and crosslinking may occur indiscriminately. If the polymers are immiscible, most of the action will be in the form of crosslinks, with grafting occurring only at the polymer–polymer interfaces. Important cases to be distinguished include the presence or absence of air or oxygen during the irradiation. In the former, oxidation of the activated polymer results in a different series of reactions than if the polymer is irradiated in the presence of vacuum or nitrogen. If polymer I is dissolved in monomer II, a mixture of homopolymer and graft copolymer results.

A similar co-crosslinking phenomenon is brought about by the sulfur vulcanization of rubber blends, as is usually practiced in the manufacture of automotive tires (1). For example, *cis*-polybutadiene and SBR elastomers may be convulcanized. The SBR provides low cost, while the *cis*-polybutadiene provides a measure of self-reinforcement, since it crystallizes under strain. (In today's terminology, it is a *smart material*.)

1.3.3. Mechanochemical Methods

Mechanochemical methods (16) often involve the mastication of either two polymers or the mastication of a polymer I and a monomer II. The mechanical action degrades the polymers, often literally ripping the polymer chain in two, producing free radicals, anions, and cations. A series of graft and block copolymers results, along with ordinary crosslinking of a single species, and homopolymer formation from monomer II.

1.3.4. AB-Crosslinked Copolymers

Bamford and Eastmond (17) describe the synthesis of AB-crosslinked copoly-mers, which are materials wherein polymer II is grafted to polymer I on both ends of the chain. This is sometimes referred to as conterminous grafting. In the case of grafting from, there are two possibilities: The chains may terminate by combination or by disproportionation.

$$\{ +(M)_x\cdot + \cdot(M)_y\} \begin{cases} \text{combination} \quad \{ -(M)_{x+y}\} \\ \\ \text{disproportionation} \quad \{ -(M)_x-H + (M)_y\} \end{cases} \tag{1.12}$$

AB-crosslinked polymers result if termination by combination predominates. Under many conditions, polystyrene terminates 10 to 1 in favor of com-bination, while poly(methyl methacrylate) terminates 4 to 1 in favor of disproportionation.

A series of systems investigated by Bamford and Eastmond emphasize polymers containing trichloro groups, such as the polycarbonate,

$$\{O-\bigcirc-\underset{CCl_3}{\overset{\overset{\displaystyle H}{|}}{C}}-\bigcirc-O-\overset{\overset{\displaystyle O}{\|}}{C}\}_n \tag{1.13}$$

The trichloro groups are thermally activated by $Mo(CO)_6$ or similar chemistry.

A very important class of AB-crosslinked polymers are formed between the reaction of unsaturated polyesters and styrene monomer (18). Polyesters are synthesized containing maleic anhydride and/or fumaric acid mers, which provide unsaturation in the backbone. This provides the reactivity with co-reactant monomers such as styrene. A redox system may be used as a source of free radicals. Initially, the free radicals produced by the redox system are neutralized by stabilizers such as quinone present in the resin system. After an induction period, the free radicals initiate a crosslinking polymerization invol-ving primarily the fumarate groups and the styrene. Commercial polyester resins usually contain 35–45% styrene monomer, corresponding to a stoichiometric ratio of approximately two styrene mers to each maleic or fumarate mer. The resulting product is, of course, thermoset.

The unsaturated polyester–styrene syrup is often added to fiber glass cloth, and additional cloth, laid down repeatedly, to form a series of layups. Such composite materials are strong, light, and puncture resistant. These materials are used for the hulls of pleasure craft boats, for example.

1.3.5. Polymer Blend Complexes

Polymer blend complexes are combinations of two polymers, where the two different polymers are bonded directly to one another along a considerable length of chain. Generally, acid–base, hydrogen bonding, or other interactions are important.

For example, poly(ethylene oxide) forms complexes with poly(acrylic acid) and poly(methacrylic acid) (19). Polyamines form complexes with polyacids. In apolar solvents, these materials show a critical concentration above which the solution gels. This is because each chain complexes with two or more different chains, causing crosslinking.

1.4. METHODS OF DETERMINING DOMAIN SIZE AND STRUCTURE

1.4.1. Electron Microscopy and Morphology

Microscopy, particularly electron microscopy, has been a powerful tool for the elucidation of the morphology of multicomponent polymer materials (20–25). Both scanning electron microscopy (SEM), and transmission electron microscopy (TEM) have been useful, with a variety of staining techniques applied.

In 1966, Kato (20) discovered that osmium tetroxide, OsO_4, was a powerful staining agent for polymers containing double bonds, such as the various diene polymers. The staining is brought about by the high concentration of electrons in the osmium atom. Additionally, osmium tetroxide hardens diene elastomers and other polymers through a crosslinking reaction,

$$OsO_4 + 2 \text{\tiny{ww}}CH{=}CH\text{\tiny{ww}} \rightarrow \begin{array}{c} \text{\tiny{ww}}CH{-}CH\text{\tiny{ww}} \\ | \qquad | \\ O \quad\; O \\ \diagdown \diagup \\ Os \\ \diagup \diagdown \\ O \quad\; O \\ | \qquad | \\ \text{\tiny{ww}}CH{-}CH\text{\tiny{ww}} \end{array} \qquad (1.14)$$

Figure 1.4 (26) illustrates the sample preparation technique, showing first a coarse cut and then an ultramicrotome sectioning. The samples must be cut thinner than the morphology to be observed, otherwise two or more domains may be seen to be overlapping. Since many morphologies are of the order of $0.1–10\,\mu m$, the ultramicrotome second section is often of the order of $0.06\,\mu m$.

It is relatively easy to overstain with osmium tetroxide. The investigator should be on the alert for domains that appear too large by this method. Notwithstanding, Berney et al. (27), measuring small-angle neutron scattering (SANS), diffraction peaks of polybutadiene-*block*-polystyrene diblock

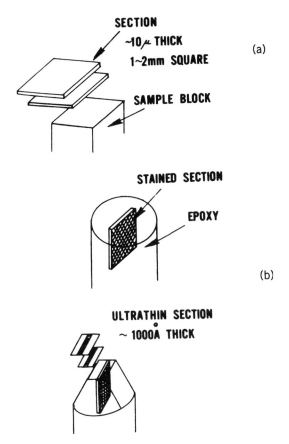

Figure 1.4. Sectioning plastic-diene rubber samples for transmission electron microscopy. (a) Section with hand microtome at low temperature. Section is exposed to OsO$_4$ (vapor or solution) for 1–2 weeks and then embedded in epoxy. A second section involves ultramicrotoming at room temperature. (b) Cross-sectional view of such a section after staining, showing the two-phased nature of such polymer blends. The TEM view normally is from the top.

copolymers, found that the polybutadiene spherical domains were *smaller* by electron microscopy than by SANS.

Acrylic phases can also be stained with osmium tetroxide. For example, Schulze et al. (28) stained polymethacrylates in the presence of polyethylene to study the morphology of a number of materials. Another method involves staining carboxyl groups (29) and is especially effective for esters such as poly(*n*-butyl acrylate). First, the polymer is treated with hydrazine or hydroxylamine, then with osmium tetroxide in the absence of water. A hydrazide forms,

$$
\overset{\overset{\displaystyle O}{\|}}{R-C}-O-C_4H_9 + NH_2-NH_2 \rightarrow \overset{\overset{\displaystyle O}{\|}}{R-C}-NH-NH_2 + C_4H_9OH \tag{1.15}
$$

which, on addition of osmium tetroxide, metallic osmium is deposited through the reducing reaction of the hydrazide.

Another staining agent is ruthenium tetroxide, RuO_4, which stains many polymers containing aromatic, ether, alcohol, amide, olefin, ester, nitrile, sulfone, halogen, carbonyl, or unsaturated moieties (20,21). In one recent study, Gonzalez-Montiel et al. (30) studied the nylon-6-*blend*-polypropylene system, using either an aqueous solution of phosphotunstic acid to stain the nylon-6 phase, or RuO_4 to stain the maleated polypropylene elastomer.

It must be emphasized that both osmium tetroxide and ruthenium tetroxide are highly toxic, in the vapor state, in solution, and in the solid state. Investigators are urged to practice a high level of safety precautions when handling these materials.

If the two polymers making up the polymer blend differ in electron density by about 15–20% or more, the contrast obtained without any staining may be sufficient for transmission electron microscopy. Thus, Voigt-Martin (31) found that the high scattering power of the bromo groups in poly(bromostyrene) were sufficient to provide contrast to polystyrene in a blend of the two. Similar research was done by Tao et al. (32) on blends of polystyrene and poly(2-chlorostyrene). With the use of very high contrast photographic film, even modest electron density differences can be distinguished.

Scanning electron microscopy (SEM) uses a focused electron beam to scan the surface of the sample. Images provide a great depth of field, allowing a three-dimensional morphology to be constructed, in contrast to TEM, which requires an imagination to reconstruct the actual three-dimensional morphology. Scanning electron microscopy is particularly useful for composite systems and for the study of fracture surfaces.

For example, the effects of static and fatigue loading on the fracture surfaces of poly(aryl ether ether ketone) reinforced with short carbon fibers were investigated by Friedrich et al. (33). Under static loading conditions, the polymer pulls away from the fibers much more than under fatigue loading conditions, see Figure 1.5 (33). Recently, the fracture surfaces of an epoxy–phenoxy system were studied by Guo (34) using SEM, finding that the extent of phase separation depended on the curing agent employed.

Many other techniques have been used on occasion. These include etching by acids, swelling by solvents, dissolving out one or the other component, and so forth. The literature is replete with studies of polymer blends and composites via microscopic methods.

1.4.2. Small- and Wide-Angle X-Ray Scattering

X-ray analysis of materials is one of the oldest methods available, with Bragg's equation being memorized by all students of science and engineering. Beyond the determination of crystal structure, however, X-ray analysis can be used to characterize the size of domains and provide a range of important information. There is a reciprocity between the scattering angle and the size

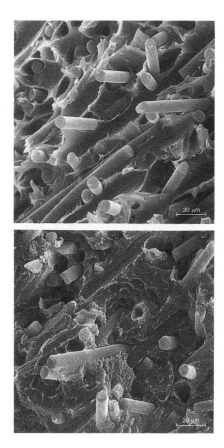

Figure 1.5. Scanning electron micrograph of a short carbon fiber reinforced poly(aryl ether ether ketone), showing the fracture surface. Note polymer still bonded to the fibers after fracture, suggesting good bonding.

scale of the structure probed. The technique of wide-angle X-ray scattering (WAXS) probes distances of 0.1–2 nm. The theory of WAXS is based on the scattering of radiation by electrons where interference effects can be correlated to atomic dimensions. In contrast, small-angle X-ray scattering (SAXS) is used in regions with different electron densities. Subsequently, the correlations are longer ranged, useful for studying phase domain sizes in the range of 5–100 nm.

Figure 1.6 (35) illustrates both WAXS and SAXS methods, where I is scattering intensity at angle 2θ, and $q = (4\pi \sin\theta)/\lambda$, where λ represents the X-ray wavelength. The maximum in q, q^*, can be used to determine the domain spacings via the relation (from Bragg's law) $d = 2\pi/q^*$. This method works if there is a (nearly) uniform domain size in the material. Alternately, the slopes of $1/I$ vs. q can be used to determine the z-average domain sizes, as is done by

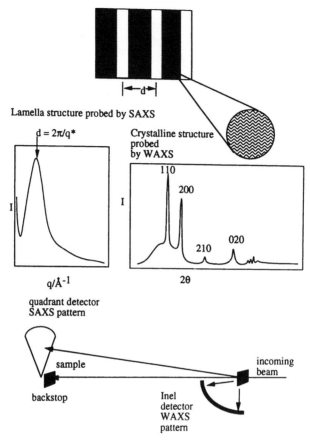

Figure 1.6. SAXS and WAXS experiments on semicrystalline polymers. While the SAXS experiments allow determination of the lamella spacing from the peak maximum, the crystal structure may be deduced from the peaks in the WAXS pattern via the Bragg equation, etc.

neutron scattering (Section 4.6), or by light scattering from polymer solutions, for the determination of the chain radius of gyration.

1.5. POLYMER TRANSITIONS

Classically, scientists and engineers learn that there are three states of matter—solids, liquids, and gases—and that two transitions connect them: melting and boiling (or freezing and condensing). Polymers behave somewhat differently. First, no polymer has been observed in the gas state: All polymers degrade before their boiling temperatures because of their high molecular weights. Many polymers do have a melting transition, T_m, such as polyamides and polyesters. However, many other polymers are amorphous, such as polystyrene and silicates such as window glass, and hence do not undergo melting. Classically, amorphous polymers were considered liquids, since they were obviously in the condensed state, but noncrystalline. Today, we would call them amorphous solids.

Virtually all polymers exhibit a glass–rubber transition, wherein the polymer softens. This transition is often abbreviated as the glass transition, and the glass transition temperature is often written T_g. The glass transition is associated with the onset of long-range, coordinated molecular motion in the material. There are a number of analogs to the glass transition in nonpolymeric materials. For example, at about 90 K, ethane undergoes a second-order transition as detected by nuclear magnetic resonance (NMR) absorption (36), where the two CH_3 units begin to rotate freely relative to one another. Another example is the Curie temperature, the temperature beyond which a ferromagnetic substance exhibits paramagnetism. Since the melting and glass transitions dominate the properties and processing of polymers, a brief review will be undertaken (37,38).

1.5.1. Melting Transition

Polymer melting may be observed by X-ray and electron diffraction, in which the sharp lines or spots obtained from crystalline materials give way to amorphous halos characteristic of the liquid state. Crystal structure and melting may also be studied with a polarizing microscope, where spherulitic or other textures give way to nearly structureless fields of view. The melting transition can also be studied quantitatively by differential scanning calorimetry (DSC), dilatometric measurements, and others.

In DSC studies, an endotherm characteristic of the enthalpy of melting of the crystalline portion of the material will be observed, noting that no high polymer is 100% crystalline. Figure 1.7 (39) illustrates the behavior of a polymer

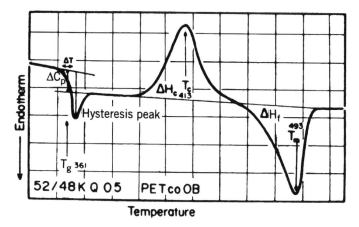

Figure 1.7. Differential scanning calorimetry heating trace of quenched poly(ethylene terephthalate-*co-p*-oxybenzoate), showing first a glass transition, then a crystallization transition, and lastly a melting transition. Also shown is a hysteresis peak, probably caused by residual orientation of the polymer.

undergoing both crystallization and melting in a DSC study, obtained by heating a polymer rapidly quenched from the melt state. The areas under the two curves are identical, showing that the heat of crystallization equals the heat of melting. Also shown in the figure is the glass transition, to be discussed later.

In dilatometric studies, the change in volume on melting is recorded. As with most organic materials, the volume increases on melting indicative of the space required for greater molecular motion. However, different from simple compounds, melting takes place over a range of temperatures, rather than sharply at one temperature. This is caused by the small size and irregular shape of the crystallites making up most polymers.

Another method of studying polymer melting involves changes in the modulus. The polymer goes from a stiff crystalline solid to a much softer melt, assuming it is a linear polymer. However, the glass–rubber transition behaves similarly, going from a glass to a liquid, so that the behavior is not unambiguous. In the old days, both transitions were sometimes referred to as the softening point of the polymer.

1.5.2. Glass Transition

1.5.2.1. Modulus and Mechanical Terms   Such terms as glassy, rubbery, and flow imply a knowledge of simple mechanical relationships. Hook's law assumes a perfect elastic material body, Young's modulus, E, is defined as

$$\sigma = E\varepsilon \tag{1.16}$$

where σ and ε represent the tensile stress and strain, respectively. The tensile stress is defined as the force applied per unit area of the material, while the strain is measured by the change in fractional length of the sample, ΔL, over the initial length, L_0

$$\varepsilon = \frac{(L - L_0)}{L_0} = \frac{\Delta L}{L_0} \tag{1.17}$$

Modulus may be reported in terms of pounds per square inch, dynes per square centimeter, or Pascals.

Young's modulus is measured in terms of stretching or compressing a sample. However, a sample may be exposed to shearing or twisting motions. The ratio of the shear stress, f, to the shear strain, s, defines the shear modulus, G,

$$G = \frac{f}{s} \tag{1.18}$$

A perfect liquid follows Newton's law for viscosity, η,

$$f = \eta \left(\frac{ds}{dt} \right) \tag{1.19}$$

Viscosity is expressed in terms of poises or pascal-seconds.

When a material undergoes any mode of deformation, in general the volume changes, usually increasing when elongational strains are applied. Poisson's ratio, ν, is defined as

$$-\nu \varepsilon_x = \varepsilon_y = \varepsilon_x \tag{1.20}$$

where the strain is applied in the x direction and the response strains are in the y and z directions, respectively. The bulk modulus, B, is given by

$$B = -V \left(\frac{\partial P}{\partial V} \right)_T \tag{1.21}$$

where P represents the hydrostatic pressure. The inverse of the bulk modulus is called the compressibility, β.

A three-way equation related the four basic mechanical properties,

$$E = 3B(1 - 2\nu) = 2(1 + \nu)G \tag{1.22}$$

where any two of these properties may be varied independently. As an especially important relationship when $\nu = 0.5$,

$$E \cong 3G \tag{1.23}$$

which is a particularly good approximation for elastomers.

The quantities E and G are measures of the stiffness of the material or how hard it is. The quantity B is a measure of its resistance to shrinkage under hydrostatic pressure. (A hooked fish actually gets a bit bigger as it is raised out of the depths.)

The viscosity is a measure of its resistance to flow. Thus, a high-viscosity liquid flows slower (molasses) than a low-viscosity liquid (water). Poisson's ratio is a measure of the change in volume on deformation. A value of exactly 0.5 (impossible physically!) means no volume change. A value of 0.0 means no lateral contraction during extension. Values of 0.3–0.4 are common for plastics, while elastomers frequently have values of about 0.49.

1.5.2.2. Glass Transition Behavior   At T_g, a linear amorphous polymer softens to a melt over about a 20–30°C temperature range. The corresponding crosslinked polymer softens to an elastomer. Quantitatively, Young's modulus drops from approximately 3×10^9 Pa to about 2×10^6 Pa, as illustrated in

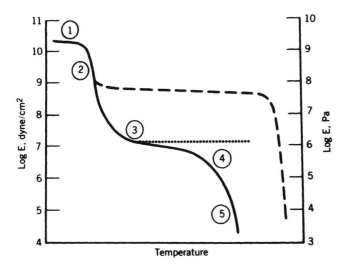

Figure 1.8. Five regions of viscoelastic behavior for a linear amorphous polymer. Also illustrated are the regions of crystallinity (dashed line) and cross-linking (dotted line).

Figure 1.8 (37). Below the T_g region, the polymer is in the glassy state, region 1 in Figure 1.8. The glass transition is shown as region 2. Immediately above T_g, the polymer is in the rubbery plateau, region 3. When the polymer begins to soften again at still higher temperatures, it is said to exhibit rubbery flow, region 4. At still higher temperatures, the polymer exhibits liquid flow, region 5.

At T_g, a polymer exhibits a change in slope in volume–temperature plots, as illustrated in Figure 1.9. At its melting temperature, a discontinuity in its volume–temperature plot occurs, as also illustrated in Figure 1.9. At T_g these materials also undergo a discontinuity in their heat capacity (Figure 1.7). A polymer always has a higher heat capacity above T_g, indicative of its greater freedom of molecular motion.

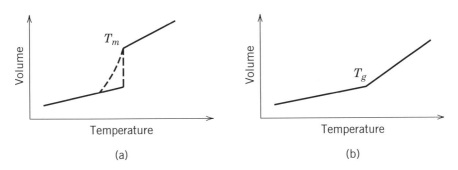

Figure 1.9. Comparison of the changes in volume with temperature for (a) melting and (b) glass transition. The dashed line in (a) represents the ideal state, while the dotted line is closer to reality. The change in slopes in (b) are explained by the theory of the glass transition.

If the polymer is crosslinked, then the rubbery plateau may extend up to the degradation temperature of the polymer (Figure 1.8). Ideally, Young's modulus of the elastomer in the rubbery plateau region is given by

$$E = 3nRT \tag{1.24}$$

where RT represents the gas constant times the absolute temperature. The quantity n represents the number of active chain segments bound on both ends by crosslinks to other chains, per unit volume. Thus, the modulus actually may increase a bit as the temperature is increased. The modulus increase with temperature suggested by equation (1.24) is mitigated by a more rapid relaxation process of the polymer chains as the temperature is increased. Hence, for many systems, the rubbery plateau modulus for elastomers appears roughly constant. A good example of a material in the rubbery plateau region is an ordinary rubber band.

If the polymer is semicrystalline, only the amorphous portion of the polymer undergoes a glass transition. Another kind of plateau appears as the temperature is raised, as illustrated in Figure 1.8. This plateau has a higher modulus than would be characteristic of rubbery materials. Frequently, the material behaves leathery. Good examples are low- and high-density polyethylene, the latter being the stiffer because it is more highly crystalline.

While a semicrystalline polymer (ideally) exhibits a discontinuity of volume on melting, a polymer going through the glass transition exhibits a change in expansion coefficient; see Figure 1.8. For polystyrene with a glass transition temperature of about 100°C, the expansion coefficient in the glassy state, α_G, is about $2.5 \times 10^{-4} \, \mathrm{K}^{-1}$, while that in the rubbery state is about $5.5 \times 10^{-4} \, \mathrm{K}^{-1}$. The theory of the glass transition, especially its relation to the underlying second-order transition, is discussed elsewhere (37).

1.5.2.3. Dynamic Mechanical Behavior   So far, the study of the glass transition has been based on quasi-static measurements. The sample is usually strained in some manner, such as elongation, and a period of time is allowed to elapse before the stress is measured, 10 s being standard. However, the properties of all materials are dependent on the time frame of the measurement. The usual case is that amorphous polymers behave softer if longer and longer time frames are used. According to the WLF (Williams–Landel–Ferry) theory (40), the glass transition increases about 6–7°C for every decade decrease in the time frame of measurement. Thus, polystyrene has a T_g of 100°C at 10 s, but a T_g of 106°C if only 1 s is allowed to elapse before measurement.

An important method of measuring the mechanical behavior of polymers utilizes dynamic mechanical spectroscopy (DMS). In DMS, the sample is deformed in a cyclic manner, often in a sine wave fashion. One then speaks of the frequency of the experiment, usually expressed in hertz (Hz). The WLF equation then suggests that for every decade increase in frequency, the glass transition temperature will increase 6–7°C, similar to the above. For DMS, 0.1 Hz corresponds roughly to the 10 s quasi-static measurement for the estimation of T_g. (In volume–temperature experiments, raising the temperature 1°C/min also yields approximately the same glass transition temperature.)

Much more information can be obtained from a polymer through the use of dynamic measurements, however. A new set of mechanical properties needs to be defined. The storage modulus, a measure of the energy stored during the deformation, is given by E'. The loss modulus, a measure of the energy lost during the deformation (usually as heat) is given by E''. The energetics can be more easily visualized by a transformation of modulus units from dyn/cm^2 to ergs/cm^3 (dyn-cm = erg), or from Pa(N/m^2) to J/m^3 (N-m = J). The loss tangent is defined by

$$\tan \delta = \frac{E''}{E'} \tag{1.25}$$

where δ is the angle between the in-phase and out-of-phase components in the cyclic motion.

While E' behaves similarly to Young's modulus, especially far from transitions, the E'' spectra looks much more like an infrared spectra, going through a series of maxima at the various polymer transitions. The largest maximum is usually associated with the glass transition temperature.

The dynamic modulus are also related to the complex Young's modulus, E^*

$$E^* = E' + iE'' \tag{1.26}$$

where i is the square root of minus one. Note that $E = |E^*|$.

Similar definitions hold for E^*, J^*, and other mechanical quantities. The complex quantities are particularly important as they are measured directly by mechanical instrumentation.

Figure 1.10 (41) illustrates the results of a typical dynamic mechanical study. Crosslinked polystyrene was measured at 110 Hz through the glass transition range. Note that the tan δ peak is about 15°C higher than the peak in E'', the usual case. The value of E' in the glassy region is about 3×10^9 Pa and about 2×10^6 Pa in the rubbery plateau region, while the peak in tan δ is about 0.4, all typical of homopolymers. Values of tan δ higher than 1.0 or 1.5 are unusual.

1.5.3. Combinations of Two Polymers

1.5.3.1. Storage Modulus Behavior   The mechanical behavior of polymer blends, block, grafts, IPNs, and so forth depends on the miscibility of a polymer pair, see Figure 1.11. If the polymer pair are miscible, forming one phase, then one sharp glass transition will be observed at a temperature governed by the Fox

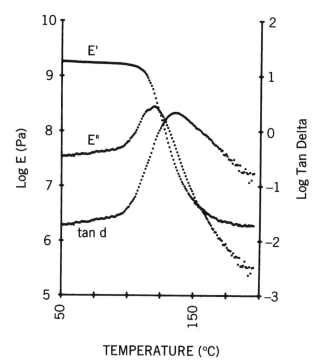

Figure 1.10. Dynamic mechanical behavior of polystyrene cross-linked with 2% divinyl benzene, at 110 Hz. The tan δ peak is usually higher than the E'' peak by 10–15°C.

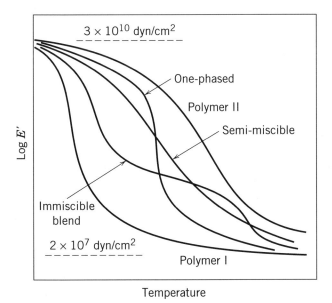

Figure 1.11. Miscibility and mechanical behavior. Polymer I may be immiscible, miscible, or semi-miscible with polymer II.

equation,

$$\frac{1}{T_g} = \frac{W_1}{T_{g1}} + \frac{W_2}{T_{g2}} \tag{1.27}$$

where T_{g1} and T_{g2} represents the glass transitions of the component polymers, and W_1 and W_2 are weight fractions. For the simple binary case, $W_1 + W_2 = 1$. If the two polymers are totally immiscible, two glass transitions will be observed at the glass transitions of the homopolymers. Usually, at least some mixing is observed at the polymer–polymer interface, resulting in a broadening of the transitions. Slight actual solubility of the components within the phases results in an inward shifting of the two glass transitions.

In the case where there is extensive but incomplete mixing, current nomenclature has a problem. If the phase domains are very small, of the same size range as the polymer chains themselves, 0.1–$0.2\,\mu$m, then the system is sometimes called microheterogeneous. The resulting $E' - T$ plot, shown in Figure 1.11, spans the entire temperature range between the two homopolymer T_gs. It is as if the entire system behaves as an interface, with all possible compositions appearing. Modern theories of the glass transition suggest that a volume occupied by about 50 mers is required for an independent contribution to the mechanical spectra.

Other nomenclature for this region of extensive mixing includes semimiscible or semicompatible. The term *compatible* in modern technology refers to a combination of two or more polymers that are useful. The term *incompatible* refers to a combination of two or more polymers that are not useful. The term *miscible* refers to the concept of complete solubility, as defined by thermodynamics and phase diagrams. The term *immiscible* refers to combinations of two or more polymers that do not mix. Partial miscibility results in an inward shifting or broadening of the two glass transitions, as mentioned above. The term *semimiscible* is in limbo, but does seem to describe the general case of one very broad transition resulting from the mixing of two or more polymers.

The reason why the current literature (42) has moved to adopt the term *miscible* rather than soluble, compatible, or some other term needs further explanation. The term *soluble* is more descriptive and exact than other terms, but with polymer–polymer blends, ideal or random molecular mixing may not adequately describe the true nature of the material even though the physical parameters of the blend would suggest true solubility. For example, the glass transitions described above are often a bit broadened even for blends exhibiting true thermodynamic solubility, because the very large molecular weights allow regions of greater than 50 mers to be nonrandom in composition. As described above, *compatibility* is used to describe the usefulness of the system and not really its thermodynamic state. It must be added that while thermodynamics in the form of negative free energies of mixing and phase diagram determination are the ultimate test of solubility, these are not always known for the particular blend under consideration at the time.

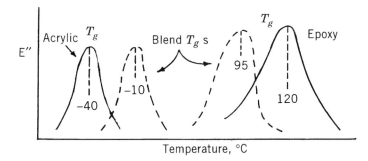

Figure 1.12. Slightly miscible compositions have glass transitions shifted inward.

While temperature is used as the x axis here, the time–temperature superposition principle shows that two other x axes will produce equivalent, although not identical, results. These are the use of log time for relaxation experiments, and the negative of log frequency, both at constant temperature. The time–temperature superposition principle states that with viscoelastic materials, time and temperature are equivalent to the extent that data at one temperature can be superimposed on data at another temperature by shifting the curves along the log time axis.

1.5.3.2. Loss Modulus Behavior   Figure 1.12 illustrates the behavior of the loss modulus, corresponding to the case of slight or partial miscibility in Figure 1.11. Here, the two transitions are shown to be broadened and shifted inward (Figure 1.12).

1.5.3.3. Example Calculation of Phase Composition   Let us assume a blend of an epoxy polymer with an acrylate polymer, 60/40 w/w overall composition. The loss moduli were determined on the blend and on the pure components, with the following results:

Composition	T_g, °C	T_g, K
Pure acrylate	−40	233
Acrylate rich blend phase	−10	263
Pure epoxy	120	393
Epoxy rich blend phase	95	368

The diagram given in Figure 1.12 can be assumed to apply. The questions are: (a) What is the composition of each phase? (b) What is the volume fraction of each phase?

The temperature of each of the two glass transitions is governed by the overall composition within each phase, which is presumed to be homogeneous. Thus, the Fox equation may be applied to each of these two transitions. The phase diagram illustrated in Figure 1.13 shows the relationship of w_1 to w_2 for

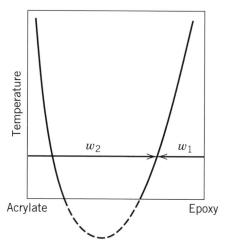

Figure 1.13. Values of w_1 and w_2 for the epoxy-rich phase, illustrated on a phase diagram. Theoretically at some lower temperature, the blend may become miscible.

the composition within each phase. The calculation starts with the Fox equation. For the epoxy-rich phase:

$$1/368 = w_1/233 + (1 - w_1)/393$$

Simple algebra yields,

$$w_1 = 0.097 \text{ acrylate}$$
$$w_2 = 0.903 \text{ epoxy}$$

Similarly for the acrylate-rich phase,

$$1/263 = w_1/233 + (1 - w_1)/393$$

and

$$w_1 = 0.72 \text{ acrylate}$$
$$w_2 = 0.28 \text{ epoxy}$$

These are the compositions of the two phases. For the volume fraction of the acrylate phase, first X is assumed to be the acrylate phase volume. Noting that the entire blend was 40% acrylate,

$$0.72X + 0.097(1 - X) = 0.40$$

and $X = 0.48$.

Similarly for the epoxy-rich phase, of volume fraction Y,

$$0.903\,Y + 0.28(1 - Y) = 0.60$$

and $Y = 0.52$. Of course, $X + Y$ must equal one here. It must be emphasized that in reality, all polymer blends must mix to at least a slight extent.

In summary, the glass transitions in a polymer blend depend on the extent of miscibility of the two components. For substantially total immiscibility, the individual glass transitions will be observed. Slight miscibility results in an inward shift of the two T_gs. Greater miscibility yields one broad transition, which narrows for thermodynamic miscibility. Using simple equations and/or analysis of the dynamic mechanical behavior as a function of temperature often yields a quantitative evaluation of the system.

APPENDIX 1.1

SELECTED ABBREVIATED TERMS RELATING TO PLASTICS

Term	Abbreviation
Acrylonitrile–butadiene–acrylate	ABA
Acrylonitrile–butadiene–styrene plastics	ABS
Acrylonitrile/ethylene–propylene–diene/styrene	AEPDM
Acrylonitrile–ethylene–styrene	AES
Acrylonitrile/methyl methacrylate	AMMA
Acrylonitrile–styrene–acrylate	ASA
Aromatic polyester	ARP
Carboxymethyl cellulose	CMC
Casein	CS
Cellulose acetate	CA
Cellulose acetate butyrate	CAB
Cellulose acetate propionate	CAP
Cellulose nitrate	CN
Cellulose plastics, general	CE
Cellulose propionate	CP
Cellulose triacetate	CTA
Chlorinated polyethylene	CPE
Chlorinated poly(vinyl chloride)	CPVC
Cresol–formaldehyde	CF
Epoxy, epoxide	EP
Ethyl cellulose	EC
Ethylene/ethyl acrylate	EEA
Ethylene/methacrylic acid	EMA
Ethylene–propylene polymer	EPM
Ethylene–propylene–diene	EPD
Ethylene–tetrafluoroethylene copolymer	ETFE
Ethylene–vinyl alcohol	EVAL
Ethylene/vinyl acetate	EVA

High-density polyethlene plastics	HDPE
High-impact polystyrene	HIPS
Linear low-density polyethylene plastics	LLDPE
Linear medium-density polyethylene plastics	LMDPE
Liquid crystal polymer	LCP
Low-density polyethylene plastics	LDPE
Medium-density polyethylene plastics	MDPE
Methacrylate–butadiene–styrene	MBS
Methyl cellulose	MC
Nylon (see also polyamide)	PA
Perfluoro(alkoxy alkane)	PFA
Perfluoro(ethylene-propylene) copolymer	PEP
Phenol–formaldehyde resin	PF
Phenol–furfural resin	PFF
Poly(acrylic acid)	PAA
Polyacrylonitrile	PAN
Poly(α-methylstyrene)	PMS
Polyamide (nylon)	PA
Polyamide–imide	PAI
Polyaryl amide	PARA
Polyarylate [poly(aryl terephthalate)]–liquid crystal polymer	PAT
Polyarylether	PAE
Polyaryletherketone	PAEK
Polyarylsulfone	PASU
Polybutadiene–acrylonitrile	PBAN
Polybutadiene–styrene	SBR
Poly(butyl acrylate)	PBA
Poly(butylene terephthalate)	PBT
Polycarbonate	PC
Poly(ester urethane)	PAUR
Polyether ketone	PEK
Poly(ether urethane)	PEUR
Polyether block amide	PEBA
Polyetheretherketone	PEEK
Poly(etherimide)	PEI
Poly(ether sulfone)	PES
Polyethylene	PE
Poly(ethylene oxide)	PEO
Poly(ethylene terephthalate)	PET
Polyimide	PI
Polyimidesulfone	PISU
Polyisobutylene	PIB
Poly(methyl-α-chloroacrylate)	PMCA
Poly(methyl methacrylate)	PMMA
Poly(4-methylpentene-1)	PMP

Polychlorotrifluoroethylene	PCTFE
Polyoxymethylene, polyacetal	POM
Poly(phenylene ether)	PPE
Poly(phenylene oxide) deprecated term and acronym, see preferred term poly(phenylene ether) and acronym	PPO
Poly(phenylene sulfide)	PPS
Poly(phenylene sulfone)	PPSU
Polyphthalamide	PPA
Polypropylene	PP
Poly(propylene oxide)	PPO
Polystyrene	PS
Polysulfone	PSU
Polytetrafluoroethylene	PTFE
Polyurethane	PUR
Polyvinylcarbazole	PVK
Polyvinylpyrrolidone	PVP
Poly(vinyl acetate)	PVAC
Poly(vinyl alcohol)	PVAL
Poly(vinyl butyral)	PVB
Poly(vinyl chloride)	PVC
Poly(vinyl chloride–acetate)	PVCA
Poly(vinyl fluoride)	PVF
Poly(vinyl formal)	PVFM
Poly(vinylidene chloride)	PVDC
Poly(vinylidene flouride)	PVDF
Saturated polyester plastic	SP
Silicone plastics	SI
Styrene–acrylonitrile plastic	SAN
Styrene–butadiene plastic	SB
Styrene–maleic anhydride plastic	S/MA
Styrene/α-methylstyrene plastic	SMS
Styrene–rubber plastics	SRP
Thermoplastic elastomer	TPE
Thermoplastic elastomer, olefinic	TEO
Thermoplastic elastomer polyether block amide	PEBA
Thermoplastic polyurethane	TPUR
Thermoset polyurethane	TSUR
Ultra-high molecular weight polyethylene	UHMWPE
Unsaturated polyester	UP
Urea–formaldehyde resin	UF
Vinyl chloride–ethylene–methyl acrylate resin	VCEMA
Vinyl chloride–ethylene–vinyl acetate resin	VCEV
Vinyl chloride–ethylene resin	VCE
Vinyl chloride–methyl acrylate resin	VCMA
Vinyl chloride–methyl methacrylate resin	VCMMA

Vinyl chloride–vinyl acetate resin   VCVAC
Vinyl chloride–vinylidene chloride resin   VCVDC

Examples:
For blends and alloys, the standard terminology is the sum of the abbreviated
terms of two components. For example: Acrylonitrile–butadiene–styrene +
polyamide ABS + PA or Poly(ethylene terephthalate) + rubber PET + RBR
where an unspecified rubber is used.

REFERENCES

1. J. A. Manson and L. H. Sperling, *Polymer Blends and Composites*, Plenum, New York, 1976.
2. L. H. Sperling and W. V. Metanomski, *Polym. Mater. Sci. Eng. (Prepr.)*, **68**, 341 (1993).
3. L. H. Sperling and W. V. Metanomski, *Polym. Mater. Sci. Eng. (Prepr.)*, **69**, 575 (1993).
4. W. V. Metanomski, Ed., IUPAC *Compendium of Macromolecular Nomenclature*, Blackwell Scientific, Oxford, 1991; available in the U.S. from CRC Press, Boca Raton, FL.
5. J. Kahovec, P. Kratochvil, A. D. Jenkins, I. Mita, I. M. Papisov, L. H. Sperling, and R. F. T. Stepto, accepted, *Pure and Appl. Chem.*, (1997).
6. IUPAC, *Pure App. Chem.*, **57**, 1427 (1985).
7. L. H. Sperling, in *Recent Advances in Polymer Blends, Grafts, and Blocks*, L. H. Sperling, Ed., Plenum, New York, 1974, p. 93.
8. E. A. Coleman, *Plast. Eng.*, **49(6)**, 47 (1993).
9. L. H. Sperling, *Polym. Prepr.*, **14**, 431 (1973).
10. L. H. Sperling, *Fiber Sci. Technol.*, **7**, 199 (1974).
11. R. J. Ehrig, Ed., *Plastics Recycling Products and Processes*, Hanser, Munich, 1992.
12. G. D. Andrews and P. M. Subramanian, Eds., *Emerging Technologies in Plastics Recycling*, ACS Symp. Ser. No. 513, ACS Books, Washington, DC, 1992.
13. L. A. Utracki, *Polymer Alloys and Blends*, Hanser, New York, 1990.
14. J. P. Kennedy, in *Recent Advances in Polymer Blends, Grafts, and Blocks*, L. H. Sperling, Ed., Plenum, New York, 1974.
15. G. Odian, *Principles of Polymerization*, 3rd ed., Wiley, New York, 1991, pp. 715–716.
16. R. S. Porter and A. Casale, in *Encyclopedia of Polymer Science and Engineering*, 2nd ed., J. I. Kroschwitz, Ed., Vol. 9, p. 467, Wiley, New York, 1987.
17. C. H. Bamford and G. C. Eastmond, in *Recent Advances in Polymer Blends, Grafts, and Blocks*, L. H. Sperling, Ed., Plenum, New York, 1974.
18. J. Selley, in *Encyclopedia of Polymer Science and Engineering*, 2nd ed., J. I. Kroschwitz, Ed., Wiley, Vol. 12, p. 256, 1988.
19. E. J. Goethals, presented at the IUPAC Macro Seoul '96, Seoul, Korea, Aug. 4–9, 1996; *Abstracts*, p. 221.

20. K. Kato, *Polym. Lett.*, **4** 35 (1966).
21. R. Vitali and E. Montani, *Polymer*, **21**, 1220 (1980).
22. J. S. Trent, J. I. Scheinbeim, and P. R. Couchman, *Macromolecules*, **16**, 589 (1989).
23. A. E. Woodward, *Atlas of Polymer Morphology*, Hanser, Munich, 1989.
24. D. R. Paul, J. W. Barlow, and H, Keskkula, in *Encyclopedia of Polymer Science and Engineering*, J. I. Kroschwitz, Ed., Wiley, New York, Vol. 12, p. 461 1988.
25. L. H. Sperling, in *Encyclopedia of Polymer Science and Engineering*, J. I. Kroschwitz, Ed., Wiley, Vol. 9, p. 789 1987.
26. D. A. Thomas, *J. Polym. Sci. Polym. Symp.*, **60**, 189 (1977).
27. C. F. Berney, R. E. Cohn, and F. S. Bates, *Polymer*, **23**, 1222 (1982).
28. U. Schulze, G. Pompe, E. Meyer, A. Janke, J. Pointeck, A. Fiedlerova, and E. Borsig, *Polymer*, **36**, 3393 (1995).
29. G. Kanig, *Prog. Coll. Polym. Sci.*, **57**, 176 (1975).
30. A. Gonzalez-Montiel, H. Keskkula, and D. R. Paul, *Polymer*, **36**, 4587 (1995).
31. I. G. Voigt-Martin, *Adv. Polym. Sci.*, **67**, 196 (1985).
32. J. Tao, M. Okada, T. Nose, and T. Chiba, *Polymer*, **36**, 3909 (1995).
33. K. Friedrich, R. Walter, H. Voss, and J. Karger-Kocis, *Composites*, **17**, 205 (1986).
34. Q. Guo, *Polymer*, **36**, 4753 (1995).
35. A. J. Ryan, S. Naylor, B. Komanschek, W. Bras, G. R. Mant, and G. E. Derbyshire, in *Hypernated Techniques in Polymer Characterization*, T. Provder, M. W. Urban, and H. G. Barth, Eds., ACS Symposium Series No. 581, American Chemical Society, Washington, DC, 1994.
36. H. S. Guawasky, G. B. Kistiakowsky, G. E. Pake, and E. M. Purcell, *J. Chem. Phys.*, **17**, 942 (1949).
37. L. H. Sperling, *Introduction to Physical Polymer Science*, 2nd ed., Wiley-Interscience, New York, 1992.
38. P. Munk, *Introduction to Macromolecular Science*, Wiley-Interscience, New York, 1989, Chapter 4.
39. W. Messiri, J. Menczel, U. Guar, and B. Wunderlich, *J. Polym. Sci., Polym. Phys. Ed.*, **20**, 719 (1982).
40. M. L. Williams, R. F. Landel, and J. D. Ferry, *J. Am. Chem. Soc.*, **77**, 3701 (1955).
41. L. H. Sperling, in *Sound and Vibration Damping in Polymers*, R. D. Corsaro and L. H. Sperling, Eds., ACS Books Symposium Ser. 424, American Chemical Society, Washington, DC, 1990.
42. O. Olabisi, L. M. Robeson, and M. T. Shaw, *Polymer–Polymer Miscibility*, Academic, New York, 1979.

MONOGRAPHS, EDITED WORKS, AND GENERAL REVIEWS

Arends, C. B., Ed., *Polymer Toughening*, Marcel Dekker, New York, 1996.

Bucknall, C. B., *Toughened Plastics*, Applied Science, London, 1977.

Coleman, M. M., J. F. Graf, and P. C. Painter, *Specific Interactions and the Miscibility of Polymer Blends*, Technomic, Lancaster, PA, 1991.

Corsaro, R. D. and L. H. Sperling, Eds., *Sound and Vibration Damping with Polymers*, ACS Symposium Series 424, American Chemical Society, Washington, DC, 1990.

Eisenberg, A. and M. King, *Ion-Containing Polymers Physical Properties and Structure*, Academic, New York, 1977.

Feast, W. J., H. M. Munro, and R. W. Richards, Eds., *Polymer Surfaces and Interfaces II*, Wiley, 1993.

Folkes, M. J., Ed., *Processing, Structure and Properties of Block Copolymers*, Elsevier Applied Science, London, 1985.

Goodman, I., Ed., *Developments in Block Copolymers-1*, Applied Science, Essex, England, 1982.

Israelachvili, J., *Intermolecular and Surface Forces*, 2nd ed., Academic, New York, 1991.

Jang, B. Z., *Advanced Polymer Composites: Principles and Applications*, ASM International, Materials Park, OH, 1994.

Klempner, D., L. H. Sperling, and L. A. Utracki, Eds., *Interpenetrating Polymer Networks*, American Chemical Society, Washington, DC, 1994.

Legge, N. R., G. Holden, and H. E. Schroeder, Eds., *Thermoplastic Elastomers: A Comprehensive Review*, Hanser, Munich, Germany, 1987.

Lewin, M., Ed., *Polymers for Advanced Technologies*, VCH, Weinheim, Germany, 1988.

Lipatov, Y. S., *Polymer Reinforcement*, ChemTec, Toronto, 1994.

Manson, J. A. and L. H. Sperling, *Polymer Blends and Composites*, Plenum, New York, 1976.

Meier, D. J., Ed., *Block Copolymers Science and Technology*, Harwood Academic, New York, 1983.

Miles, I. S. and S. Rostami, Eds., *Multicomponent Polymer Systems*, Longman Scientific & Technical, Avon, UK, 1992.

Nielsen, L. E. and R. L. Landel, *Mechanical Properties of Polymers and Composites*, 2nd ed., Marcel Dekker, New York, 1994.

Noda I., and D. N. Rubingh, Eds., *Polymer Solutions, Blends, and Interfaces*, Elsevier Science, Amsterdam, 1992.

Noshay, A. and J. E. McGrath, *Block Copolymers Overview and Critical Survey*, Academic, New York, 1977.

Olabisi, O., L. M. Robeson, and M. T. Shaw, *Polymer-Polymer Miscibility*, Academic, New York, 1979.

Paul, D. R. and L. H. Sperling, Eds., *Multicomponent Polymer Materials*, Advances in Chemistry Series No. 211, American Chemical Society, Washington, DC, 1986.

Paul, D. R. and S. Newman, Eds., *Polymer Blends*, Vol. I and II, Academic, New York, 1978.

Riew, C. K., Ed., *Rubber Toughened Plastics*, Adv. Chem. Ser. 222, American Chemical Society, Washington, DC, 1989.

Seferis, J. C. and L. Nicolais, Eds., *The Role of the Polymeric Matrix in the Processing and Structural Properties of Composite Materials*, Plenum, New York, 1983.

Solc, K., Ed., *Polymer Compatibility and Incompatibility: Principles and Practices*, Harwood Academic, Chur, Switzerland, 1982.

Utracki, L. A. *Polymer Alloys and Blends*, Hanser, New York, 1990.

Utracki, L. A. and R. A. Weiss, Eds., *Multiphase Polymers: Blends and Ionomers*, ACS Symposium Series No. 395, American Chemical Society, Washington, DC, 1989.

Vigo, T. L. and B. J. Kinzig, Eds., *Composite Applications: The Role of Matrix, Fiber, and Interface*, VCH Publishers, New York, 1992.

Walsh, D. J., J. S. Higgins, and A. Maconnachie, Eds., *Polymer Blends and Mixtures*, NATO ASI Series, Martinus Nijhoff Publishers, Dordrecht, Netherlands, 1985.

Wool, R. P., *Polymer Interfaces: Structure and Strength*, Hanser, Munich, 1995.

Wu, S., *Polymer Interface and Adhesion*, Marcel Dekker, New York, 1982.

HANDBOOKS AND ENCYCLOPEDIAS

Allen, G., Ed., *Comprehensive Polymer Science*, Pergamon, Oxford, 1989.

Brandrup, J. and E. H. Immergut, Eds, *Polymer Handbook*, 3rd ed., Wiley-Interscience, New York, 1990.

Kaplan, W. A., Ed, *Modern Plastics Encyclopedia*, McGraw-Hill, New York, 1997.

Kroschwitz, J. I., Ed., *Encyclopedia of Polymer Science and Engineering*, 2nd dd., Wiley, New York, 1985–1990.

Mark, J. E., Ed., *Physical Properties of Polymers Handbook*, AIP, Woodbury, NY, 1996.

Metanomski, W. V., Ed., *Compendium of Macromolecular Nomenclature*, IUPAC, CRC Press, Boca Raton, FL, 1991.

Salamone, J. C., Ed., *Polymeric Materials Encyclopedia*, CRC Press, Boca Raton, FL, 1996.

2

MODELING
MULTICOMPONENT
POLYMER BEHAVIOR

There are a number of basic concepts for polymer blends and composites that are more or less held in common. Among these are several that model the physical and mechanical behavior of multicomponent polymer materials: modulus, viscosity, phase continuity, and so forth. This chapter will describe these concepts and show how they can be applied.

2.1. TAKAYANAGI MODELS

All composite materials and most polymer blends, blocks, grafts, and inter-penetrating polymer networks (IPNs) are phase separated. Frequently the two phases differ in modulus, indeed, behave synergistically because they differ in modulus. Of course, there are several types of modulus; Young's modulus, shear modulus, bulk modulus, and so forth. In addition, creep, stress relaxation, and melt viscosity may vary widely from one phase to another. The mechanical and rheological behavior of combinations of such systems can be modeled by the Takayanagi models (1).

In the original Takayanagi models, two phases were indicated, rubbery, R, and plastic, P, see Figure 2.1 (2). These may be generalized in the reader's mind to read any two phases. The models progress from simple parallel and series models (Figures 2.1a and 2.1b, respectively) to quite complex, represent-ing a phase within a phase (Figures 2.1c and 2.1d). Three-dimensional models have also been constructed. These models bear a direct resemblance to arrays of springs and dashpots widely used to express viscoelastic behavior of polymers (3,4), where the Maxwell element represents a spring and dashpot in series, and the Voigt–Kelvin element represents a spring and dashpot in parallel.

In the Takayanagi models, the quantities λ and ϕ or their indicated multi-plications represent volume fractions of the materials. For example, in Figures 2.1c and 2.1d, the volume fraction of the rubber phase is given by $\lambda\phi$. The power

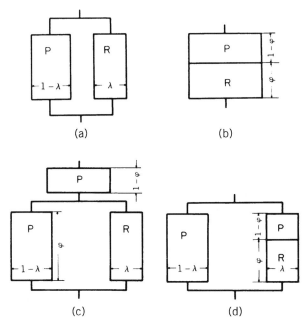

Figure 2.1. The Takayanagi models for two-phased systems. P and R represent plastic and rubber phases. The quantities λ and ϕ are used to calculate volume fractions. (a) An isostrain model; (b) isostress model; (c) and (d) combinations.

of the Takayanagi models lies in their ability to express mechanical and rheological behavior analytically. For Figure 2.1a, the horizontal bars connecting the two elements are constrained to remain parallel and horizontal, yielding an isostrain condition, $\varepsilon_P = \varepsilon_R$. The total stress on the system is the sum of the stresses,

$$\sigma = \sigma_R + \sigma_P \tag{2.1}$$

and Young's modulus for the system is easily derived,

$$\sigma = \varepsilon(E_R + E_P) \tag{2.2}$$

Noting volume fractions,

$$E = (1 - \lambda)E_P + \lambda E_R \tag{2.3}$$

For the series model (Figure 2.1b) an isostress condition exists, Then, the strains are additive,

$$\varepsilon = \varepsilon_P + \varepsilon_R \tag{2.4}$$

and Young's modulus is given by

$$E = \left(\frac{\phi}{E_R} + \frac{1 - \phi}{E_P} \right)^{-1} \tag{2.5}$$

Similarly, Young's moduli for Figures 2.1c and 2.1d are given, respectively, by

$$E = \left[\frac{\phi}{\lambda E_R + (1 - \lambda)E_P} + \frac{1 - \phi}{E_P} \right]^{-1} \tag{2.6}$$

and

$$E = \lambda \left[\frac{\phi}{E_R} + \frac{1 - \phi}{E_P} \right]^{-1} + (1 - \lambda)E_P \tag{2.7}$$

The Takayanagi models assume different morphological structures in the blend or composite. Figure 2.1a assumes that both phases are continuous in space. An example is a continuous or long fiber composite, stretched in the direction of the fibers. Figure 2.1b assumes neither phase is continuous in space. While many materials have phases continuous in space, their morphology is grossly oversimplified by Figure 2.1a. One must go to extremes to make a morphology with neither phases continuous, such as making a multilayered film, and stretching it perpendicular to the surfaces. More complex morphologies are illustrated by Figures 2.1c and 2.1d. In both cases a continuous plastic phase and a discontinuous rubber phase are assumed. The inverse morphology is easy to imagine, with the plastic phase discontinuous. As will be illustrated, these models can also be applied to composite materials.

2.1.1. Example Takayanagi Calculations with Polymer Blends

Assume a blend of a plastic and an elastomer (rubber), a system originally envisioned by Takayanagi. A typical Young's modulus for a glassy, amorphous plastic such as polystyrene or poly(methyl methacrylate) is close to 3×10^9 Pa. Young's modulus of a typical elastomer such as polybutadiene or poly(dimethyl siloxane) is close to 2×10^6 Pa. An ordinary rubber band has a modulus close to this value. Suppose polystyrene and polybutadiene are blended together in 50/50 proportions. What ranges of Young's moduli might be expected?

According to Figure 2.1a and equation (2.3),

$$E = 0.50 \times 3 \times 10^9 \,\text{Pa} + 0.50 \times 2 \times 10^6 \,\text{Pa}$$

$$E \simeq 1.5 \times 10^9 \,\text{Pa}$$

Thus, a Young's modulus of approximately half of the plastic modulus is found. Figure 2.1a is an upper bound model, that is, no modulus higher than derived from equation (2.3) is attainable. (However, other models might predict different upper bound moduli.) For such materials as rubber-toughened plastics, the rubber modulus term in equation (2.3) can be dropped because it contributes a value two orders of magnitude lower than the plastic term. Since for most rubber-toughened plastics the rubber content is approximately 10–15% rather than 50% in the example, the contribution of the rubber modulus term is even smaller.

However, accorded to Figure 2.1b, and equation (2.5), Young's modulus for the above system is given by

$$E = (0.50/2 \times 10^6 \, \text{Pa} + 0.50/3 \times 10^9 \, \text{Pa})^{-1}$$

$$E \cong 4 \times 10^6 \, \text{Pa}$$

which is seen as a value about twice that of the rubber modulus. Thus, Figure 2.1b is a lower bound model because no modulus lower than that predicted by equation (2.5) can be found. Of course, intermediate moduli may exist and, in fact, is the usual case. These may be derived via Figures 2.1c and 2.1d and their corresponding equations. The same homopolymer data will be employed, using equation (2.6).

Since the rubber volume fraction is 0.5,

$$\phi\lambda = 0.5$$

However, ϕ and λ need specific numerical values to solve equation (2.6). Assume $\phi = \lambda$; then $\phi = \lambda = \sqrt{0.5} = 0.707$.

Plugging into equation (2.6) similar to the above examples yields

$$E = 1.11 \times 10^9 \, \text{Pa}$$

Thus, equation (2.6) predicts a value for E slightly lower than equation (2.3). Note that it was necessary to assume that $\phi = \lambda$ or some other relationship to obtain a numerical answer in this case.

2.1.2. Example Takayanagi Calculations for Composites

Let us assume 30% by volume continuous parallel glass fibers in a plastic matrix. As before, Young's modulus of the plastic will be assumed to be $3 \times 10^9 \, \text{Pa}$. Glass fibers have Young's moduli in the range of about $85 \times 10^9 \, \text{Pa}$ (S-glass fibers assumed). Note that the glass has a Young's modulus some 30 times higher than the polymer. Two cases may be assumed. First, that Young's modulus is being measured in the fiber direction. In this case, both phases are actually continuous. Figure 2.1a and equation (2.3) apply.

$$E = 0.30 \times 85 \times 10^9 \, \text{Pa} + 0.70 \times 3 \times 10^9 \, \text{Pa}$$

$$E \cong 27.6 \times 10^9 \, \text{Pa}$$

which is approximately 10 times stiffer than the plastic alone.

Alternately, Young's modulus may be calculated in the transverse direction. Note that the fibers are now discontinuous. Let us assume Figure 2.1d and equation (2.7) apply. The geometry of a circular cross section of the cylindrical fibers is isometric, so $\phi = \lambda$. Again, $\phi\lambda = 0.3$, therefore, $\phi = \lambda = 0.548$.

$$E = 0.548(0.548/85 \times 10^9 \, \text{Pa} + 0.452/3 \times 10^9 \, \text{Pa})^{-1} + 0.452 \times 3 \times 10^9 \, \text{Pa}$$

$$E = 4.85 \times 10^9 \, \text{Pa}$$

As might be expected, Young's modulus is only slightly increased. If the nonpolymer phase is discontinuous in space, as with glass spheres, calcium carbonate, and so forth, lower bound modulus increases will be obtained, unless the concentration of the nonpolymer phase is so high that the particles are touching one another.

Of course, other moduli may be calculated as well. More complex properties may also be treated. For example, elastomeric Young's moduli obey equation (1.24), where the modulus is proportional to the temperature. Similarly, the Williams–Landel–Ferry (WLF) equation may be inserted into the model. The various equations for springs and dashpots, alluded to above, may also be inserted as appropriate, to describe the time dependence of one or both phases.

2.2. KERNER EQUATION AND OTHER RELATIONSHIPS

There are a number of algebraic relationships in the literature from which one can calculate the modulus or other properties of polymer blends and composites. While many of these models are quite old, they are still valid and provide not only numerical solutions to problems but provide insight as to the organization of multiphase matter.

Perhaps the most well known of these relationships is the Kerner equation (5):

$$\frac{G_c}{G_p} = \frac{G_f v_f / [(7 - 5\nu)G_p + (8 - 10\nu)G_f] + v_p / [15(1 - \nu)]}{G_p v_f / [7 - 5\nu)G_p + (8 - 10\nu)G_f] + v_p / [15(1 - \nu)]} \tag{2.8}$$

where G_c is the shear modulus of the composite, G_p and G_f represent the shear moduli of the polymer and the filler, continuous and discontinuous phases, respectively, ν is Poisson's ratio of the polymer, v_f is the volume fraction of the filler, and v_p is the volume fraction of the polymer. In most applications, the polymer is in its plastic state, and the filler is spheroidal in shape. For most plastics, Poisson's ratio is in the range of 0.30–0.35. The Kerner equation predicts a lower bound increase in the modulus, assuming that the filler has a higher modulus than the polymer.

The Guth–Smallwood (6,7) equation is widely used to predict the increase in modulus in carbon-black-reinforced elastomers:

$$\frac{G_c}{G_p} = 1 + 2.5v_f + 14.1v_f^2 \tag{2.9}$$

The coefficient 2.5 arises from the Einstein viscosity relationship, the origin of the Guth–Smallwood equation. The last term on the right is important for more concentrated dispersions of fillers, taking into account their interaction. This equation also arises from a consideration of a dispersed filler in a polymeric matrix and is also of the lower bound modulus increase type.

There are also a series of models helpful in evaluating features such as phase continuity and phase inversion. They are useful across the entire composition range of polymer blends.

The series–parallel model (8) establishes an upper and lower bound modulus of the Takayanagi type:

$$[(v_1/G_1) + (v_2/G_2)]^{-1} \leq G_c \leq v_1 G_1 + v_2 G_2 \tag{2.10}$$

The Hashin–Shtrikman model (9) sets rigorous bounds without any specification of morphology except that one phase is continuous and the other discontinuous:

$$G_1 \frac{v_1 G_1 + (\alpha_1 + v_2)G_2}{(1 + \alpha_2 v_2)G_1 + \alpha_1 v_1 G_2} < G_c < G_2 \frac{v_2 G_2 + (\alpha_2 + v_1)G_1}{(1 + \alpha_1 v_1)G_2 + \alpha_2 v_2 G_1} \tag{2.11}$$

The Budiansky model (10,11), a self-consistent approximate model, considers a macroscopically homogeneous and isotropic composite or blend. The model illustrates a phase inversion at the midrange composition:

$$\frac{v_1(G_1 - G_c)}{G_c + \alpha G_1} + \frac{v_2(G_2 - G_c)}{G_c - \alpha G_2} = 0 \tag{2.12}$$

where v_1 and v_2 represent the volume fractions of the two phases. A dual-phase continuity model was derived by Davies (12), showing the moduli raised to the one-fifth power:

$$G_c^{1/5} = v_1 G_1^{1/5} + v_2 G_2^{1/5} \tag{2.13}$$

Coran and Patel (13) derived an equation that represents a phenomenological interplay between the series and parallel models and is characterized as a one-parameter model:

$$G_c = v_2^n(nv_1 + 1)(G_H - H_L) + G_L \tag{2.14}$$

where $G_H = v_1G_1 + v_2G_2$ and $G_L = [(v_1/G_1) + (v_2/G_2)]$, and G_c, G_1, G_2 represent the modulus of the composite and the constituents 1 and 2, where $G_2 > G_1$, and v_1 and v_2 are the volume fractions of the constituents 1 and 2, respectively, and $\alpha = 2(4 - 5\nu)/(7 - 5\nu)$, a function of Poisson's ratio. In these equations, the terms *blend* and *composite* are essentially synonymous.

For comparison of the behavior of several of these models, the modulus–composition behavior of a poly(*n*-butyl acrylate)–polystyrene IPN is illustrated in Figure 2.2 (14). The IPN undergoes a phase inversion near its midpoint of composition, going from an elastomer to a plastic as the polystyrene component is increased. The Budiansky model fits best, showing the phase inversion. The Coran–Patel model fits fair, depending on the numerical value of the quantity *n*.

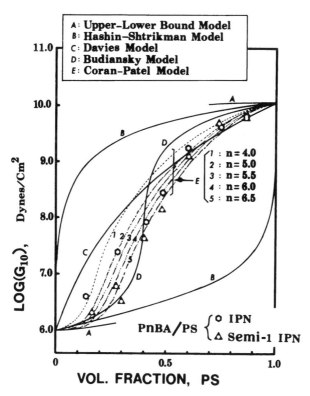

Figure 2.2. Modulus–composition behavior of poly(*n*-butyl acrylate)–polystyrene IPNs and semi-I IPNs at 25°C. The quantity *n* represents the fitting parameter of the Coran–Patel model.

2.3. PHASE CONTINUITY AND INVERSION

2.3.1. Basic Morphologies in Polymer Blends

2.3.1.1. Dual Phase Continuity In phase-separated systems, the physical arrangement of the phase domains may be simple or complex. One of the simplest involves spheres of phase II dispersed in phase I or vice versa. The notation of phase I and phase II rather than polymer I and polymer II is used because some degree of mixing of the two polymers always exists. Polymer blends exhibit a number of morphologies, many of them exhibiting some degree of dual-phase continuity; see Figure 2.3 (15). Some of the simpler, or better defined morphologies, are as follows (16):

1. *Spheres of Phase I in Phase II.* This may follow from nucleation and growth phase separation kinetics in blends or from block copolymers having much longer polymer II chains than polymer I chains (17,18).
2. *Cylinders of Phase I in Phase II.* This is also a well-known block copolymer morphology, where polymer I contains roughly 25–40% of the volume (19). Several variations on this theme might exist: Simple parallel cylinders (well-known for long fiber composites), randomly placed cylinders (typical of short fiber composites), or layers of cylinders with different orientations (typical of large-body layup composites). Alternately, high shear flows can turn polymer blend spherical domains into elongated, cylindrical structures.
3. In the case of polymer blends, interconnected cylinders typically arise through spinodal decomposition phase separation kinetics (20).

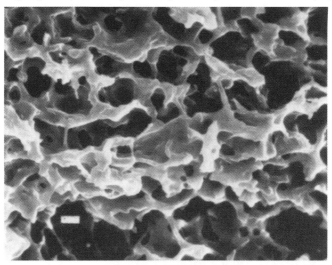

Figure 2.3. Scanning electron micrograph of a 70/30 blend of EPM and PP. The EPM phase was extracted with heptane prior to photography, leaving the PP. Note the suggestion of dual-phase continuity. Bar equals 1 μm

4. Ordered bicontinuous, double-diamond interconnected cylinders of phase I in phase II, as have documented for star block copolymers (21–23).

5. Alternating lamellae of the two phases, well-known for diblock copolymers of nearly equal chain length. Block copolymer morphologies will be treated further in Chapter 9. Certain composites have alternating lamellae. Some polymer films have very many multilayers, contributing either permeability resistance or toughness.

6. Onion rings of two phases have been noted in AB-crosslinked polymers (24). There are two possible kinds of onion rings, which have never been fully distinguished in the laboratory: (a) true rings, forming nested spherical shells, and (b) spirals, where the two phases spiral outwards, to form pseudo-spherical shells. Although spheres have been found, it is also possible to find the cylindrical analogs. Of course, these are all special modifications of the alternating lamellae morphology.

7. *Phase Within a Phase Within a Phase.* High-impact polystyrenes and acrylonitrite–butadine–styrene (ABS) materials, that exhibit a secondary phase separation, have this morphology (25).

Of course, multiples of these morphologies might occur simultaneously. For example, spheres and cylinders might appear in the same sample. The morphologies may be the result of the thermodynamics of phase separation, kinetics of phase separation, method of mixing such as shear rate, or combinations of these.

There are several morphologies that have not yet been observed in polymer blends that might be found in the future (16):

8. *Nested Tepee Cylinders.* Here the cylinder density in the dispersed phase will vary significantly.

9. *Catenated Ring Morphologies.* A number of parallel chains of catenated ring morphologies are envisioned. Here, it is theoretically possible for reasonably equal spacings of the two morphologies to exist, as needed for block copolymers.

10. *Interlocking Helices of Various Kinds.*

11. *Single-Diamond Morphology.* Note that the double diamond has already been found.

The reader will observe that all of the above except simple spheres dispersed in a continuous medium have some degree of dual phase continuity, albeit that in most of these materials it may be said that one phase is more continuous than another.

In a material exhibiting dual-phase continuity, a Maxwell demon can, in principle, traverse the entire sample while staying in either phase. Returning to Figure 2.3, the dual-phase continuity can be modeled as an interlocking three-dimensional network of phases. Surely, our Maxwell demon can traverse the entire sample in either phase in this morphology. An unknown factor, however, is the degree of tortuosity of the individual phases. In a random walk, how much further, on average, must the Maxwell demon travel in phase I than in phase II to traverse the sample?

2.3.1.2. Basic Rules of Phase Continuity The above known and still unknown morphologies can be generalized by the formulation of some basic rules:

1. To be continuous, a phase must traverse the macroscopic material in at least one direction.
2. A continuous phase may traverse the material in one, two, or three dimensions.
3. The directions that a multidimensional continuous phase traverses may be not connected, connected in two dimensions, or connected in all three dimensions. (It is possible that a phase exhibiting continuity in two or even three dimensions may be composed of elements that have no continuity between the directions of travel. An example of this latter might be cylinders running at right angles to laminates of a chicken wire structure.)
4. In general, the more dimensions a phase traverses, and the more connected it is, the greater its continuity is likely to be.
5. While one phase may be discontinuous, the other phase must be continuous.
6. Both phases may have discontinuous portions, as in phase-within-a-phase-within-a-phase morphologies.
7. Both phases may exhibit equal continuity. Two examples include: (a) alternating lamellar structures and (b) a version of the single-diamond morphology.

2.3.2. Requirements for Phase Inversion

Phase inversion is defined as a phenomenon in which the more continuous phase becomes the less continuous phase, and vice versa. Usually, it is brought about by shearing conditions during a polymerization or crosslinking stage in a polymer blend or graft. Important examples of phase inversion occur during the polymerization of high-impact polystyrene, where first the polybutadiene-rich phase is continuous, then the polystyrene-rich phase becomes continuous (see Section 8.2), and the polypropylene–ethylene-propylene-diene monomer (EPDM) system, where the EPDM phase is first continuous, then the system is brought to dual-phase continuity (see Section 10.12.1).

It was known for many years that the larger volume phase tended to be more continuous in space. It was also known that the lower viscosity phase tended to be more continuous in space. These concepts were formulated quantitatively by Paul and Barlow (26) and Jordhamo et al. (27):

$$\frac{V_{\mathrm{I}}}{V_{\mathrm{II}}} \times \frac{\eta_{\mathrm{II}}}{\eta_{\mathrm{I}}} = X \qquad (2.15)$$

where if

$X > 1$, phase I continuous

$X \approx 1$, dual phase continuity or phase inversion

$X < 1$, phase II continuous

In equation (2.15), V represents the volume fraction and η is the melt viscosity of phase I or phase II, as per subscript.

Equation (2.15) is true in the limiting case for zero shear. (Note, however, that phase inversions of the type being described here only occur under shearing conditions.) Utracki (20) has proposed somewhat more complex relationships for finite shear rates.

2.3.3. Phase Inversion Experimental Results

The phase continuity of a number of polymer blends was reviewed by Jordhamo et al. (27); see Figure 2.4. The results generally support equation (2.15). It was

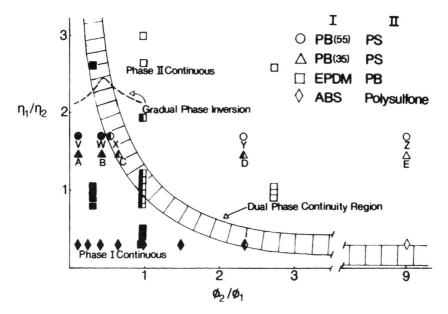

Figure 2.4. Phase continuity diagram for blends and IPNs undergoing phase inversion during polymerization or under shear. Filled points, phase I continuous, open points, phase II continuous. Half-filled points, dual-phase continuity. ϕ is the volume fraction.

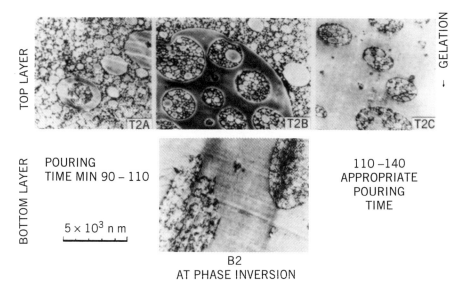

Figure 2.5. Castor oil–polyester SIN phase inversion. Osmium tetroxide staining of the castor-oil-rich phase.

noted that dual-phase continuity and/or phase inversion took place over about a factor of 4 in the data, that is, equation (2.15) could range from about 0.5 to about 2 and the sample still exhibit some degree of dual-phase continuity.

The morphological changes during phase inversion were observed by Devia et al.; see Figure 2.5 (28). Here, a simultaneous interpenetrating network was being formed by the simultaneous polymerization of a castor oil–polyester (stained) and polystyrene. Initially, the oil phase was continuous, but then at the end the polystyrene phase is continuous. The reader should note the phase-within-a-phase-within-a-phase morphology after phase inversion, reminiscent of high-impact polystyrene (Chapter 8). Since both the phase volumes and the phase viscosities were changing with time, the exact path during the phase inversion is not known.

2.3.3.1. Example Calculations of Phase Inversion Suppose we mix a blend having the characteristics of Table 2.1. As a first approximation, the phases will be considered totally immiscible, so that neither the viscosity nor the volume fractions of the phases changes during the mixing process. Assuming a modest

Table 2.1 Polymer blend characteristics

Component	Component Viscosity, Poise	Volume Fraction
Polybutadiene	3×10^5 P	0.4
Polystyrene	5×10^6 P	0.6

shear rate, with polybutadiene being phase I:

$$(0.4/0.6) \times (5 \times 10^6 \, P/3 \times 10^5 \, P) = 1.11 \times 10^1$$

Thus, the polybutadiene will be the continuous phase.

Somewhat more complex calculations can also be undertaken using equation (2.15). For example, assume a new commercial ternary blend of polypropylene (PP), crosslinkable ethylene–propylene rubber (EPDM), and polyethylene (PE). Because of the low price of PE compared to the other two, especially EPDM, it is important to have PE replace EPDM as far as possible. The desired morphology is to have drops of EPDM within the PP. The PE is to be dispersed in the EPDM, to create a phase within a phase within a phase, all three phases being different. If the ratio of PP/EPDM/PE is 75/15/10, and the melt viscosity of PP is $4 \times 10^5 \, P$, what possible values of viscosity can the other two components have? Conditions are assumed to be in a sheared melt at 200°C.

Again, for simplicity, we assume that there is no mixing between the components. Solution:

$$\eta_{PP} = 4 \times 10^5 \, P \ @ \ 200°C \qquad \eta_{EPDM} = ? \qquad \eta_{PE} = ?$$

First, it must be assumed that η_{EPDM} represents the EPDM-rich phase, PE included. (A minor dispersed phase does not make a major contribution to the melt viscosity.) Then, with PP being phase I,

$$(\eta_{EPDM}/4 \times 10^5 \, P) \times (75/25) > 1$$

$$\eta_{EPDM} > 1.33 \times 10^5 \, P$$

Then, the PE must be dispersed in the EPDM. With EPDM now phase I:

$$(\eta_{PE}/1.33 \times 10^5 \, P) \times (15/10) > 1$$

$$\eta_{PE} > 8.9 \times 10^4 \, P$$

Thus, both the EPDM and PE components must have viscosities higher than the indicated quantities. The value of these relationships is that it gives the scientist or engineer analytical equations with which to estimate numerical quantities to obtain the desired morphology.

2.3.3.2. Blending with Commodity Polymers

Polyethylene, polypropylene, and EPDM are all polyolefin polymers. Several of these polyolefins are in the *commodity polymer* class, so-called because of their relatively huge tonnages in commerce. In general, many important commercial examples exist of the blending of two (or more) polyolefins. These include blends of low-density polyethylene (LDPE), high-density polyethylene (HDPE), ultra-high-molecular-weight polyethylene (UHMWPE), linear low-density polyethylene, polybutene-1, ethylene proplylene monomer (copolymer) (EPM), and its cross-linkable counterpart, ethylene propylene diene monomer (copolymer) (EPDM), and PP. The latter exists in both the isotactic form and the atactic form. At room temperature, all but the EPM, EPDM, and the atactic PP are semicrystalline. Of Course, if EPM or EPDM have long ethylene sequences, some crystallinity will develop. These various blends are used for improved processibility, to increase tear strength, to reduce permeability (particularly oxygen and water), for rubber toughening, and a host of other reasons, including cost.

2.3.4. Morphology of Polymer Composites

2.3.4.1. Glass Fibers and Spheres and Mica Platelets The structure of the composite phase may take many shapes, but the usual assumption is that the nonpolymer (filler or reinforcing) phase does not deform during the formation stage. Common shapes for the nonpolymer phase include spheres or spheroidal shapes (such as glass beads or calcium carbonate), cylindrical (such as glass or graphite fibers), or platelets (such as mica). The nonpolymer phase may be oriented in various ways. For example, long glass fibers may be parallel, or in the form of a cloth (29). The use of ribbon composites to make windings is useful, particularly for large, hollow structures. By winding back and forth, alternate layers will have different orientations. Short glass fibers are also in common use. Sometimes the short glass fibers appears in the form of tufts dispersed in the polymer. Mica platelets tend to take parallel positions relative to one another, especially at high volume loadings, when the platelets begin to touch each other. The filler particles or fibers may touch each other, especially desirable for semiconductor applications or be nontouching, as in high-performance glass fiber–epoxy materials. In the latter case, the fibers might damage each other under high strain conditions.

2.3.4.2. Polymer-Impregnated Concretes The polymer may be dispersed in the nonpolymer phase as well. The literature recognizes three kinds of mixtures of polymers and concretes, for example (30), (a) polymer-impregnated concrete (PIC) in which fully cured (hydrated) cement concrete is impregnated with a monomer and subsequently polymerized in situ. A typical impregnating monomer is methyl methacrylate, and the product can be used on concrete highways to protect both against freeze/thaw cycles of water and against salt corrosion of the reinforcing rods. (b) Polymer cement concrete (PCC) in which either a monomer or a polymer is added to the fresh, uncured concrete while mixing, and subsequently simultaneously cured and polymerized. While most monomers interfere with the concrete curing, an important commercial product is latex-impregnated concretes, where poly(vinyl acetate) or similar latexes are incorporated into the concrete mix. This toughens the product, making it useful for repair projects such as concrete step edges, which perennially fail. (c) The third type is polymer concrete (PC) in which the entire binder is polymeric in replacement of the cement, which is recognizable as a composite material containing stones and the like.

2.3.4.3. Foams   One case where the nonpolymer phase is clearly deformable involves foams. Here, the gas phase is dispersed in a polymer phase. Two cases must be distinguished: The foam bubble cells may be closed or open. In a closed-cell foam, each gas bubble is isolated. In an open-cell foam, the gas phase forms a continuous phase, as in air filter systems. Seat cushions are often partly open cell and partly closed cell, providing the most comfortable elastic compressive behavior.

2.3.4.4. Wood   The structure of wood provides an example of a complex morphology, natures's best. Here, cells of cellulose are dispersed in lignin, a crosslinked polymer resembling a phenolic. The cellulose itself is formed on oriented fibrils, which wind around a porous (hollow) open space, called the lumen. During growth, the lumina contain the organic structure of life. In mature wood, the lumina are interconnected for transport of water and nutrients from the roots to the leaves, and vice versa. These lumina can be filled with polymer for the manufacture of very wear resistant structures, such as knife handles and gymnasium flooring. In this case, the wood serves as the filled phase.

Polymer-impregnated wood is sometimes called a wood–polymer composite (WPC) (31). The monomer, frequently methyl methacrylate, is forced into evacuated wood, and polymerized using heat or irradiation. In general, compressive strength, impact resistance, hardness, and abrasion resistance can be improved. Clearly, such materials are more water and rot resistant.

2.4. POLYMER–POLYMER MISCIBILITY

Up until the early 1970s, it was thought that most polymer blends were substantially immiscible. Further, if there were any mixing taking place, the kinetics of interdiffusion were thought to be so slow as to discourage measurement. Today, we know that polymers and/or their blends must only be above T_g for interdiffusion to be fast enough for measurement. Beginning with the works of McMaster (32,33), Patterson and Robard (34), and others (35–38), well-defined phase diagrams for polymer blends were worked out. To the surprise of most workers in the field, lower critical solution temperatures (LCST) were found in nearly all cases of a blend of two polymers; see Figure 2.6 (39). Thus, early workers attempting to increase miscibility by raising the temperature were going the wrong way, doomed to failure.

Of course, thermodynamic theories all point to some extent of mixing of two components, polymeric or not. This mixing may be great or small, but is always finite.

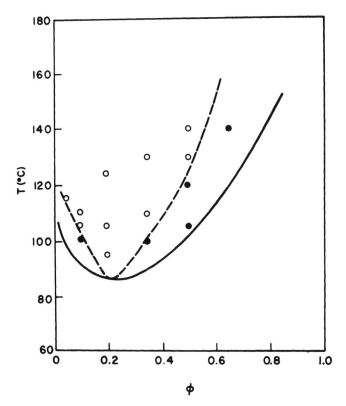

Figure 2.6. Phase separation by spinodal decomposition (open circles) and nucleation and growth (solid circles), as observed under an optical microscope. Solid line, binodal, and dashed line, spinodal.

2.4.1. Phase Diagrams

In general, the shapes of phase diagrams may take many forms. The temperature–volume fraction of several common types of phase diagrams are illustrated in Figure 2.7 (40). Figure 2.7a illustrates total miscibility; Figure 2.7b an upper critical solution temperature (UCST), that is, a temperature above which all compositions are miscible (Figure 2.7c), an LCST; Figure 2.7e illustrates both UCST and LCST conditions; and Figure 2.7d and 2.7f show two cases where the mixture exhibits regions of partial immiscibility.

Phase diagrams are useful for analyzing polymer blend compositions. Consider Figure 2.8, a mixture of polymer I and polymer II, prepared with volume fraction x of polymer II at temperature T. Then, the relative volumes or volume fractions of the two phases can be calculated. Consider that the system has a total volume V. The volumes of the two phases are V_I and V_{II}, and these phases contain volume fractions of polymer II of x_I and x_{II}. Assuming that the

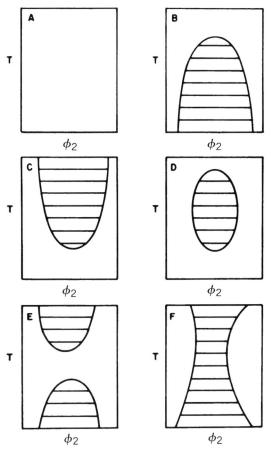

Figure 2.7. Schematics of several types of liquid–liquid temperature–composition phase diagrams. Shaded areas indicate phase-separated regions. $\phi_2 = v_{\mathrm{II}}$.

volume changes but little on mixing,

$$V = V_1 + V_{\mathrm{II}} \tag{2.16}$$

Most polymer phase diagrams are reported in volumes or volume fractions. Weights and weight fractions can easily be substituted. For polymer II in both phases,

$$xV = x_{\mathrm{I}} V_{\mathrm{I}} + x_{\mathrm{II}} V_{\mathrm{II}} \tag{2.17}$$

Substitution of V from equation (2.16) into equation (2.17) yields

$$x(V_{\mathrm{I}} + V_{\mathrm{II}}) = x_{\mathrm{I}} V_{\mathrm{I}} + x_{\mathrm{II}} V_{\mathrm{II}} \tag{2.18}$$

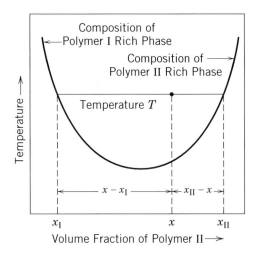

Figure 2.8. Schematic of a phase diagram, useful for calculating phase volumes.

which rearranges to

$$V_I(x - x_I) = V_{II}(x_{II} - x) \tag{2.19}$$

and thus,

$$\frac{V_I}{V_{II}} = \frac{(x_{II} - x)}{(x - x_I)} \tag{2.20}$$

Interestingly, the volumes of the two phases are in the proportion of the two line lengths shown in Figure 2.8.

2.4.2. Model Calculation of Phase Volume Fraction

Using Figure 2.6, let us assume a mix of 60% polystyrene and 40% poly(vinyl methyl ether) (PVHE) at 100°C, and determine the volume fraction of each phase. Then, $x = 0.40$ for polymer II. Using the binodal curve (the outer curve), $x_I = 0.04$ and $x_{II} = 0.48$ from vertical constructions. From equation (2.20),

$$V_I/V_{II} = (0.48 - 0.40)/(0.40 - 0.04)$$

or

$$V_I = 0.22V_{II}$$

Assuming unit volume in equation (2.16),

$$V_I + V_{II} = 1$$

Then, V_I and V_{II} become volume fractions. Accordingly,

$$0.22V_{II} + V_{II} = 1$$

and

$$V_{II} = 0.82$$
$$V_I = 0.18$$

Of course, the extreme case is when x approaches V_{II}, then V_{II} approaches unity.

The volume fractions of each phase are important in determining phase continuity relationships as well as mechanical behavior. For the latter, note the applicability of the Takayanagi models for modulus calculations, for example. An important further assumption, however, is that the composition of each phase does not change appreciably on (rapid) cooling from the melt. Note that under these conditions, the system might not be quite in thermodynamic equilibrium, a very common case.

2.4.3. Optical Clarity and Glass Transition Miscibility Criteria

Phase diagrams may be determined by many different kinds of experiments. Among the more important are optical clarity, appearance of phase separation in the microscope, and the glass transition behavior, among others. However, in many cases, it is important to establish the multiphase nature of a system, without the labor of establishing a phase diagram. Indeed, sometimes the LCST (or UCST) may be out of experimental reach.

The amount of light scattered depends on the square of the difference in refractive index in the two phases. Thus, if a polymer blend is hazy, cloudy, or milky white to the eye, it is usually accepted as being immiscible. (A possible exception is when one or both of the polymers is crystalline.)

The amount of light scattered also depends on the sixth power of the size of the domains, for domains much smaller than the wavelength of visible light. Thus reasonably large domains will ordinarily produce a cloudy or milk white appearance. Cases where these methods fail are when the refractive indices of the two polymer is substantially identical, and/or the domains are very small (less than about 20 nm). The sample will then appear clear to the eye, but the amount of light scattered may still be detected instrumentally.

Phase domains or particles much smaller than the wavelength of light scatter as the inverse fourth power of the wavelength. Thus, backscattered light may appear bluish, while transmitted light may appear pinkish, as in the blue of the sky and of sunsets. Phases or particles approximately the same size as the light, $\sim$5000 Å, exhibit a characteristic white appearance. Larger domains or particles in a plastic may give a chalk white appearance, the wavelength dependence diminishing toward zero. If the particles are uniform-sized spheres larger than the wavelength of the light, characteristic lobes in the scattering pattern may produce brilliant colors at specific angles, like the rainbow.

In the qualitative sense, of course, modulus is a measure of the stiffness of the material. By ordinary feeling of samples, it is easy to determine if the sample is glassy, rubbery, or leathery. For many polymer blends, blocks, grafts, or IPNs composed of a glassy and a rubbery polymer a leathery plateau exists, with moduli in the 10^8–10^9 dyn/cm^2 range. Sometimes, if a glass transition of either polymer is near room temperature, the sample can be made to soften appreciably even by warming between the hands. Thus, a leathery polymer with the upper glass transition near 25°C may become rubbery on such gentle warming.

Glass transitions were already discussed in Section 1.5. If the polymer blend exhibits two glass transitions, it is usually considered to be immiscible. However, the two glass transitions might nearly be the same, as is the case with polystyrene and poly(methyl methacrylate).

Both clarity and glass transition methods have been used traditionally. Together with other experiments such as refractive index determination of the homopolymers, glass transition measurement of the homopolymers, microscopy, and so forth, these methods are usually reliable. It must be emphasized that clarity and/or one glass transition temperature are of the *necessary* condition for miscibility, not the *sufficient* condition. A multicomponent polymer blend must be clear and exhibit one glass transition, and all other evidence for miscibility, but these alone are not sufficient. Again, the best method for determining phase relationships always involves phase diagrams.

2.4.4. Immiscible and Miscible Polymer Pairs

Very many polymer blends, grafts, blocks, and IPNs have been prepared to date. While temperature, composition, molecular weight, and so forth, all play a role in miscibility, many important polymer pairs are classified as being either immiscible or miscible, see Tables 2.2 and 2.3, respectively. To be considered miscible, a polymer pair usually must exhibit a range of compositions and temperatures where total miscibility occurs. Otherwise, the polymer pair are usually considered immiscible.

Most of the pairs shown in both tables are presently commercial. In Table 2.2, polymer I is typically a plastic and polymer II an elastomer. An example is polybutadiene-toughened polystyrene, to make it more impact resistant. However, there are many more kinds of blends besides plastic–rubber. There are many examples of both plastic–plastic and rubber–rubber blends. For a number of years, poly(ethylene terephthalate) was used to reduce flat spotting in nylon-6 tire cords, a blend of two semicrystalline plastics. As an example of a rubber–rubber blend, polybutadiene or natural rubber, by reason of their ability to crystallize on elongation, is used to increase tear resistance of the more inexpensive styrene–butadiene rubber (SBR) elastomers in automotive tires. One of the more interesting miscible polymer blends is polystyrene and poly(2,6-dimethylphenylene oxide) (PPO). This blend is usually further mixed

Table 2.2 Selected immiscible polymer pairs

Polymer I	Polymer II
Polystyrene	Polybutadiene
Polystyrene	Poly(methyl methacrylate)
Polystyrene	Poly(dimethyl siloxane)
Polypropylene	EPDM
Polypropylene	Polyethylene
Epoxies	CTBN
Polycarbonate	ABS
Nylon 6,6	EPDM
Nylon 6	Poly(ethylene terephthalate)
Polybutadiene	SBR

^{a} Abbreviations: EPDM, ethylene–propylene–diene monomer (a copolymer elastomer); CTBN, carboxy-terminated butadiene nitrile (a copolymer elastomer); ABS, acrylonitrile–butadiene–styrene (a complex latex structure used both independently, and to toughen other polymers); SBR, styrene–butadiene rubber (an elastomer).

with high-impact polystyrene or part or all of the polystyrene replaced by high-impact polystyrene, to make an extremely tough plastic, as will be described in Chapter 8. Poly(vinylidene fluoride) contributes both light stability and gasoline resistance to poly(methyl methacrylate), making it useful for coating automotive parts.

Again, polymer blend miscibility must be distinguished from compatibility. The latter term is used to describe polymer pairs with acceptable or desired engineering behavior and should not convey the notion of miscibility. Many of the pairs shown in Tables 2.2 and 2.3 will be the subject of repeated discussion throughout this monograph.

2.4.5. Methods of Causing Phase Separation

In Section 2.4.1, the concept of a phase diagram was introduced, along with the experimental finding that most polymer I–polymer II blends, grafts, blocks, and

Table 2.3 Selected miscible polymer pairs

Polymer I	Polymer II
Polystyrene	Poly (2,6-dimethylphenylene oxide)
Polystyrene	Poly(vinyl methyl ether)
Poly(vinyl chloride)	Poly(butylene terephthalate)
Poly(methyl methacrylate)	Poly(vinylidene fluoride)
Poly(ethylene oxide)	Poly(acrylic acid)

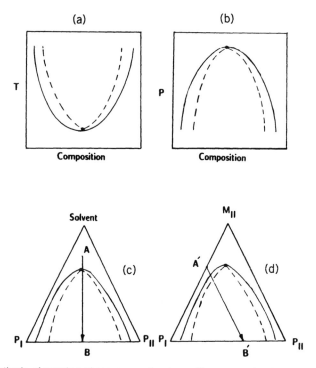

Figure 2.9. Methods of causing phase separation in multicomponent polymer materials. The solid lines represent the binodal, while the dashed lines the spinodals.

IPNs are immiscibility to a greater or lesser extent. Figure 2.6 showed a lower critical solution temperature, for example.

However, simple blending of two polymers is not the only way that phase separation can be brought about or studied; see Figure 2.9 (3). While temperature–composition diagrams are the most common, thermodynamics teaches us that pressure is also a variable. Pressure is important because all materials, including polymers, have molecular-sized holes and attractive and repulsive forces. If the pressure is increased, generally the system will become more dense, shifting the phase diagram significantly. Figure 2.9b shows that increasing the pressure in a polymer blend generally increases miscibility, the system exhibiting an upper critical solution pressure. Research on polyurethane-poly(methyl methacrylate) simultaneous interpenetrating networks synthesized at 140°C under 20,000 bars of pressure by Lee and Kim (41) showed much reduced domain size, one broad glass transition, and became optically transparent. Those synthesized at atmospheric pressure exhibited two distinct glass transitions. Poly(vinyl acetate)-*blend*-poly(methyl methacrylate) became miscible when synthesized at 10,000 bars at 300°C (42). Both of these studies depended on cooling the simultaneous interpenetrating network (SIN) or blend to room temperature before reptation diffusion motions caused demixing. Thus, the materials may not actually be thermodynamically miscible at one atmosphere and 25°C.

Figure 2.9c shows the effect of removing a solvent, perhaps by evaporation. At point *A*, the mix is homogeneous. On drying, phase separation ensues. By drying slowly enough, phase equilibrium can be achieved. At point *B*, all of the solvent is removed, leaving the dry blend. A point of caution for experimentalists trying to make polymer blend films by this method: Many such materials will form layers rather than microscopic or submicroscopic domain morphologies.

A very important case involves the polymerization of monomer II in the presence of polymer I (Figure 2.9d). The mixture is homogeneous at A', where polymer I may be considered dissolved in monomer II. After partial polymerization, phase separation sets in, the blend being fully synthesized at B'. As will be discussed in Sections 8.2.1 and 8.3.1, in reality such syntheses are rarely carried out under equilibrium conditions. Sections 8.2.1 and 8.3.1 outline the syntheses of high-impact polystyrene (HIPS) and ABS polymers, respectively, important commercial materials. In both of these cases, polymerization is going on simultaneously with phase separation processes.

2.4.6. Statistical Thermodynamics of Mixing

The general problem that thermodynamic theories of mixing and phase separation need to solve is simple: They should be able to predict the coexistence curves demarking miscible regions of the phase diagram from immiscible regions. Along the coexistence curve, the free energy of mixing is zero by definition. Especially interesting are the upper and lower critical temperatures, defining the highest and lowest temperatures of phase separation, respectively.

2.4.6.1. Types of Thermodynamic Theories   There are three basic types of thermodynamic theories of mixing: the classical theory (43), the statistical theory (44–46), and the equation of state theory (47–49). Each has many particular approaches by individual authors. Greatly oversimplifying the theories, the classical theory makes few assumptions about the size, shape, and relative positions of the molecules, relying on such concepts as heats of mixing and phase transitions. The great general equation for mixing of two components arising out of the classical theory can be written

$$\Delta G_M = \Delta H_M - T\Delta S_M \tag{2.21}$$

where ΔG_M is the change in the Gibbs free energy on mixing, ΔH_M is the enthalpy of mixing, and $T\Delta S_M$ is the absolute temperature times the entropy of mixing.

The statistical thermodynamics theory puts all of the molecules on a lattice and counts the number of molecular arrangements in space in determining the entropy of mixing. This theory assumes incompressibility of the system, that is, no mer-sized holes are allowed in the lattice. The equation-of-state theories remove this assumption, that is, the polymers and/or blends are assumed compressible. The equation-of-state theory will be developed in Section 2.4.7.

2.4.6.2. Entropic Changes on Mixing Two Polymers   The statistical thermo-dynamic theory of mixing starts with Boltzmann's relation,

$$\Delta S_M = k \ln \Omega \tag{2.22}$$

where Ω is the number of possible arrangements in space that the molecules may assume. Boltzmann did not derive equation (2.22), but rather it is a postulate, stating that the entropy of a system is a measure of its disorder, proportional to the natural logarithm of the number of possible ways that matter can be arranged in space on a molecular level, counting each such arrangement equally. The constant of proportionality, k, is the famous Boltzmann constant. It must be noted here that k times Avogadro's number give the gas constant, R, thus serving as the basic definition of R. (The appearance of the quantities k or R indicate that the origin of any such equation is traceable to a counting of arrangements in space. The ideal gas equation, $PV = nRT$, is a simple example.)

The quantity ΔS_M is calculated by counting the total number of ways of arranging N_1 identical molecules on a lattice comprising (50)

$$N_0 = N_1 + N_2 \tag{2.23}$$

cells, see Figure 2.10 (3). The total number of such arrangements is given by

$$\Omega = \frac{N_0}{N_1! N_2!} \tag{2.24}$$

Applying Stirling's approximation for large N,

$$\ln N! = N \ln N - N \tag{2.25}$$

yields the entropy of mixing,

$$\Delta S_M = k[(N_1 + N_2)\ln(N_1 + N_2) - N_1 \ln N_1 - N_2 \ln N_2] \tag{2.26}$$

Strictly speaking, the above holds only for small molecules. Consider that the N_0 cells together occupy a volume V_0. If each of the N_0 cells have a volume V_r, and each of the two chain species occupy V_1 and V_2 volume per molecule (or x_1 and x_2 mers each occupying V_r), then rearrangement of equation (2.26) yields

$$\Delta S_M = -k(N_1 \ln v_1 + N_2 \ln v_2) \tag{2.27}$$

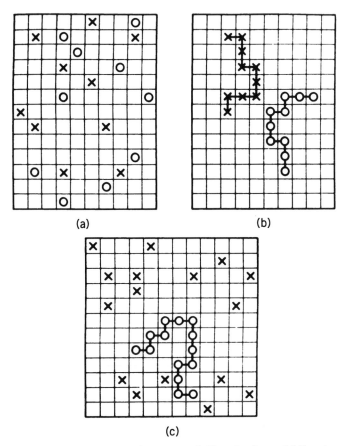

Figure 2.10. Mixing of types of molecules on quasilattice structures. (a) Two types of small molecules, (b) a blend of two types of polymer molecules, and (c) a polymer and a small molecule. The entropy of mixing decreases from (a) to (c) to (b) because of the decrease in the number of ways of arranging the molecules in space. Note that the mers of the polymer chains are constrained to remain in juxtaposition with their neighbors.

where v_1 and v_2 represent the volume fractions of the two polymers,

$$v_1 = \frac{V_1 N_1}{V_1 N_1 + V_2 N_2} \tag{2.28}$$

and

$$v_2 = \frac{V_2 N_2}{V_1 N_1 + V_2 N_2} \tag{2.29}$$

Thus, ΔS_M represents the combinatorial entropy computed by considering the possible arrangements of the molecules in Figure 2.10.

2.4.6.3. Free Energy of Mixing   The Flory–Huggins theory (44) introduces the quantity χ_1, proportional to the heat of mixing:

$$\chi_1 = \frac{\Delta H_M}{kTN_1 v_2} \tag{2.30}$$

leading to the free energy of mixing:

$$\Delta G_M = kT(N_1 \ln v_1 + N_2 \ln v_2 + \chi_1 N_1 v_2) \tag{2.31}$$

Equation (2.31) can be rearranged to the following form:

$$\frac{\Delta G_M}{V_0 kT} = \frac{v_1 v_2 \chi_1}{V_1} + \frac{v_1 \ln v_1}{V_1} + \frac{v_2 \ln v_2}{V_2} \tag{2.32}$$

Taking into account the lattice coordination number, z, the following approximate equation results:

$$\frac{\Delta G_M}{V_0 kT} = \frac{v_1 v_2 \chi_1}{V_x} + \frac{v_1 \ln v_1}{V_1} + \frac{v_2 \ln v_2}{V_2} \tag{2.33}$$

where $1/V_x = (1 - 2/z)/V_1$, z commonly ranging from 6 to 12. It is important to note that for high-molecular-weight polymers, V_1 and V_2 are large compared to the volumes obtained for small molecules. Thus, the contribution of the entropy of mixing in equation (2.33) drops orders of magnitude on going from mixing two small molecules (Figure 2.10a) to mixing two long molecules (Figure 2.10b). [Dissolving a polymer in a solvent (Figure 2.10c) represents an intermediate case.] The reduced entropy of mixing arises because the long chains are required to maintain connected configurations.

The quantity V_x in equation (2.33) is somewhat indeterminate (40,51). It is supposed to be the volume of an interacting segment, which conveniently preserves the value of χ_1 and the notion of a lattice. Although its value is probably larger than V_r, substitution of the latter for V_x is suggested for numerical calculations.

If the free energy of mixing, ΔG_M, is positive, phase separation will occur spontaneously. The quantity χ_1 is a dimensionless number and may be positive or negative. Its values range from about -1 to $+2$. If the value of χ_1 is negative, equation (2.33) will always yield a negative value of ΔG_M, since the entropic portion is always negative. Then, the blend is usually miscible. [A second requirement for miscibility is that the quantity $(\partial \Delta G^2 / \partial v_2^2)_{P,T}$ be negative. This ensures that some other phase-separated state does not have a still more thermodynamically stable state.] If the value of χ_1 is positive, then the system may phase separate. Because of the very small entropy of mixing of two high polymers, even a small positive value of $+0.1$ Kcal/mol for ΔH_M may ensure phase separation.

Table 2.4   Solubility parameters and densities of common polymers

Polymer	δ (cal/cm^3)$^{1/2}$	Density (g/cm^3)
Polybutadiene	8.4[a]	1.01
Polyethylene	7.9	0.85 (amorphous)
Poly(methyl methacrylate)	9.45	1.188
Polytetrafluorethylene	6.2	2.00 (amorphous, estimated)
Polyisobutene	7.85	0.917
Polystryene	9.10	1.06
Cellulose triacetate	13.60	1.28[b]
Cellulose tributyrate	—	1.16[b]
Nylon 66	13.6	1.24
Poly(ethylene oxide)	9.9	1.20
Poly(vinyl chloride)	9.6	1.39

[a] Note: $1(\text{cal/cm}^3)^{1/2} = 2.046 \times 10^3 \, (\text{J/m}^3)^{1/2}$.
[b] C. J. Malm, C. R. Fordyce, and J. A. Tanner, *Ind. Eng. Chem.*, **34**, 430 (1942).

2.4.6.4. Solubility Parameter   The value of χ_1 may be calculated through the solubility parameter δ:

$$\chi_1 = \frac{V_r(\delta_1 - \delta_2)^2}{RT} \tag{2.34}$$

where V_r is the molar volume of the solvent. The solubility parameter has the units of $(\text{cal/cm}^3)^{1/2}$, being the square root of the energy of evaporation per cubic centimeter. More recently, the units of $(\text{J/m}^3)^{1/2}$ have come into widespread use. For polymers, which ordinarily do not vaporize, the solubility parameter is determined by measuring swelling of crosslinked polymers in various solvents, among other methods. The solvent that swells it the most is accepted as having the same solubility parameter as the polymer.

Solubility parameters of some common polymers are shown in Table 2.4, selected values from Brandrup and Immergut (52). If the solubility parameter of the polymer is not known, it can be estimated though the use of the group molar attraction constants, which have been calculated by Small (53), Hoy (54), and others. These tables are based on a group contribution analysis. Small's table, perhaps the most widely used, is shown in Table 2.5 (53).

For completion, the solubility parameters of some common solvents are shown in Table 2.6, selected values from Brandrup and Immergut (52), since the more ordinary use of the solubility parameter has been to estimate solubility of polymers in various solvents. In this case, if $(\delta_1 - \delta_2)^{1/2}$ has a value of less than one $(\text{cal/cm}^3)^{1/2}$, the polymer will probably dissolve in the solvent. No such rule of thumb exists for polymer–polymer blends. Unfortunately, equation (2.34) will always predict a positive value for χ_1, a frequent case, but a real limitation

Table 2.5 Group molar attraction constants at 25° according to Small

Group		G^a	Group		G^a	Group		G^a
-CH₃	Single-bonded	214	Ring	5-membered	105–115	Br	Single	340
-CH₂-		133	Ring	6-membered	95–105	I	Single	425
-CH<		28	Conjugation		20–30	CF₂	n-fluorocarbons only	150
>C<		−93				CF₃		274
CH₂=	Double-bonded	190	H	Variable	80–100	S	Sulfides	225
-CH=		111	O	Ethers	70	SH	Thiols	315
>C=		19	CO	Ketones	275	ONO	Nitrates	~440
-CH=C<		285	COO	Esters	310	NO₂	Aliphatic nitro-compounds	~440
>C=C<		222	CN		410	PO₄	Organic phosphates	~500
Phenyl		735	Cl	Mean	260	Si	In silicones	−38
Phenylene o,m,p		658	Cl	Single	270			
Naphthyl		1146	Cl	Twinned as in >CCl₂	260			
			Cl	Triple as in -CCl₃	250			

[a] G has units of $\mathrm{cal}^{1/2}\text{-}\mathrm{cm}^{3/2}/\mathrm{mol}$
Source: P. A. Small, *J. Appl. Chem.*, **3**, 71 (1953).

Table 2.6   Solubility parameters of some common solvents

Solvent	$\delta\ (cal/cm^3)^{1/2}$	H-bonding[a] Group	Specific Gravity 20°C (g/cm^3)
Acetone	9.9	m	0.7899
Benzene	9.2	p	0.87865
n-Butyl acetate	8.3	m	0.8825
Carbon tetrachloride	8.6	p	1.5940
Cyclohexane	8.2	p	0.7785
n-Decane	6.6	p	—
Dibutyl amine	8.1	s	—
Difluorodichloro methane	5.1	p	—
1,4-Dioxane	10.0	m	1.0337
Low-odor mineral spirits	6.9	p	—
Methanol	14.5	s	0.7914
Styrene	9.3	p	0.9060
Toluene	8.9	p	0.8669
Turpentine	8.1	p	—
Water	23.4	s	0.99823
Xylene	8.8	p	0.8611

[a] Hydrogen bonding is an important secondary parameter in predicting solubility. *p*, Poorly H-bonded; *m*, moderately H-bonded; and *s*, stongly H-bonded.
Source: J. Brandrup and E. H. Immergut, *Polymer Handbook*, 3rd ed., Wiley-Interscience, New York, 1989.

in the use of the theory. The principal value, however, is the ease in making numerical calculations with equations (2.33) and (2.34).

2.4.6.5. Example Solubility Parameter Calculations   Suppose that poly(methyl methacrylate) was a new polymer, with the just determined structure,

$$
\begin{array}{c}
CH_3 \\
| \\
{+\!(}CH_2{-}C{)\!}_n \\
| \\
O{=}C{-}O{-}CH_3
\end{array}
\qquad (2.35)
$$

What is its solubility parameter?

The solubility parameter is calculated from

$$
\delta = \frac{\rho \sum G}{M}
\qquad (2.36)
$$

where G is the group molar attraction constant (see Table 2.5), ρ is the density, and M is the mer molecular weight. Then,

$$\delta = [1.188\,\text{g/cm}^3(133 - 93 + 2 \times 214 + 310)(\text{cal}^{1/2}\text{-cm}^{3/2}/\text{mol})]/100\,\text{g/mol}$$

since the molecular weight of the poly(methyl methacrylate) mer is exactly 100 g/mol. Then,

$$\delta = 9.24\,\text{cal}^{1/2}/\text{cm}^{3/2}$$

Table 2.4 shows a value of $9.45\,\text{cal}^{1/2}/\text{cm}^{3/2}$ by comparison. Usually, the calculated value and the value derived from experiment agree within $\pm 0.5\,\text{cal}^{1/2}/\text{cm}^{3/2}$.

2.4.6.6. Example Calculation of Phase Separation   Suppose one wants to make a mutually soluble blend of polybutadiene and polystyrene, 50/50 volume/volume at 125°C. What is the maximum molecular weight of the two polymers before phase separation? For simplicity, let's take the two molar volumes to be equal, and the cell (mer) sizes ($V_r = V_x$) to be equal to that of a styrene mer. The mer molar volume is calculated from the ratio of the mer molecular weight to the density:

$$V_r = (104\,\text{g/mol})/(1.06\,\text{g/cm}^3) = 98\,\text{cm}^3/\text{mol}$$

where the density was obtained from Table 2.4. Then, solving equation (2.34) for χ_1, using δ's from Table 2.4:

$$\chi_1 = [(98\,\text{cm}^3/\text{mol})(9.1 - 8.4)^2\,\text{cal/cm}^3]/[(1.987\text{cal/mol-K}) \times 398\,\text{K}]$$

$$\chi_1 = 0.0607^*$$

This value is needed for the solution of equation (2.33). At the point of phase separation, which is the critical point, $\Delta G_M = 0$, and V_0 is conveniently taken at $1\,\text{cm}^3$. Given also are that $V_1 = V_2$, and $v_1 = v_2 = \frac{1}{2}$. Since Boltzmann's constant is used, V_1 and V_2 are the volumes per polymer molecule. (If the gas constant R was used, then all quantities would be on a per mole basis.) Division of the mer molar volume by Avogadro's number yields the volume per mer:

$$(98\,\text{cm}^3/\text{mol})/6.02 \times 10^{23}\text{mers/mol} = 1.63 \times 10^{-22}\,\text{cm}^3/\text{mer}$$

The unknown for equation (2.33) is V_1:

$$0 = (0.50 \times 0.50 \times 0.0607)/(1.63 \times 10^{-22}\,\text{cm}^3/\text{mer}) + 2(0.5\ln 0.5)/V_1$$

$$V_1 = 7.44 \times 10^{-21}\,\text{cm}^3/\text{molecule}$$

---

* Experimentally, $\chi_1 = 0.01$–0.03; see H. S. Lee, W. N. Kim, and C. McBurns, *J. Appl. Polym. Sci.*, **64**, 1301 (1997), references cited therein, and Table 5.5.

Multiplication of V_1 by Avogadro's number puts us on a molar basis:

$$(7.44 \times 10^{-21} \, \text{cm}^3/\text{molecule}) \times (6.02 \times 10^{23} \, \text{mers/mol}) = 4.478 \times 10^3 \, \text{cm}^3/\text{mol}$$

and multiplication by the density yields the molecular weight:

$$(4.478 \times 10^3 \, \text{cm}^3/\text{mol}) \times (1.06 \, \text{g/cm}^3) = 4.74 \times 10^3 \, \text{g/mol}$$

for the polystyrene. Since the density of polybutadiene is $1.01 \, \text{g/cm}^3$, its molecular weight is $4.52 \times 10^3 \, \text{g/mol}$. Thus, at 125°C, this polymer pair is predicted to phase separate at somewhere between 4000 and 5000 g/mol, a very low molecular weight. This blend is known to be immiscible for all useful molecular weights.

Calculations of this nature, while simple, are only approximate. As the polymers become more similar, χ_1 becomes smaller, and its exact value more difficult to determine (55). When the incompressible limitation is removed (Section 2.4.7), the calculation is more complex, but the reliability of the results are improved.

Both the statistical theory, above, and the equation of theory, below, assume no specific interactions between the two polymers. As will be developed in many places further on in this book, attractive forces such as hydrogen bonding or acid–base interactions, the use of compatibilizers, or grafting, and so forth, all tend to change the thermodynamic picture dramatically. Usually, the need and the result is to make ΔG_M smaller or actually negative.

2.4.7. Equation-of-State Theories

The equation-of-state theories take into account the interrelationships among temperature, pressure, and volume (47–49). The equation-of-state theories grew out of a need to explain the appearance of a lower critical solution temperature, rather than an upper critical solution temperature, which the statistical thermodynamic theory cannot do, and also to provide both a better understanding of the physical nature of phase separation and improved prediction of conditions for miscibility and phase separation.

Frequently, density, ρ, is substituted for volume because density measurements are easier to make precisely in polymeric materials than volume measurements. In these theories, the polymers are considered to have mer-sized holes that can be partially squeezed out by an increase in the hydrostatic pressure. This contrasts with the statistical thermodynamic theory (Section 2.4.6), which assumes incompressibility, or the near equivalent, a constant number of holes per unit volume. Actually, the statistical thermodynamic theory of mixing ignores holes completely; thus the number of holes considered by the theory is really zero.

The reality of holes may be seen by considering dynamic experiments such as rheological deformation and flow on a macroscopic and microscopic scale. Matter, and polymers in particular, flow by the movement of molecules or mers into holes. When the molecule or mer has moved into the hole, the hole is observed to have moved also, into the site vacated by the moving molecule or mer. The only states of matter that do not have holes are perfect crystals. They do not flow in the ordinary sense of the term, but rather only exhibit short-range elasticity.

The equation-of-state theories are expressed in terms of the reduced temperature, $\tilde{T} = T/T^*$, the reduced pressure, $\tilde{P} = P/P^*$, and reduced density, $\tilde{\rho} = \rho/\rho^* = V^*/V$. The starred quantities represent characteristic values for particular polymers, often referred to as the *hard-core* or *close-packed* values, that is, with no free volume.

Consider a mixture of N_0 holes of volume fraction v_0. Following equation (2.27), the entropy of mixing is given by

$$\Delta S_M = -N_0 \ln v_0 - \sum_i N_i \ln v_i \tag{2.37}$$

Since the number of holes can be varied, this allows for compressibility and a concomitant change in entropy. Noting that the free volume is given by $1 - \tilde{\rho}$, and the entropy of mixing vacant sites with the molecules is given by

$$\Delta S_M = -\frac{k}{\tilde{\rho}}[(1 - \tilde{\rho}) \ln(1 - \tilde{\rho}) + \tilde{\rho} \ln \tilde{\rho}] \tag{2.38}$$

where $\tilde{\rho}$ is less than unity. If all of the sites are occupied, $\tilde{\rho} = 1$, and the right-hand side of equation (2.38) is zero. In reality, the cell size also increases with temperature, to accommodate increased vibrational amplitudes.

In equation-of-state notation, the Gibbs free energy of mixing is

$$\Delta G_M = \tilde{\rho}\varepsilon^* - T\Delta S_M \tag{2.39}$$

where the quantity ε^* represents the van der Waals energy of interaction. It must be noted that $\tilde{\rho} = \tilde{\rho}(P, T)$; $\tilde{\rho} \to 1$ as $T \to 0$; $\tilde{\rho} \to 1$ as $P \to \infty$.

It must be remembered that equations of state, in general, express relationships among volume, temperature, and pressure. The Sanchez equation of state is obtained by taking $\partial \Delta G_M / \partial \tilde{\rho} = 0$:

$$\tilde{\rho}^2 + \tilde{P} + T[\ln(1 - \tilde{\rho}) + (1 - 1/r)\tilde{\rho}] = 0 \tag{2.40}$$

where r represents the number of sites (mers, if each occupies a site) in the chain, and

$$\frac{1}{r} = \frac{v_1}{r_1} + \frac{v_2}{r_2} \tag{2.41}$$

The quantity r substantially goes to infinity for high polymers, yielding a general

Table 2.7   Equation of state parameters for some common polymers

Polymer	T^* (K)	P^* (bars)	ρ^* (g/cm^3)
Poly(dimethyl siloxane), PDMS	6050	3020	1.104
Poly(vinyl acetate), PVAc	6720	5090	1.283
Poly(n-butyl methacrylate), PnBMA	7000	4310	1.125
Polyisobutylene, PIB	8030	3540	0.974
Polyethylene (low density), LDPE	7010	4250	0.887
Poly(methyl methacrylate), PMMA	8440	5030	1.269
Polystyrene (atactic), PS	7950	3570	1.105
Poly(2,6-dimethylphenylene oxide), PPO	8260	5170	1.161
Polyethylene (high density)[a]	6812	4720	0.995
Polypropylene[a]	6802	4080	0.992
Polybutadiene[a]	4665	5400	0.932

[a] D. J. Walsh, W. W. Graessley, S. Datta, D. J. Lohse, and L. J. Fetters, *Macromolecules*, **25**, 5236 (1992).

equation of state for both homopolymers and miscible polymer blends:

$$\tilde{\rho}^2 + \tilde{P} + \tilde{T}[\ln(1 - \tilde{\rho}) + \tilde{\rho}] = 0 \tag{2.42}$$

The corresponding Flory equation of state reads

$$\tilde{\rho}^2 + \tilde{P} - \tilde{T}\tilde{\rho}(1 - \tilde{\rho}^{1/3})^{-1} = 0 \tag{2.43}$$

Data for several polymers are shown in Table 2.7 (56). The quantity P^* scales with cohesive energy density, and the quantity T^* scales with the energy divided by the density. Qualitatively, the criteria for miscibility are as follows: The T^* values must be similar for miscibility. If $T_1^* > T_2^*$, then it is desirable to have $P_1^* > P_2^*$. Polymers must have similar coefficients of expansion for miscibility, that is, the ratio of densities ρ_1^*/ρ_2^* must remain similar as the temperature is changed. Thus, from Table 2.7, it is easy to predict that the system polystyrene-*blend*-poly(2,6-dimethylphenylene oxide), with T^* (K) values of 735 and 739, respectively, should be miscible, which it is. Values of T^* for poly(dimethyl siloxane), however, are so different from the other polymers, immiscibility with all of the other members of Table 2.7 are predicted.

In general, the Sanchez theory predicts a better solubility of polymer systems than both the statistical thermodynamic theory and the Flory theory (20). Other theories, developed by Koningsveld and Kleindjens (57), Simha and co-workers (58,59), Walker and Vause (60), and Hong and Noolandi (61,62) provide additional insight into polymer blend mixing. A major problem is finding the best expression for χ_1 or its equivalent.

The equation-of-state theories predict a lower critical solution temperature: When two polymers are mixed, negative heats of mixing cause a negative volume change. Reducing the number of holes causes the entropy of mixing to be negative. At sufficiently high temperatures, the unfavorable entropy change associated with the densification of the blend becomes prohibitive, that is, $T\Delta S > \Delta H$, and the blend phase separates. The two-phased mixture has a larger volume as the holes are reintroduced, increasing the entropy. This scenario leads directly to the prediction of LCST behavior. Significantly differing coefficients of expansion also contributes to phase separation.

As the theories continue to evolve, improved predictions of miscibility are anticipated. However, the reader will remember that most of the above applies to noninteracting systems. More recently, both theory and experiment has turned to systems that interact at the interfaces, and/or within the domains, as treated in Chapters 4–7.

2.4.8. Some Specific Approaches for Mixing Thermodynamics

Polymer–polymer interactions can be measured via analog calorimetry (63,64). The method utilizes a direct measure of the heat of mixing of model pairs of low-molecular-weight compounds whose chemical structure closely resemble the mers they represent. Hydrogen bonding and other specific interactions in polymer blends can often be characterized by an increase in the glass transition temperature (65,66). Evidence for hydrogen-bonding interaction was detected by infrared spectroscopy in mixtures of poly(styrene-*stat*-vinylphenylhexafluor-odimethyl carbinol) and bisphenol A polycarbonate (66). Kwei et al. (67) took the enthalpy–infrared frequency shift correlation for simple acids and bases and extended these studies to hydrogen bonding in polymer systems. This led to a series of simple quantitative relationships.

Further quantitative infrared methods of the vibrational spectra of hydrogen bonded systems and related concepts were developed by Coleman et al. (68). This book specifically outlines a computer approach to determining the heats of mixing, χ_1, and so forth.

2.5. MECHANISMS OF PHASE SEPARATION

There are two major kinetic mechanisms of phase separation that have been identified and treated analytically: nucleation and growth (NG) and spinodal decomposition (SD). These occur in different parts of the phase diagram, as shown in Figure 2.6.

Nucleation and growth is the more familiar mechanism to most scientists and engineers. A supersaturated salt solution in water provides a common example: Subcritical-sized nuclei form and reform, until by chance a critical-sized nucleus occurs. The crystal grows by addition of material to its surface. Similar phenomena frequently occur with liquid–liquid type of separations. Nucleation and growth requires activation, and the region between the binodal and the spinodal is considered a metastable region, where NG occurs. Often, the domains formed are spheroidal in shape. The chemical composition of this phase does not change with time.

Spinodal decomposition also constitutes an important part of nature, but partly because of its somewhat greater complexity, it is not taught to students as often. The mixture becomes unstable with temperature or pressure changes, and activation energies are not required. Concentration fluctuations of a given wavelength grow in amplitude with time and eventually saturate. These may form large domains. Phase separation by SD kinetics often takes place in the form of nearly equally spaced interconnected cylinders, especially for phase separation involving melts of liquid phases. During SD, phase separation takes place under conditions in which the energy barrier is negligible, so that even small fluctuations in composition initiate phase separation.

Nucleation and growth kinetics are marked by having a positive diffusion coefficient [Figure 2.11 (69)], because matter is considered to flow from the supersaturated solution into a region partly depleted of that component. Of course, the component phase separating continuously diffuses into the growing phase surface, causing the depletion. On the other hand, SD has a negative diffusion coefficient arising from the flow of each component from regions of low concentration to regions of higher concentration. Often, SD is described by a wavelength, Λ, marking the distance between the cylinders (Figure 2.11). Figure 2.12 (70) shows the interconnected appearance of a spinoidally decomposed system via microscopy. The amplitude of the waves increases during the early stages of phase separation through an exchange of mass across the boundary, purer phases (i.e., more like equilibrium) forming with time. However, during the early stages of phase separation, the wavelength usually remains constant. Table 2.8 (3) summarizes the mechanisms that take place for both NG and SD. It must be emphasized that the more phase separation in polymer blends is researched, the more important SD seems to be. Spinodal decomposition is the commonly observed phenomenon in polymer blend phase separation, not NG.

Both NG and SD phases tend to coarsen with time, increasing the domain sizes and sometimes, especially in spinodal decomposition, changing the domain shapes. Figure 2.12 shows the original cylindrical morphology coarsening and approaching a more spheroidal morphology. Thus, if an investigator studies phase separation via microscopy, the element of time must be a factor, and observations made starting the earliest possible times.

This coarsening is known as Ostwald ripening and often follows the relation (40,71)

$$d^3 = d_0^3 + 64\gamma_{1,2}X_e V_m Dt/9RT \tag{2.44}$$

where d_0 is the initial drop diameter, $\gamma_{1,2}$ is the interfacial tension between the droplet and matrix phases, X_e represents the equilibrium mole fraction of the

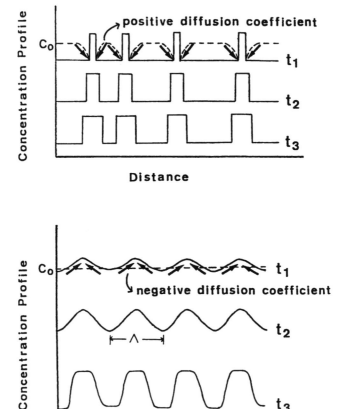

Figure 2.11. Illustration of nucleation and growth and spinodal decomposition kinetic mechanisms of phase separation.

droplet-rich component in the matrix phase, V_m represents the molar volume of the droplet phase, and D represents the diffusion coefficient for the matrix phase. Equation (2.44) shows that the volume of the coarsening phase increases linearly with time. In Ostwald ripening, the larger droplets grow at the expense of the smaller ones, favoring a smaller interfacial area. Of course, at equilibrium a layer effect should dominate, where the less dense phase lies on top of the more dense phase.

Ostwald ripening assumes a quiescent system. If the material is being sheared, as in a stirred reactor or in an extruder, the morphology for both NG and SD kinetics may appear as oriented ellipsoids of revolution or as oriented cylinders. Also, no chemical reactions are assumed. If phase separation takes place during polymerization, a phase-within-a-phase-within-a-phase

90×90μm

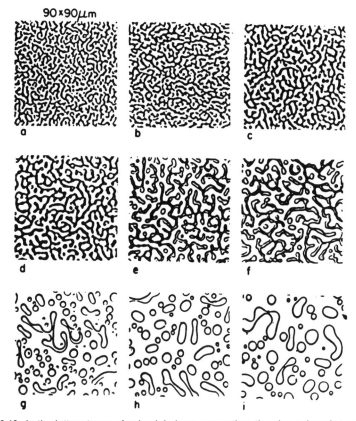

Figure 2.12. In the latter stages of spinodal phase separation, the phase domains coarsen.

morphology may result. For example, there is some evidence that the inner polymer styrene phase of HIPS materials (see Chapter 8) may be the result of a spinodal decomposition. The initially identified spheroidal structures obtained through salami slicing of the HIPS may really have a wormy, cylindrical morphology.

More detail on the kinetics of phase separation is treated in Appendix 2.1 (72–75).

2.6. EXPERIMENTAL SHAPES OF COEXISTENCE CURVES IN POLYMER BLENDS

Section 2.4 and other places in this chapter delineated phase separation, phase diagrams, and related thermodynamic theories as per polymer blends. This section will summarize some of the experimentally known phase diagrams.

Historically, polymer–polymer phase diagrams for high-molecular-weight polymers were unknown until about 1974. Up until then, it was thought that

Table 2.8 Mechanisms of phase separation

1. Nucleation and growth
   a. Initial fragment of more stable phase forms.
   b. Two contributions to free energy: (i) work spent in forming the surface and (ii) work gained in forming the interior.
   c. Concentration in immediate vicinity of nucleus is reduced; hence diffusion into this region is *downhill*. (Diffusion coefficient is positive.)
   d. Droplet size increases by growth initially.
   e. Requires activation energy.

2. Spinodal decomposition
   a. Initial small-amplitude composition fluctuations.
   b. Amplitude of wavelike composition fluctuations increases with time.
   c. Diffusion is *uphill* from the low concentration region into the domain. (Diffusion coefficient is negative.)
   d. Unstable process: no activation energy required.
   e. Phases tend to be interconnected.

either substantially all polymer pairs were immiscibile or that the kinetics of phase mixing would be too slow to be measured. In the late 1970s and early 1980s, there was a wave of experimental and theoretical results that transformed the field of polymer blends. A rather large number of polymer blends were found to exhibit lower critical solution temperatures. The early results were reviewed by Paul and Barlow (76). Up until the middle 1980s, only miscible, immiscible, or LCST-forming polymer pairs were known. Now, the path to a more general understanding is coming forth.

Table 2.9 shows selected polymer pair behavior. While several of the polymer pairs shown indeed show the expected LCST behavior, the poly(*o*-chlorostyrene-*co-p*-chlorostyrene)-*blend*-poly(2,6-dimethyl-1,4-phenylene oxide) blend exhibits both LCST and a miscibility window. The latter is defined as a range of compositions within which one phase exists and outside of which two phases exist. The poly(4-vinyl phenol)-*blend*-poly(*n*-butyl methacrylate) system is known to be more miscible than otherwise predicted because of the hydrogen-bonded carbonyl groups. The coexistence curves of model polyolefins based on different copolymer compositions of deuterated and protonated poly(ethylene-*co*-butene-1), synthesized by deuterating or hydrogenating anionic 1,2-rich poly-butadienes, reveals a UCST. This latter is thought to arise because the very small heats of interaction between the copolymers allows the enthalpic contribution to be positive and the entropic contribution (because of $-T\Delta S$) to be negative.

While the system polystyrene-*blend*-poly(2,6-dimethyl-1,4-phenylene oxide) is miscible at all attainable temperatures, because of the chlorinated polystyrene results and other experiments, the coexistence curves for LCST behavior are only thought to be above the degradation temperature of the polymers.

Table 2.9   Selected systems illustrating various phase diagram behaviors

Polymer I	Polymer II	Phase Behavior	Reference
Polystyrene	Poly(vinyl methyl ether)	LCST	a
Polystyrene	Poly(2,6-dimethyl-1,4-phenylene oxide)	Completely miscible	b
Poly(o-chlorostyrene-co-p-chlorostyrene)	Poly(2,6-dimethyl-1,4-phenylene oxide)	LCST and miscibility window	c
Poly(vinyl chloride)	Poly[ethylene-stat-(vinyl acetate)]	LCST	d
Poly(vinylidene fluoride)	Poly(methyl methacrylate)	LCST	e
Poly(ε-caprolactone)	Polycarbonate	LCST	f
Poly(4-vinyl phenol)	Poly(n-butyl methacrylate)	LCST	g
Poly(ethylene -co-52% butene-1)	Poly(ethylene-co-66% butene-1)	UCST	h

[a] T. Nishi, T. T. Wang, and T. K. Kwei, *Macromolecules*, **8**, 227 (1975).

[b] J. Stoelting, F. E. Karasz, and W. J. MacKnight, *Polym. Eng. Sci.*, **10**, 133 (1970).

[c] P. Alexandrovich, F. E. Karasz, and W. J. MacKnight, *Polymer*, **17**, 1022 (1977).

[d] E. Nolley, D. R. Paul, and J. W. Barlow, *J. Appl. Polym. Sci.*, **23**, 623 (1979).

[e] R. E. Bernstein, C. A. Cruz, D. R. Paul, and J. W. Barlow, *Macromolecules*, **10**, 681 (1977).

[f] C. A. Cruz, D. R. Paul, and J. W. Barlow, *J. Appl. Polym. Sci.*, **23**, 589 (1979).

[g] C. J. Serman, P. C. Painter, and M. M. Coleman, *Polymer*, **32**, 1049 (1991).

[h] R. Krishnamoorti, W. W. Graessley, N. P. Balasara, and D. J. Lohse, *J. Chem. Phys.*, **100**, 3894 (1994).

Again, as developed in Chapter 8, the use of HIPS blended with poly(2,6-dimethyl-1,4-phenylene oxide) results in a very impact-resistant composition. The rubber portion, however, is completely immiscible with both plastic-forming polymers.

Phase diagrams for block copolymers will be explored in Chapter 9, and those for interpenetrating polymer networks will be explored in Chapter 10.

APPENDIX **2.1**

SUMMARY OF SPINODAL DECOMPOSITION KINETICS

A2.1.1. CAHN–HILLIARD RELATIONS

Cahn and Hilliard (72–74) developed a theory of spinodal decomposition, in part to explain the phase separation behavior of certain types of glasses. In a homogenous binary system, the local chemical potential difference caused by concentration fluctuations is given by the difference of the chemical potential, μ, of the two components:

$$\mu_1 - \mu_2 = \partial G / \partial v_1 \qquad (A2.1)$$

where G represents the free energy of the homogeneous system, and v_1 represents the volume fraction of component 1.

If the system is nonhomogeneous, a gradient energy term must be added, describing the extra energy associated with departures from uniformity:

$$\mu_1 - \mu_2 = \partial G / \partial v_1 - K \nabla^2 v_1 \qquad (A2.2)$$

where K is the energy coefficient. The local free energy difference causes an interdiffusion flux, $\mathbf{j}$. The associated thermodynamic driving force is given by

$$\Omega_1 \nabla (\mu_1 - \mu_2) = -\mathbf{j} = \Omega_1 \left(\frac{\partial^2 G}{\partial v_1^2} \nabla v_1 - 2K \nabla^3 v_1 \right) \qquad (A2.3)$$

where Ω_1 represents the Onsager coefficient. This dynamic flux equation yields the diffusion equation

$$\frac{\partial v_1}{\partial t} = \Omega_1 \left[\left(\frac{\partial^2 G}{\partial v_1^2} \right) \nabla v_1 - K \nabla^4 v_1 + \cdots \right] \qquad (A2.4)$$

By comparison, the conventional diffusion equation is,

$$\frac{\partial v_1}{\partial t} = D\nabla^2 v_1 \tag{A2.5}$$

where D is the diffusion coefficient.

The general solution of the Cahn–Hilliard relation is expressed in trigonometric terms:

$$v_1 - v_{1,0} = \sum_\beta [\exp[R(\beta)t]][A\cos(\beta x) + B\sin(\beta x)] \tag{A2.6}$$

where the wavelength Λ is determined from $\Lambda = 2\pi/\beta$, and A and B are wave vector, β, dependent parameters. The Rayleigh growth factor is given by

$$R(\beta) = -\Omega_1 \beta^2 \left(\frac{\partial^2 G}{\partial v_1^2} + 2K\beta^2\right) \tag{A2.7}$$

and x is the position variable.

The phase separation process is dominated by the fluctuations in β. Setting the first derivative of equation (A2.7) to zero yields

$$\Gamma_m = 2\sqrt{2}\pi \left[\left(-\frac{1}{2K}\right)\frac{\partial^2 G}{\partial v_1^2}\right]^{-1/2} \tag{A2.8}$$

where Λ_m is the most rapidly growing wavelength.

From Debye's thermodynamic theory of nonhomogeneous solutions (69),

$$K = \frac{RT\chi_1}{V_1}\frac{r^2}{2} \tag{A2.9}$$

where r is the end-to-end distance of the polymer chain.

A2.1.2. NUMERICAL VALUES FROM EXPERIMENT

Nishi et al. (39), in their study of the spinodal decomposition of the poly(vinyl methyl ether)-*blend*-polystyrene system solved these equations via optical microscopy and NMR experiments, obtaining the following:

$$\Lambda_m = 1.98 \times 10^{-4}\,\text{m}$$

Thus, the domains are about $2\,\mu\text{m}$ apart just after formation, and

$$D = -2.8 \times 10^{-17}\,\text{m}^2/\text{s}$$

where the negative value is typical of spinodal decomposition, and

$$r = 3.34 \times 10^{-8}\,\text{m}$$
$$K = 3.73 \times 10^{-8}\,\text{J-m}^2$$
$$\partial^2 G/\partial v_1^2 = -4.06\,\text{J}$$

The latter two are given in per cubic meter of poly(vinyl methyl ether). Additionally, χ_1 is approximately -0.6 at 30°C, and approaches zero near 100°C, the phase separation temperature.

REFERENCES

1. M. Takayanagi, *Mem. Fac. Eng. Kyushu Univ.*, **23**, 11 (1963).
2. J. A. Manson and L. H. Sperling, *Polymer Blends and Composites*, Plenum, New York, 1976.
3. L. H. Sperling, *Introduction to Physical Polymer Science*, 2nd ed., Wiley-Interscience, New York, 1992.
4. J. J. Aklonis and W. J. Macknight, *Introduction to Polymer Viscoelasticity*, 2nd Ed., Wiley, NY, 1983.
5. E. H. Kerner, *Proc. Phys. Soc. London*, **69B**, 802, 808 (1956).
6. E. Guth and H. Smallwood, *J. Appl. Phys.*, **15**, 758 (1944).
7. E. Guth, *J. Appl. Phys.*, **16**, 20 (1945).
8. B. Paul, *Trans. AIME*, **218**, 36 (1960).
9. Z. Hashin and S. Shtrikman, *J. Mech. Phys. Solids*, **11**, 127 (1963).
10. B. Budiansky, *J. Mech. Phys. Solids*, **13**, 223 (1965).
11. B. Budiansky, *J. Compos. Mater.*, **4**, 286 (1970).
12. W. E. A. Davies, *J. Phys. (D)*, **4**, 318 (1971).
13. A. Y. Coran and R. Patel, *J. Appl. Polym. Sci.*, **20**, 3005 (1976).
14. J. K. Yeo, L. H. Sperling, and D. A. Thomas, *Polym. Eng. Sci.*, **21**, 696 (1981).
15. E. N. Kresge, in *Polymer Blends*, Vol. 2, D. R. Paul and S. Newman, Eds., Academic, New York, 1978.
16. L. H. Sperling, *Polym. Eng. Sci.*, **16**, 87 (1976).
17. N. R. Legge, G. Holden, and H. E. Schroeder, Eds., *Thermoplastic Elastomers: A Comprehensive Review*, Hanser, Munich, 1987.
18. S. Sakurai, *TRIP (Trends in Polymer Science)*, **3**, 90 (1995).
19. M. Matsuo, *Jpn. Plastics*, **2** (July), 6 (1968).
20. L. A. Utracki, *Polymer Alloys and Blends: Thermodynamics and Rheology*, Hanser, Munich, 1990.
21. D. B. Alward, D. J. Kinning, E. L. Thomas, and L. J. Fetters, *Macromolecules*, **19**, 215 (1986).
22. E. L. Martin, D. B. Alward, D. J. Kinning, D. C. Martin, D. L. Handlin, Jr., and L. J. Fetters, *Macromolecules*, **19**, 2197 (1986).

23. D. M. Anderson and E. L. Thomas, *Macromolecules*, **21**, 3221 (1988).

24. G. C. Eastmond and D. G. Phillips, *Polymer*, **20**, 1501 (1979).

25. C. B. Arends, Ed., *Polymer Toughening*, Marcel Dekker, New York, 1996.

26. D. R. Paul and J. W. Barlow, J. Macromol. *Sci. Rev. Macromol. Chem.*, **C18**, 109 (1980).

27. G. M. Jordhamo, J. A. Manson, and L. H. Sperling, *Polym. Eng. Sci.*, **26**, 518 (1986).

28. N. Devia, J. A. Manson, L. H. Sperling, and A. Conde, *Macromolecules*, **12**, 360 (1979).

29. H. G. Kia, in *Composite Applications: The Role of the Matrix, Fiber, and Interface*, T. L. Vigo and B. J. Kinzig, Eds., VCH Publishers, New York, 1992.

30. S. Saccubai, M. Sarojadevi, and A. Raghavan, *J. Appl. Polym. Sci.*, **61**, 577 (1996).

31. H. D. Rozman, R. N. Kumar, and A. Abusamah, *J. Appl. Polym. Sci.*, **57**, 1291 (1995).

32. L. P. McMaster, *Macromolecules*, **6**, 760 (1973).

33. L. P. McMaster, in *Copolymers, Polyblends, and Composites*, N. A. J. Platzer, Ed., Adv. Chem. Ser. No. 142, American Chemical Society, Washington, DC, 1975.

34. D. Patterson and A. Robard, *Macromolecules*, **11**, 691 (1978).

35. T. K. Kwei and T. T. Wang, in *Polymer Blends*, Vol. I, D. R. Paul and S. Newman, Eds., Academic, New York, 1978.

36. R. E. Bernstein, C. A. Cruz, D. R. Paul, and J. W. Barlow, *Macromolecules*, **10**, 681 (1977).

37. A. Robard and D. Patterson, *Macromolecules*, **10**, 1021 (1977).

38. F. E. Karasz and W. J. MacKnight, in *Contemporary Topics in Polymer Science*, Vol. 2, E. M. Pearce and J. R. Schaefgen, Eds., Plenum, New York, 1977.

39. T. Nishi, T. T. Wang, and T. K. Kwei, *Macromolecules*, **8**, 227 (1975).

40. O. Olabisi, L. M. Robeson, and M. T. Shaw, *Polymer–Polymer Miscibility*, Academic, New York, 1979.

41. D. S. Lee and S. C. Kim, *Macromolecules*, **17**, 268 (1984).

42. S. Miata and T. Hata, *Rep. Prog. Polym. Phys. Jpn.*, **12**, 313 (1969).

43. J. Hildebrand and R. Scott, *The Solubility of Nonelectrolytes*, 3rd ed., Reinhold, New York, 1949.

44. P. J. Flory, *Principles of Polymer Chemistry*, Cornell University Press, Ithaca, NY, 1953.

45. R. L. Scott, *J. Chem. Phys.*, **17**, 279 (1949).

46. S. Krause, *Macromolecules*, **3**, 84 (1970).

47. P. J. Flory, *J. Am. Chem. Soc.*, **87**, 1833 (1965).

48. I. C. Sanchez, in *Encyclopedia of Physical Science and Technology*, 2nd ed., R. A. Meyers, Ed., Academic, San Diego, Vol. 13, 1992, p. 153.

49. D. Patterson, *Polym. Eng. Sci.*, 22, 64 (1982).

50. U. W. Gedde, *Polymer Physics*, Chapman & Hall, London, 1995.

51. D. D. Patterson, *J. Paint Technol.*, **45**, 37 (1973).

52. J. Brandrup and E. H. Immergut, Eds., *Polymer Handbook*, 2nd ed., Wiley-Interscience, New York, 1975, Sect. IV.

53. P. A. Small, *J. Appl. Chem.*, **3**, 71 (1953).

54. K. L. Hoy, *J. Paint Technol.*, **46**, 76 (1970).

55. D. J. Dunn and S. Krause, *J. Polym. Sci., Polym. Lett. Ed.*, **12**, 591 (1974).

56. I. C. Sanchez, in *Polymer Blends*, Vol. I, D. R. Paul and S. Newman, Eds., Academic, Orlando, FL 1978, Chapter 3.

57. R. Konigsveld and L. A. Kleintjens, *J. Polym. Sci., Polym. Symp.*, **61**, 221 (1977).

58. O. Olabisi and R. Simha, *J. Appl. Polym. Sci.*, **21**, 149 (1977).

59. R. Simha and R. K. Jain, *Polym. Eng. Sci.*, **24**, 1284 (1984).

60. J. S. Walker and C. A. Vouse, *Am. Soc. Mech. Eng.*, **1**, 411 (1982).

61. K. M. Hong and J. Noolandi, *Macromolecules*, **14**, 727, 736, 1229 (1981).

62. J. Noolandi and K. M. Hong, *Macromolecules*, **15**, 482 (1982).

63. S. Ziaee and D. R. Paul, *Polym. Mater. Sci. Eng. (Prepr.)*, **75**, 14 (1996).

64. D. R. Paul and J. W. Barlow, *Polymer*, **25**, 487 (1984).

65. T. K. Ahn, M. Kim, and S. Choe, *Polym. Mater. Sci. Eng. (Prepr.)*, **75**, 22 (1996).

66. S. P. Ting, E. M. Pearce, and T. K. Kwei, *J. Polym. Sci., Polym. Lett. Ed.*, **18**, 201 (1980).

67. T. K. Kwei, E. M. Pearce, F. Ren, and J. P. Chen, *J. Polym. Sci., Part B: Polym. Phys.*, **24**, 1597 (1986).

68. M. M. Coleman, J. F. Graf, and P. C. Painter, *Specific Interactions and the Miscibility of Polymer Blends*, Technomic, Lancaster, PA, 1991.

69. J. H. An and L. H. Sperling, in *Cross-Linked Polymers: Chemistry, Properties and Applications*, R. A. Dickie, S. S. Labana and R. A. Bauer, Eds., American Chemical Society, Washington, DC, 1988.

70. S. Reich, *Phys. Lett.*, **114A**, 90 (1986).

71. I. M. Lifshitz and V. V. Slyozov, *J. Phys. Chem. Solids, Lett. Sect.*, **19**, 35 (1961).

72. J. W. Cahn and J. E. Hilliard, *Chem. Phys.*, **28**, 258 (1958).

73. J. W. Cahn and J. E. Hilliard, *Chem. Phys.*, **31**, 688 (1959).

74. J. W. Cahn, *J. Chem. Phys.*, **42**, 93 (1965).

75. P. J. Debye, *J. Chem. Phys.*, **31**, 680 (1959).

76. D. R. Paul and J. W. Barlow, *J. Macromol. Sci. Rev. Macromol. Chem.*, **C18**, 109 (1980).

MONOGRAPHS, EDITED WORKS, AND REVIEWS

Collyer, A. A., Ed., *Rubber Toughened Engineering Plastics*, Chapman & Hall, London, 1994.

Finlayson, K., Ed., *Advances in Polymer Blends and Alloys Technology*, Vol. 4, Technomic, Lancaster, PA, 1993.

Privalko, V. P. and V. V. Novikov, *The Science of Heterogeneous Polymers: Structure and Thermophysical Properties*, Wiley, Chichester, England, 1995.

3

FRACTURE BEHAVIOR

3.1. BASICS OF MECHANICAL AND FRACTURE BEHAVIOR

This chapter is concerned with the strength of polymeric materials, mechanisms of fracture, and improvements that may result from the addition of a second component. In Section 1.5, the concepts of modulus, stress and strain, and dynamic mechanical behavior were introduced. In all of those cases, the assumption of small strains was implicit, that is, the polymer was never strained to the breaking point. Now, this restriction is removed, and indeed, the stresses and strains imposed on the sample will be expected to cause failure. Much current research relates to how failure actually occurs, and what can be done to toughen plastics, elastomers, blends, and composites.

Fracture behavior may be considered both on the micromechanical level and the molecular level. Both of these approaches are governed by the thermodynamics and kinetics of motion, albeit they may superficially appear quite different. The bottom line is that it takes work to break a material; the laws of thermodynamics apply. It takes time and motion for the fracture processes to occur, therefore the laws of kinetics apply.

3.1.1. Energy Balance

Figure 3.1 (1) illustrates a closed system undergoing stress. An increment of energy, δU_1, is transferred into the system, increasing the stress. While the nature of δU_1 is general, and might be mechanical, thermal, electrical, chemical, and the like, the present case involves mostly mechanical work, although chemical work in the sense of bond breakage is also important. The quantity δU_1 is divided into three types of energy inside the closed system:

δU_2 = change in irreversibly dissipated energy
δU_3 = change in stored or potential energy
δU_4 = change in the kinetic energy

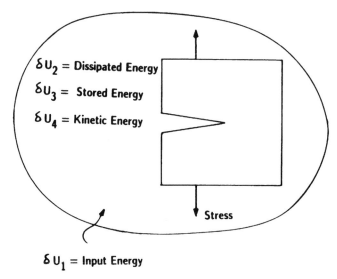

δU_1 = Input Energy

Figure 3.1. Thermodynamic modeling of crack growth, shown as a closed system.

The quantity δU_2 represents the energy dissipated by plastic or viscous flow, being converted into heat (1, 2). On a molecular scale, chain scission or chain pull-out absorbs large amounts of energy (3–8); on a micromechanical scale, rheological flow and deformation are important. Specific mechanisms include cavitation, cracking, crazing, shear yielding and shear band formation, fatigue, and debonding between phases (9, 10).

Two general subcases must be considered: adiabatic systems and isothermal systems. In the adiabatic case, there is a temperature rise in the system, but in the isothermal case, the thermal energy is all transferred to the surroundings across the closed system's boundary. Which one predominates depends on the rates of the deformation and fracture processes relative to the heat transfer rate. However, new information now forthcoming strongly suggests that because of the adiabatic nature of crazing and chain pull-out mechanisms, crazing and crack growth are more and more thought to occur in the *melt state* in polymeric materials, including plastics and composites. This will be developed further below.

The quantity δU_3 represents the elastically stored energy in the system, but also includes stored thermal energy. The quantity δU_4 represents changes in velocity; for example, when a sample is fractured, a portion of the sample is given kinetic energy of motion relative to some other portion.

In the isothermal case where the energy stored is entirely elastic, the net input of energy is

$$\delta U_1 - \delta U_2 \tag{3.1}$$

The first law of thermodynamics provides an energy balance for the conservation of energy:

$$\delta U_1 - \delta U_2 = \delta U_3 + \delta U_4 \tag{3.2}$$

If the body undergoes a displacement δu,

$$\frac{\delta U_1}{\delta u} - \frac{\delta U_2}{\delta u} = \frac{\delta U_3}{\delta u} + \frac{\delta U_4}{\delta u} \tag{3.3}$$

An incremental increase in energy of the system may be written

$$\delta U_1 = f \cdot \delta u \tag{3.4}$$

where f is the force in the direction of the deformation or displacement δu. This follows from the basic consideration that force times distance equals work.

Consider the case of the body in Figure 3.1 undergoing fracture, where the area of the crack increases by δA. Then the energy changes are

$$\frac{\delta U_1}{\delta A} - \frac{\delta U_2}{\delta A} = \frac{\delta U_3}{\delta A} + \frac{\delta U_4}{\delta A} \tag{3.5}$$

The term $\delta U_2/\delta A$ is called the *fracture resistance*, $\bar{R}$:

$$\bar{R} = \frac{\partial U_2}{\partial A} \tag{3.6}$$

The quantity $\bar{R}$ indicates the energy dissipated in propagating a fracture over an increment of crack area, δA. Larger values of $\bar{R}$ indicate tougher material. For homopolymers, $\bar{R}$ is associated with chain scission, chain pull-out, chain orientation at the crack surface, and the creation of new surfaces associated with surface tensions. For multicomponent materials, new mechanisms, associated with rubber toughening, fiber reinforcement, and the like all play important roles in addition.

Another quantity of interest is the *strain energy release rate*, G:

$$G = \frac{dU_1}{dA} - \frac{dU_3}{dA} \tag{3.7}$$

If $G > \bar{R}$, then the system is unstable, and the crack grows. The quantity G is also called the fracture energy, with units of joules per metre squared (J/m^2), where the area under consideration is the crack area.

3.1.2. Griffith Equation

One of the first quantitative analyses of the energy balance between the energy applied and the energy released in crack growth was due to Griffith (11). For the simplest case, imagine a large sheet, panel, or plate of glassy material that exhibits only simple elastic behavior, that is, no viscoelastic deformation. A crack of length $2a$ is inserted into the plate; see Figure 3.2 (12). Griffith noted

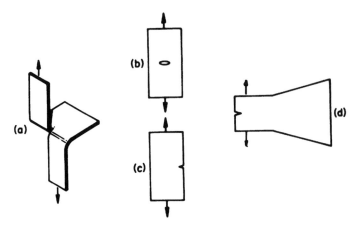

Figure 3.2. Four specimen shapes used in fracture mechanics for the determination of the fracture energy and stress intensity parameters: (a) trouser-leg design, often used with elastomers, (b) center-notched panel, (c) edge-notched panel, and (d) cantilever-beam type. Specimen shapes (b)–(d) are used for plastics.

that when such a cracked plate is stressed, a balance must be struck between the decrease in potential energy, δU_3, and the increase in surface energy resulting from the growth of the crack. This surface energy is expressed by the surface tension of the material, γ. Griffith determined that the surface energy term had to be the product of the total crack surface area, $2a \times 2t$, where t is the thickness of the sample, and γ, which has the units of energy per unit area.

If σ is the applied stress and E is Young's modulus, then the Griffith equation reads (13)

$$\gamma = \frac{\pi}{2} \frac{\sigma^2 a}{E} \tag{3.8}$$

The Griffith equation works reasonably well for some inorganic glasses because viscoelastic motions are virtually nonexistent. However, for polymers, the stresses, and hence work to fracture, is 100–1000 times the value predicted by equation (3.8). The additional work is needed for chain scission, chain pull-out, chain straightening (rubber elasticity effects), crazing, and general viscoelastic deformation of the sample under stress. Thus, the Griffith equation is a lower bound relationship.

3.1.3. Strain Energy Release Rate

A more general equation was derived by Orowan (14) and Irwin (15), who accounted for the energy involved in plastic deformation,

$$\sigma = \left(\frac{EG}{\pi a}\right)^{1/2} \tag{3.9}$$

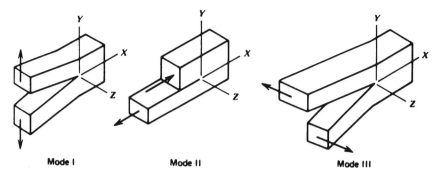

Figure 3.3. Three basic modes of loading, involving different crack surface displacements.

By replacing γ with G, the Griffith equation was broadened to include requirements for all kinds of energy, including surface energy.

3.1.4. Basic Modes of Loading

All materials are assumed to be inherently flawed. Suppose the sample has a crack as illustrated in Figure 3.2. The crack may be inserted at the edge of a sample or in the center. For many scientific studies, a crack is deliberately inserted in the sample *before* stressing for several reasons. This puts the flaw where it can be easily studied. The size and shape of the crack, particularly its edge(s) (sharp, rounded, etc.) are important both experimentally and theoretically. The downside of deliberately introducing a crack is that in fatigue and related experiments, the sample experiences internal damage before the external damage (the crack) begins to grow. The deliberately introduced crack may obviate such internal damage, thus changing the behavior of the crack during growth.

There are three different mechanical stress modes that may be applied to such cracks; see Figure 3.3 (13). Each of these modes involves different crack surface displacements. Mode I is the cleavage or tensile mode, where the crack surfaces are forced directly apart. Mode II is an in-plane shear or sliding mode, where the crack surfaces slide over one another in opposite directions in and out over the surface of the crack. Mode III is the antiplane shear or tearing mode, where the crack surfaces slide relative to one another left and right over the surface of the crack. Mode I is the one most commonly encountered, and the one that most frequently results in failure. Of course, mixed modes are often encountered in engineering practice.

3.1.5. Plane Stress and Plane Strain Conditions

In general, stresses on a sample may be uniaxial, biaxial, or triaxial. Thus, in the general triaxial case, the stresses may be written $\sigma_x + \sigma_y + \sigma_z$ where the x, y, and z subscripts refer to the directions shown in Figure 3.3. Consider a load applied in the Y direction. The plastic zone would develop a positive strain in the Y direction and in general develop corresponding negative strains in the X and Z directions, thus achieving a constant volume condition (Poisson's ratio assumed 0.5) required for a plastic deformation process. Thus, the strains $\varepsilon_x + \varepsilon_y + \varepsilon_z = 0$. The negative strain ε_z is counteracted by an induced tensile stress σ_z. However, since there can be no stress normal to a free surface, the through-thickness stress σ_z must be zero at both surfaces, but in general may maintain a large value internally.

There are two extremes to be considered. First, a thin plate where σ_z cannot increase appreciably in the thickness direction, where a condition of plane stress is said to dominate, that is, $\sigma_z \approx 0$. Second, for very thick sections, a stress σ_z may be developed that severely restricts straining in the z direction. This condition is referred to as plane strain. Then, $\sigma_z \approx \nu(\sigma_x + \sigma_y)$, where ν is Poisson's ratio; see Section 1.5.

3.1.6. Stress Intensity Factor

A key development in fracture mechanics was the analysis of the stress field at the crack tip, see Figure 3.4 (13). The radius of the damage zone is given by r_y.

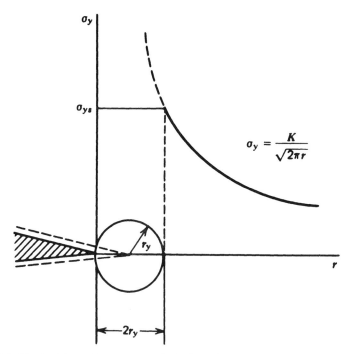

Figure 3.4. Plastic deformation at the crack tip. The *effective* crack length is the initial crack length plus the plastic zone radius, r_y.

The stress in the Y direction is given by

$$\sigma_y = \frac{K}{(2\pi x)^{1/2}} \tag{3.10}$$

where the stress in the Y direction declines with increasing distance from the crack tip. At the elastic–plastic boundary, $\sigma_y = \sigma_{ys}$, where σ_{ys} is the yield stress, and the crack-tip plastic zone radius is computed to be

$$r_y = \frac{1}{2\pi} \frac{K^2}{\sigma_{ys}^2} \tag{3.11}$$

The quantity K is called the stress intensity factor. It is the number of times that the stress is magnified at the crack tip.

For the case of plane stress, K is given by

$$K^2 = EG \tag{3.12}$$

and for the case of plain strain,

$$K^2 = \frac{EG}{1 - v^2} \tag{3.13}$$

For the simple case of Figure 3.2b,

$$K = \sigma(\pi a)^{1/2} \tag{3.14}$$

At the critical condition of crack propagation, the critical stress intensity factor, K_c, is given by

$$K_c = \sigma_c(\pi a)^{1/2} \tag{3.15}$$

Similarly, the work done per unit area of new crack surface is given by G_c. A more general case may be written

$$K_c = Y\sigma_c a^{1/2} \tag{3.16}$$

where Y is a geometrical factor that corrects for the sample shape and crack length.

Table 3.1 (2) summarizes typical values for G_{Ic}, the critical strain energy release rate (fracture energy) in mode I, and K_{Ic}, the mode I critical stress intensity factor. The quantity J_{Ic} is a material property version of G_{Ic}, more nearly independent of crack length and specimen geometry.

Table 3.1 compares typical values for epoxy homopolymers and rubber-toughened epoxies. For the rubber-toughened materials, the modulus declines slightly, roughly following the Takayanagi upper bound modulus, equation

Table 3.1 Typical values of G_{Ic} and K_{Ic} for various materials

Material	Young's Modulus E (GPa)	G_{Ic} (kJm^{-2})	K_{Ic} (MNm$^{-3/2}$)^a
Rubber	0.001	13	—
Polyethylene	0.15	20.0 (J_{Ic})	—
Polystyrene	3	0.4	1.1
High-impact polystyrene	2.1	15.8 (J_{Ic})	—
PMMA	2.5	0.5	1.1
Epoxy	2.8	0.1	0.5
Rubber-toughened epoxy	2.4	2	2.2
Glass-reinforced thermoset	7	7	7
Glass	70	0.007	0.7
Wood	2.1	0.12	0.5
Aluminum alloy	69	20	37
Steel, mild	210	12	50
Steel alloy	210	107	150

^{a} MNm$^{-3/2}$ = MPa-m$^{1/2}$

(2.3). The value of G_{Ic} increases by a factor of 20, hence the energy to create a unit of surface has increased by a factor of 20. The quantity K_{Ic} increases by a factor of 4. Thus, from equation (3.15), four times the stress is required to propagate the same sized crack.

Equation (3.15) is also important in design philosophy. The quantities K_c or K_{Ic} provide a basis for material selection, provide a minimum for the design stress, and the flaw size a provides the maximum allowable crack size without failure. The premise here is that all materials are inherently flawed, whether or not such flaws are easily detectable. Another point is that the equations in this section are derived under the assumption of linear elastic fracture mechanics deformation modes, that is, that long-range elasticity does not occur. Thus, elastomers and carbon-black-reinforced elastomers are only poorly described by these equations, since they commonly fail only after 500–1000% extension.

3.1.7. Example Calculation with Stress Intensity Factors

A center-notched polystyrene sample has a crack of 0.01 m width. If a stress of 2×10^8 Pa is imposed, will the sample fail?

From equation (3.14),

$$K = \sigma(\pi a)^{1/2}$$

$$K = 2 \times 10^8 \, \text{Pa} \, (3.14 \times 0.005 \, \text{m})^{1/2}$$

$$K = 2.5 \times 10^7 \, \text{Pa-m}^{1/2} = 2.5 \times 10^7 \, \text{N-m}^{-3/2} = 25 \, \text{MNm}^{-3/2}$$

Data are commonly reported both in MN-m$^{-3/2}$ and MPa-m$^{1/2}$, numerical values being equal. Since K_{Ic} from Table 3.1 is 1.1 MN-m$^{-3/2}$ < 25 MN-m$^{-3/2}$, the sample will fail.

3.2. EXPERIMENTAL METHODS FOR FRACTURE BEHAVIOR

The experiments that scientists and engineers utilize in testing the mechanical and fracture behavior of polymers vary from more or less fundamental to highly empirical. A question that must always be asked of engineers is: "To what stress and strain conditions will the material be exposed?"

3.2.1. Tensile Measurements

Indeed, the most common experiment is the measurement of the stress–strain relationships of polymers. In an Instron or similar instrument, a dog-bone-shaped sample is clamped and stressed to failure, providing a measure of the tensile strength.

The behavior of a polymer in a stress–strain experiment depends very much on the temperature, Figure 3.5 (16). At very low temperatures, the modulus is high, and the stress–strain curve is substantially straight. As the glass transition or melting temperature is approached, there is a yield point, and the strain to break increases very significantly. At still higher temperatures, the polymer becomes very soft, the strain to break decreasing again. Crosslinked polymers will follow rubber elasticity theory above the glass transition temperature, not shown. The curve illustrating the yield point also would be typical of rubber-toughened plastics, while the low-temperature curve is more typical of simple brittle plastics, as will be developed in Chapter 8. The areas under the curves represent the work to break of the sample. As a first approximation, the toughness is given by one half the product of the stress times the strain to fracture. Figure 3.5 is for single glass transition materials, such as homopolymers. The behavior of multiple T_g materials is more complex.

Strangely, while almost all scientific and engineering literature on stress–strain behavior is concerned with the sample in *tension*, nearly all actual load-bearing applications have the polymer in *compression*. Most materials are much stronger in compression than in tension, and may exhibit yield points in compression that are entirely absent in tension. It should also be remarked that few engineers will design parts that stress the part (either in tension or compression) more than about 10% of its strength.

Brittle plastics such as polystyrene and poly(methyl methacrylate) fail in tensile and impact experiments by a variety of mechanisms (17, 18); see Chapter 8. The most important of these is the dilational mechanism of crazing. A model of a craze is given in Figure 3.6, illustrating the presence of fibrillar strands spanning the craze, which may break as the craze moves forward, producing a true crack. Cavitation, shear band formation, and bridging of a rubber particle

STRESS – STRAIN

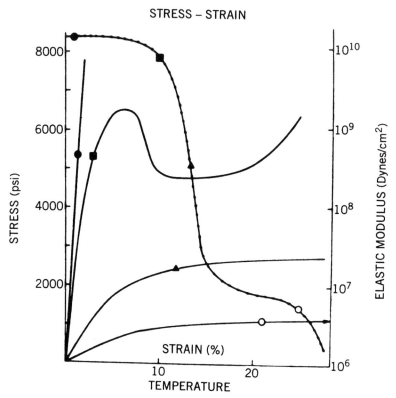

Figure 3.5. Stress–strain curves of a homopolymer or copolymer taken at various points along the modulus–temperature curve.

across a craze are all important mechanisms of energy absorption, and/or of stress relief. The fibrils spanning the craze are highly oriented material, large amounts of energy being absorbed in the orientation process. Cavitation is a major mechanism of stress relief because the triaxial stresses that cause the craze to grow are alleviated.

Tensile studies can also be used to measure improvements in fiber containing composites. Table 3.2 (19) compares the properties of plain epoxy matrix polymer with various modes of fiber reinforcement. The tensile strength increases by a factor of 10–20 depending on the nature of the fiber. The modulus also increases on the order of 100 times. This latter follows the upper bound Takayanagi model (Section 2.1).

Table 3.2 also provides the properties of specific stiffness and specific strength. In both of these cases, the property is divided by the density. Thus, per kilogram of weight, fiber-reinforced polymers are several times stiffer and stronger than steel, titanium, or aluminum. It must be pointed out that for many applications, especially aerospace, weight is the important quantity, not volume or even price. For composites, crack pinning by several closely spaced particles

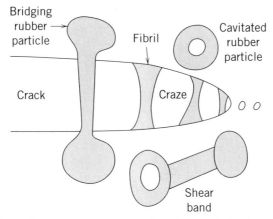

Figure 3.6. Structure of a craze in a rubber-toughened plastic is illustrated, showing several toughening mechanisms.

may prevent crack passage, microcracking via biforcation, and crack path deflection, where the particle behaves as a mirror, all contribute to toughness.

3.2.2. Impact Testing

Another popular method of testing the fracture behavior of plastics involves impact loading. Here, a sample, often with a sharp notch cut in it, is struck a

Table 3.2   Comparative fiber and unidirectional composite properties

Fiber/Composite	Elastic Modulus (GPa)[a]	Tensile Strength (GPa)[a]	Density (g/cm³)	Specific Stiffness (MJ/kg)	Specific Strength, M (J/kg)
Epoxy matrix resin	3.5	0.09	1.20	—	—
E-glass fiber	72.4	2.4	2.54	28.5	0.95
Epoxy composite	45	1.1	2.1	21.4	0.52
S-glass fiber	85.5	4.5	2.49	34.3	1.8
Epoxy composite	55	2.1	2.1	27.5	1.0
Boron fiber	400	3.5	2.45	163	1.43
Epoxy composite	207	1.6	2.1	99	0.76
High-strength graphite fiber	25.3	4.5	1.8	140	2.5
Epoxy composite	145	2.3	1.6	90.6	1.42
High-modulus graphite fiber	520	2.4	1.85	281	1.3
Epoxy composite	290	1.0	1.63	178	0.61
Aramid[b] fiber	124	3.6	1.44	86	2.5
Epoxy composite	80	2.0	1.38	58	1.45

[a] To convert GPa to psi, multiply by 145,000.
[b] Aromatic polymide.

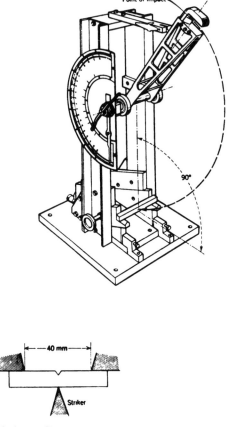

Point of impact

90°

40 mm

Striker

Figure 3.7. Simple beam Charpy-type impact instrument (ASTM D-256-56). Inset: Charpy test piece and support. Note the striker hits the sample on the opposite side from the notch.

sudden blow, causing failure. There are two basic instrument designs. In the Charpy test, the sample is supported at both ends, and struck on the back side of the notched position, see Figure 3.7 (20). In the Izod test, the sample is supported at one end only, cantilever style. In this case, the sample is struck at the unsupported end, on the side of the notch. The idea is to put an opening or mode I stress on the notch. Using a standard-sized sample, early data were reported in terms of foot pounds per inch of notch. Current data are reported in terms of joules per meter. Both refer to the energy required to create a crack on a sample of square cross section, approximately 1×1 cm (20). Figure 3.8 (21) illustrates the effect of rubber particle size and concentration in polystyrene blends determined in a Izod impact tester. While Figure 3.8 shows that the Izod impact energy increases with increasing particle size in the range shown, acrylonitrile–butadine–styrene (ABS) plastics show maximum toughness for rubber particles in the 0.5-μm size range; see Chapter 8.

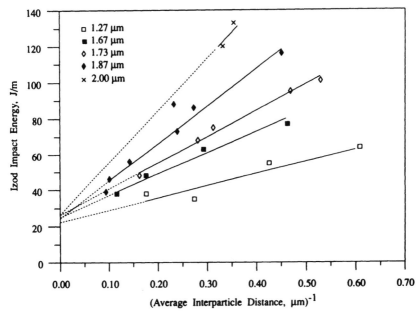

Figure 3.8. Izod impact energy for several HIPS/PS blends, showing the effects of rubber particle size and average interparticle distance. Note that the materials become tougher as the percent of rubber increases and the interparticle spacings decreases, to the right of the figure.

Impact tests are not limited to the basic Charpy and Izod methods, however. Special-purpose tests, sometimes very highly instrumented, are used to characterize polymer blends and composites. For example, Adams et al. (22) used a three-point instrumented Charpy-like method of impact, with controlled striker speeds of up to 5 m/s to study various composites. Zee et al. (23) studied the ballistic response of various fiber composites, finding that fracture toughness and frictional heat were more important than tensile strength in determining bullet-proof properties. Rodriguez (24) studied the puncture impact resistance of printed circuit boards, specifically noting the interfacial bonding characteristics of the composites.

Reed (25) summarized the characteristics of several impact tests; see Table 3.3. While impact tests are generally considered to cover the high strain rate end of the test spectrum, the flexed beam and falling weight tests do not necessarily provide particularly high strain rates.

3.2.3. Fatigue Tests

While tensile and impact tests are single motion experiments, very often cyclic loads are imposed on engineering materials in service. After many thousands of such cycles, such deformations may cause failure even at relatively low stress levels. In such engineering applications, the damage may remain hidden, until

Table 3.3   Order of magnitude characteristics of various impact tests

Test Method	Order of Magnitude of Strain Rate (s^{-1})	Impact Velocity (ms^{-1})
Flexed beam		
Charpy	10	3
Izod	60	2
Falling weight	$10^{-1}-10$	$1-4$
Conventional tensile	$10^{-3}-10^{-1}$	$10^{-5}-10^{-1}$
Pneumatic gun	$10^{2}-10^{4}$	$20-240$
Hydraulically operated	$1-10^{2}$	$0.008-4$

late, so that the apparent failure occurs in only a relatively few cycles, while in reality damage was accumulating with each cycle from the beginning.

There are three regions of fatigue response of a polymer to an imposed cyclic stress; see Figure 3.9 (26). At first, the sample appears to have no cracks or crazes. Then a crack or craze appears, which may be external or internal. Perhaps many such cracks and crazes are generated nearly simultaneously. The cracks or crazes propagate until the sample cannot withstand the load with the remaining intact material, and it fails. As illustrated in Figure 3.9, the greater the stress level, the shorter the life of the material.

In many experiments, a notch is deliberately introduced in the specimen at a convenient location. This obviates the first region of Figure 3.9, shortening the experiment considerably, and eases the crack growth study itself.

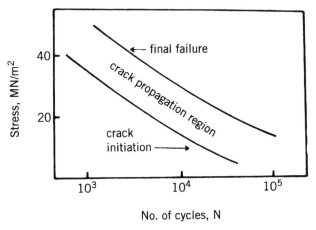

Figure 3.9. Three regions of fatigue response, showing the relationship between the applied stress, σ, and the number of stress cycles, N. Initially, there is no obvious damage. This is followed by a region where cracks are visible, leading to ultimate failure.

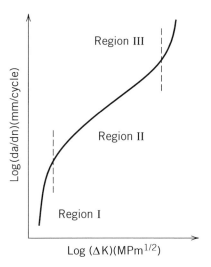

Figure 3.10. Idealized fatigue crack propagation for a plastic. Region I, very slow crack propagation. Region II, crack propagation proceeds though a quasi-linear region, thousands of cycles required to reach region III. Region III, failure usually takes place after only a few cycles.

For a crack of length a, the fatigue crack growth rate is da/dn, where n represents the number of cycles. The stress intensity factor range is given by

$$\Delta K = K_{max} - K_{min} \qquad (3.17)$$

where K_{max} and K_{min} represent the maximum and minimum values of the stress intensity factor during the cyclic loading. The crack growth rate is related to ΔK by Paris's law:

$$\frac{da}{dn} = A\Delta K^{m} \qquad (3.18)$$

where A and m depend on material variables, mean stress, environment, and frequency.

The semi-empirical Paris law can be used to express region II in Figure 3.10. Note the log–log plotting mode. Often, a large segment of the curve in region II is substantially linear (27). Displacement of the curve to the right indicates a tougher material, that is, a greater K_{max} is required to cause a given rate of crack propagation, assuming K_{min} stays constant. While K_{min} may be zero, investigators often keep it at a low level, to prevent buckling and other problems, and to better simulate actual engineering stress levels in application.

3.3. MECHANISMS OF FRACTURE RESISTANCE

Fracture must be considered on the basis of thermodynamics; see Section 3.1. External energy is put into the system in terms of some type of stress, and the system responds by absorbing the energy. Failure occurs when the energy input exceeds the mechanism by which energy can be absorbed. Energy absorption mechanisms can be considered on two levels: the molecular and the super-molecular, the latter sometimes known as the micromechanical mechanism.

3.3.1. Molecular Mechanisms in Fracture

Plastics have two main molecular mechanisms of absorbing energy, *chain scission* and *chain pull-out*; see Figure 3.11 (28). Other mechanisms, such as chain straightening and orientation, creation of new surfaces, and the like, while important conceptually, contribute only minor amounts to the total energy picture. Chain scission and chain pull-out terminology should replace such terminology as *viscoelastic deformation*, which has a similar connotation, but

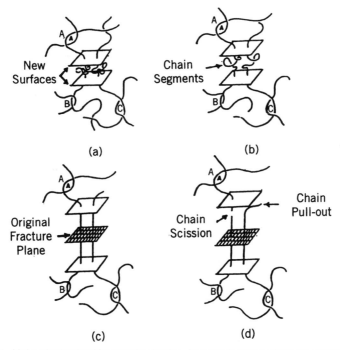

Figure 3.11. Molecular basis of fracture in plastics. As a crack opens, chain scission will occur if it is held by entanglements at both ends. If the chain lacks entanglements at one or both ends, it will pull out. While these two mechanisms are responsible for most of the energy absorption, note the creation of new surfaces and chain straightening, which together absorb about 1% or less of the total energy to fracture.

does not allow quantitative, fundamental calculations of fracture energy so easily.

The basic theory evolves from a consideration of the Lake–Thomas (29) postulate, which states that all of the main chain bonds between entanglement points on opposite sides of a growing crack are activated when the chain is stressed. When a chain is broken, all of the energy necessary to stress each of these bonds to near the breaking point is converted to heat, except for one bond, the one that breaks. For polystyrene, the carbon–carbon backbone chain energy is 70 kcal/mol, and there are about 300 bonds between entanglements.

At this time, the exact energy well characteristics of a polymer chain are not known. The hydrogen molecule model, popular in physical chemistry text-books, shows the energy levels getting closer together as the bond is excited to higher levels. The Lake–Thomas model implies that the energy levels get further apart as the bond is excited, making it cheaper for neighboring bonds to be excited, all rising in energy levels together.

For very long chains, there is mostly chain scission. For short chains, near the entanglement molecular weight, there is mostly pull-out (30–33). Portions of chains that pull out are primarily those not involved in entanglements on both sides. The fraction, F, of chain scission energy contribution scales as

$$F = \left(\frac{M - M_c}{M}\right)^3 \tag{3.19}$$

where M is the molecular weight of the whole chain, and M_c is the molecular weight between entanglements (33).

A useful instrument to determine the fracture energy on a molecular basis is the dental burr fracture apparatus (34), which is attached to a rheometer. Three basic quantities can be determined separately: the total energy to fracture a unit volume of material, the number of scissions created per unit volume [via gel permeation chromatography (GPC) of the sample parts before and after the experiment], and the size of the ground particles broken off, and hence the total surface area created per unit volume. Thus, for homopolymer polystyrene, at 32,000 g/mol, $G_{\mathrm{Ic}} = 17 \, \mathrm{J/m^2}$, being 100% pull-out energy, while at 151,000 g/mol, $G_{\mathrm{Ic}} = 230 \, \mathrm{J/m^2}$, being about 40% chain scission energy (33).

There is a growing body of evidence that the actual fracture process takes place in the melt state. Anyone who has physically examined a growing fatigue crack tip can attest to the fact that it is hot. Sambasivam et al. (32) estimated the temperature of a chain during actual pull-out for polystyrene to be between 150 and 250°C. Quite independently, Wool (35), quoting Kramer's (36) analysis of the Saffman–Taylor meniscus instability mechanism (37), arrived at 170°C for the temperature of craze formation in polystyrene. Both of these estimates are well above the glass transition temperature of polystyrene, near 100°C.

If the actual fracture process heats the sample hot enough that the crack or craze grows in the melt state, then a great deal of rethinking of the fracture processes in both homopolymers and blends needs to be done. The melt state fracture concept, however, permits the use of the Lake–Thomas theory, and the basis for equation (3.19), both of which are predicated on the polymer being in the melt state, rather than in the glassy state. The molecular mechanism of fracture model is currently being applied to multicomponent polymer materials.

3.3.2. Micromechanical Model

The micromechanical model concentrates on the development of stress fields, crazes and cracks, and cavitation, and methods of toughening plastics. The addition of a second phase to plastics, particularly rubber domains or fibers, produces very significant increases in fracture resistance over that provided by the homopolymer. Considering first the addition of rubber domains to brittle plastics, there are three main mechanisms of toughening: crazing, shear band formation, and cavitation (38). Crazing is responsible for the phenomenon known as stress whitening, where the stressed portion takes on a milk white appearance, as briefly discussed in Section 3.2.1. Stress whitening arises from light scattering caused by the many microscopic-sized void regions of the crazes. Again, bridges of oriented polymer span the opening, providing mechanical support. Shear band formation and cavitation are both illustrated in Figure 3.12 (39). In a shearing action, one portion of the polymer slips relative to another, with the chains caught in the middle becoming highly oriented. The orientation process absorbs energy. In Figure 3.12, the majority of the shear bands formed are at approximately 53° from the principal axis. Often, the average is closer to 45°.

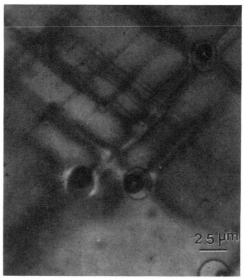

Figure 3.12. Shear bands and cavitated rubber particles of a rubber-toughened epoxy [DER 331/ Pip/CTBN (10) tensile specimen]. A quarter-wave plate inserted in the optical path of this micrograph facilitates the observation of the two effects.

Usually, the most desired situation is that on stressing, shear bands form first, and then crazes. The shear bands tend to stop the crazes from growing indefinitely, thus preventing gross failure. An example of this is in the blend of high-impact polystyrene (HIPS) with poly(2,6-dimethyl phenylene oxide) (PPO) (40, 41), which forms a very tough engineering plastic sold as Noryl. In HIPS–PPO blends as well as rubber-toughened epoxies (Figure 3.12), the shear bands run from the rubber particles. Both crazes and shear bands tend to initiate at stress concentration sites produced by the rubber. The orientation of the matrix polymer within a shear band is roughly parallel to the applied tensile stress, and therefore roughly normal to the plane of the crazes. It is for this reason that shear bands act as obstacles to craze propagation.

Cavitation usually occurs in the rubber domains; see Figure 3.12. While the microfracture inside of the rubber domains absorbs only minor amounts of energy, the local increase in volume allows an important kind of stress relief to the matrix (42). Cavitation within the rubber domains, as well as the crazing process, are the principal causes for stress whitening observed in mechanically damaged plastic specimens. While the relative importance of shear bands, cavitation, and crazing vary among the engineering polymer materials, these mechanisms are now believed to be the most important trio in toughening such materials.

Azimi et al. (38) described toughening mechanisms when both rubber particles and glass spheres are present in a brittle plastic such as an epoxy. They pointed out that the addition of a compliant rubbery phase toughens the matrix through cavitation and shear band formation. Further, glass reinforcement works through crack path deflection, crack tip pinning, bridging of particles across the crack, and associated microcracking mechanisms. These mechanisms work by shielding the crack tip from the intensity of the applied stress, that is, the stress intensity factor, K, is kept lower. Ideally, the simultaneous occurrence of multiple toughening mechanisms results in positive interactions.

These toughening mechanisms can be divided into two broad categories: namely, process zone and crack wake mechanisms. Bridging particles are an example of the crack wake mechanism, whereas cavitation and shear band formation reflect a process zone mechanism.

3.4. BRITTLE–DUCTILE TRANSITION

The words *brittle* and *ductile* have both common and specific technical meanings. In the common sense of the term, if a piece of plastic is bent, and it snaps, it is said to be brittle. If it yields, or even necks on extension, it is said to be ductile. Technically, these terms are defined in terms of specific failure mechanisms, intended to characterize the common meaning. Whether brittle or ductile behavior is observed depends on the thickness of the sample, the temperature, added components, and so forth. For example, polycarbonate fails by a ductile

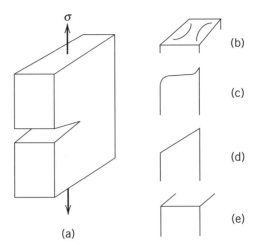

Figure 3.13. Gross fractography of a plastic. (a) Material with crack forming; (b)–(e) appearance of crack surfaces; (b) plane stress (thin sample approximation), showing lips; (c) appearance of the lips, side view; (d) another form of plane stress, showing all shear, side view; and (e) plain strain (thick sample approximation), showing smooth, flat fracture surfaces.

mechanism if the specimen is thinner than about 3 mm. Thicker sections, such as 6 mm, fail by brittle mechanisms.

Figure 3.13 shows the gross behavior of failure in mode I, failure in extension. Here, Figure 3.13 illustrates surfaces developed during plane stress and plane strain failure mechanisms. Plane stress conditions are generally desired for greater toughness, influencing engineering design of plastics. Taking polycarbonate as an example again, thin sections are much tougher per unit cross section, because they fail in a ductile mode.

Lee et al. (43) made a comparison of brittle and ductile fracture modes:

Brittle	*Ductile*
Crazing	Shear tearing
Herring bone surfaces	Shear lip formation
	Pop-in

The herring bone fracture surface is a plane strain mode characterized by repeated interception of the main crack with a series of secondary cracks resulting in a fish bone appearance. The pop-in fracture surface refers to the center of a fracture surface, where the center is depressed, and the edges are raised, creating a valley-ridge morphology. This is similar to Figure 3.13c, where both edges are either raised or depressed.

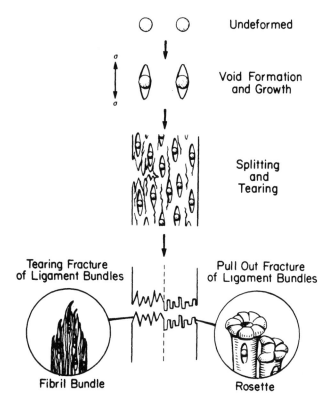

Figure 3.14. Mode A ductile fracture of rubber-toughened plastics.

Lee et al. (43) studied the fractography of the commercially successful blends of polycarbonate and ABS as a function of temperature and composition. They found that addition of 25% of ABS to polycarbonate reduced the ductile–brittle temperature from -20 to $-60°C$.

Li et al. (44) investigated the fractography of calcium-carbonate-filled thermoplastic polyesters. Several failure modes were described. Mode A ductile fracture surfaces contained two regions: a pullout region of slower crack growth and a rosette region of faster crack growth. Mode B ductile fracture surfaces, however, only contained the ductile pull-out texture. Mode C was a quasi-brittle fracture mode with herring bone patterns. Mode D quasi-brittle fracture surfaces consisted of a stress-whitened dimple region and a brittle fracture region. Mode E fracture surfaces exhibited the rough texture of brittle fracture. Figure 3.14 (44) schematically illustrates mode A fracture morphology during necking. Filler particles were debonded from the matrix polyester during yielding, forming voids around the particles as the polymer was plastically deformed. The width of the ligaments between the particles decreased during neck propagation. However, the width of the voids remained constant, equal to the particle diameter.

3.5. FRACTURE RESISTANCE IN FIBER COMPOSITES

Gerstle (45) defines continuous fiber composites as being composed of aligned fibers in a polymer matrix, the latter sometimes called the resin or binder. The fibers are typically 50 times stronger and 20–150 times stiffer than the matrix polymers. The role of the matrix is primarily that of a glue or binder that enables the fibers to support the applied loads. Thus, Gerstle (45) defines an advanced composite as bonded-fiber materials rather than reinforced plastics.

3.5.1. Mechanical Requirements

For continuous-fiber composites, the fibers must support all main loads and limit acceptable deformations. The material must contain a sufficiently high volume fraction of fibers to support these imposed loads.

The matrix must maintain the desired fiber orientation and spacing, transmit shear loads between neighboring fibers so that they can resist bending and/or compression, and protect the fiber from surface damage. By separating the fibers from each other by a polymer matrix, the stress concentration and hence the tendency to transmit a crack from a broken fiber to an intact one is reduced. If one fiber breaks, the matrix helps support the material. A good matrix system allows fibers to break at random points, the load being transmitted to remaining intact fibers. Of course, when the number of fibers remaining intact becomes smaller than the number required for maintaining the load, the material fails.

For continuous-fiber composites, the literature distinguishes between simple unidirectional fibers, usually aligned along the long direction of the sample, in the direction of the intended imposed load, and cloth layups, which usually contain two-directional woven fibers, with the polymer placed between the layers, sometimes called "layups." For the latter, on compression during molding, the polymer is forced between the fibers. Other materials use windings, where fibers can be laid down at an angle of 45° to cigar-shaped structures, running back and forth with the binder being laid down behind, in a continuous operation. This mode is useful for airplane body design, for example.

Consider the simple case of a composite consisting of unidirectionally oriented glass fibers, stressed in the direction of fiber orientation. First of all, the fibers absorb elastic energy and stretch slightly. On further extension, a crack may begin to form in the matrix, propagating into the fibers. Of course, the fibers must be bonded to the matrix. If the natural chemistry of the fiber surface does not attract or bond to the polymer sufficiently, coatings of silane bonding agents or similar may be added.

Ideally, the bonding energy of the fibers to the matrix should be slightly less than the energy necessary to fracture the fibers. Then, extensive debonding occurs first, absorbing energy. If the fiber is too well bonded to the matrix, it breaks without debonding, hence obviating an important energy absorption mechanism. This concept is developed quantitatively in Section 7.3.

Inorganic fillers such as glass, carbon, or boron fibers, short fibers, planar mica particles, or even glass spheres contribute to toughness of the polymeric system. While the advantage of unidirectional, continuous fibers is that the maximum stress resistance can be put in the direction of the expected maximum load, processing can be difficult. Thus, many materials utilize multidirectional laminates such as the layups and windings mentioned above, short, random fiber systems, and so forth. For some systems, it may be more economical to extrude short fiber containing composites. For good strength, however, a short fiber must have an aspect ratio (ratio of length to diameter) of at least 50.

While the above discussion emphasizes simple strain, composites are frequently used in compression or torsion. Each mode has its own requirements for greatest resistance to failure. (Here, the literature has a problem: While more data is presented in extension, the majority of applications involve compression!)

3.5.2. Model Experiments for Fiber Composites

Several model experiments have been developed for studying the effectiveness of fiber-reinforced plastics. Piggott (46) describes four such methods; see Figure 3.15. In the pull-out method, the fiber is embedded in a solid block of the matrix polymer. The fiber is stressed increasingly, until the fiber is pulled out. The stress at debonding is recorded to estimate the interfacial shear strength. The

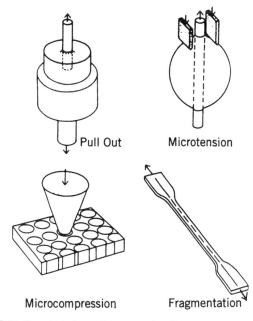

Figure 3.15. Four techniques for measuring fiber–composite interface failure.

microtension method uses a small drop of matrix polymer and is useful for thermoplastics. The microcompression test uses a small indenter to debond the fiber. It is possible to obtain quantitative results via a finite-element analysis. In the fiber fragmentation test, a single fiber is embedded in a matrix that has a breaking strain at least three times higher than the fiber. When the matrix polymer is stretched, the fiber breaks into ever shorter pieces, until finally the fiber lengths are all less than a critical length, enabling a fiber length–strength relation to be developed.

3.6. RUBBER BLENDS AND TIRES

While most of the material covered so far has a plastic as the continuous phase, there are important elastomer blends and composites. The single most important application area for such materials is rubber tires. This section will briefly treat these kinds of materials. Other elastomers treated in this book include the block copolymers of Chapter 9.

3.6.1. Historical Aspects

The term *rubber* dates from 1770, when Joseph Priestly observed that gum from the *Hevea brasilensis* tree could be used to rub out pencil marks (47). Another early term is *caoutchouc*, a word derived from the American Indian word meaning weeping wood, and widely used by the French. Of course, in those days the only rubber available was natural rubber, *cis*-1,4-polyisoprene. In 1839 Charles Goodyear discovered that rubber and sulfur combined in the presence of heat, yielding a new substance known as vulcanized rubber. This latter was the first true elastomer, defined as an amorphous crosslinked polymer above its glass transition temperature. Tires made around 1905 could go about 5000 miles at 25 miles an hour. That modern tires can go 50,000 miles at 55 miles an hour speaks to the improvements incorporated since then.

3.6.2. Blending of Rubbery Polymers

For amorphous polymers, three basic types of blends may be considered: the blending of two glassy polymers, the blending of a glassy and a rubbery polymer, and the blending of two rubbery polymers. The basic difference is in the temperature range of blending, since in each case all components must be in the melt state, that is, above T_g, and, of course, not be crosslinked. Blends involving polystyrene or poly(methyl methacrylate) are frequently mixed at 160–190°C. For elastomer (rubber) blends, temperatures are frequently in the range of 35–65°C.

Rubber–rubber blends frequently contain natural rubber (NR), styrene–butadiene rubber (SBR), ethylene–propylene diene monomer (EPDM), polybutadiene (BR), chloroprene (CR), nitrile rubber (NBR), and others, see Table 3.4 (48).

Table 3.4 Typical elastomers names and structures

Name	Abbreviation	Chemical Name	Structure
Natural rubber	NR	*cis*-1,4-Polyisoprene	$-(CH_2-C=CH-CH_2)_n-$ $\quad\quad\quad\mid$ $\quad\quad\quad CH_3$
Polyisoprene (synthetic rubber)	IR	*cis*-1,4-Polyisoprene	$-(CH_2-C=CH-CH_2)_n-$ $\quad\quad\quad\mid$ $\quad\quad\quad CH_3$
Polybutadiene	BR	Polybutadiene	$-(CH_2-CH=CH-CH_2)_n-$
Styrene–butadiene	SBR	Poly(butadiene-*co*-styrene)	$-[(CH_2-CH=CH-CH_2)_5(CH_2-CH)]_n-$ $\quad\quad\quad\quad\quad\quad\quad\quad\quad\quad\quad\quad\mid$ $\quad\quad\quad\quad\quad\quad\quad\quad\quad\quad\quad\quad C_6H_5$
Butyl rubber	IIR	Poly(isobutylene-*co*-isoprene)	$\quad\quad CH_3$ $\quad\quad\mid$ $-[(CH_2-C)_{50}(CH_2-C=CH-CH_2)]_n-$ $\quad\quad\quad\mid\quad\quad\quad\mid$ $\quad\quad\quad CH_3\quad\quad CH_3$
Ethylene–propylene–diene	EPDM	Poly(ethylene-*co*-propylene-*co*-diene)	$-[(CH_2-CH_2)_{37}(CH_2-CH)_{13}\text{-diene}]_n-$ $\quad\quad\quad\quad\quad\quad\quad\quad\quad\quad\mid$ $\quad\quad\quad\quad\quad\quad\quad\quad\quad\quad CH_3$

Source: M. Morton, Ed., *Rubber Technology*, 2nd ed., R. E. Krieger Pub. Co., Malabar, FL 1981. R. S. Bhakuni, S. K. Mowdood, W. H. Waddell, I. S. Rai, and D. L. Knight, *Encyclopedia of Polymer Science and Engineering*, 2nd ed., J. I. Kroschwitz, Ed., Wiley, New York, Vol. 16, 834, 1989.

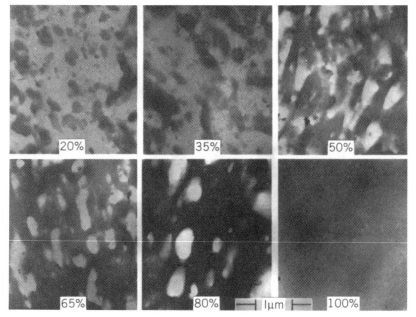

Figure 3.16. Transmission electron micrographs of pyrolytically etched sections of unfilled natural rubber/polybutadiene, with varied amounts of polybutadiene.

Figure 3.16 (49) shows transmission electron micrographs of a series of blends of natural rubber, *cis*-1,4-polyisoprene with *cis*-1,4-polybutadiene. The electron micrographs were prepared by a pyrolytic etching technique. Osmium tetroxide and similar techniques cannot be used because both NR and BR contain double bonds in each mer. The pyrolytic etching technique involves cutting the sample into sections of about 0.1 μm and mounting on 200-mesh tungsten specimen grids containing thin carbon substrates. The specimens are then heated slowly under vacuum to a temperature of about 350°C. The NR degrades at a significantly lower temperature than the BR and is thus selectively removed. Finally, the remaining BR phase appears darker relative to the etched areas representing the now removed NR. Figure 3.16 shows a phase inversion at midrange compositions.

3.6.3. Carbon Black in Tire Rubber

The carbon blacks used in toughening rubber are extremely fine colloidal structures, usually on the order of 10 nm in size, 50–150 m^2/g of surface area. These have extremely active surfaces, containing hydrogen, hydroxyls, carboxyls, esters, and aldehyde groups, among others, all capable of interacting with the rubber to produce a series of weak bonds. These weak bonds allow significant slip under stress, and recovery [see Figure 3.17 (50)] providing an important reinforcement mechanism (50a–d).

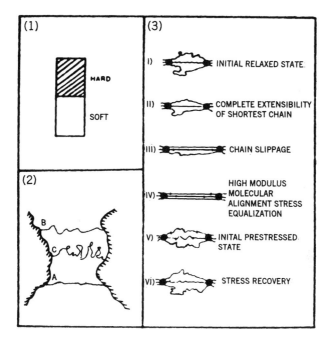

Figure 3.17. Schematic illustration of carbon-black-reinforced rubber stress softening theories. (1) Mullins and Tobin (50a) considered the material in a manner similar to the Takayanagi models (Figure 2.1), with only hard and soft regions. Deformation broke down the hard phase, but the degree of breakdown depended on the maximum extension of the sample. (2) F. Bueche (50b) attributed stress softening to the breakage of network chains attached to carbon black particles. (Molecule A breaks first.) (3) Dannenberg (50c) and Boonstra (50d) suggested that carbon black reinforcement involves chain slippage mechanisms; see chain marks. The slippage is partly reversible with the Mullins effect.

The carbon blacks are known to aggregate significantly, adding to their strength-imparting capabilities. The primary unit of carbon black is the aggregate, which is composed of spheroidal particles with some degree of coalescence. Higher structured aggregates typically have branches that form voids between them where polymer and/or oil can be occluded. This results in a higher effective loading of the carbon black in elastomeric compounds (51).

Figure 3.18 (52) illustrates the aggregated structure of a graphitized medium thermal (MT) carbon black by three methods: transmission electron microscopy (TEM), scanning electron microscopy (SEM), and scanning tunneling microscopy (STM). Graphitization tends to remove the hydrogen, oxygen, and other noncarbon elements, and emphasizes the crystalline structure of the carbon. The STM image in Figure 3.18c shows a better resolution than the SEM image in Figure 3.18b. Note the open structure and state of aggregation of the material. Table 3.5 (53) shows large amounts of carbon black blended into the rubber, GPT Black (N-299) in this case. Often, tire rubber is almost 50% carbon black by weight.

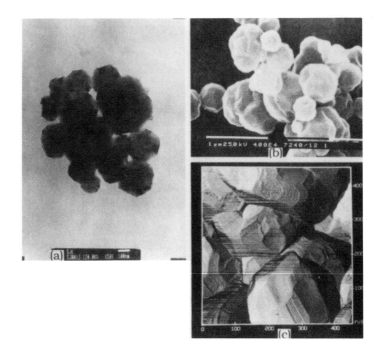

Figure 3.18. Three views of graphitized MT carbon black: (a) TEM, (b) SEM, and (c) STM images. The last shows the carbon black as polyhedrons covered with smooth facets and terraced facets.

Of course, interactions occur between the various components during tire fabrication, curing, and service life. Migration of soluble components must be limited. These involve the curatives sulfur, accelerators and activators, processing aids such as oils, peptizers, and tackifiers, and antidegradants such as antioxidants, antiozonants, and waxes.

In the early days, zinc oxide was used as an accelerator for the vulcanization reaction. In 1905, its reinforcement capabilities were recognized. Between 1910 and 1915, carbon black replaced zinc oxide as a reinforcing agent, as it proved much superior than the latter. (However, since people liked the white appearance of the tire, they still sometimes buy "whitewalls," containing zinc oxide as ornamental material.) A further discussion of carbon black in rubber will be found in Section 5.4.1.

3.6.4. Rubber Composition of Modern Pneumatic Tires

The modern pneumatic tire is a complex blend-composite material; see Figure 3.19 (48). The reader will note the presence of steel belting and nylon overlays. Normally, the rubber is a blend of elastomers; see Tables 3.4 and 3.5. For passenger tire tread compounds, an alternate to SBR and natural rubber can be SBR and cis-polybutadiene.

Table 3.5   Passenger tire tread recipe

PLIOFLEX 1502 SBR	50
TSR 20 natural rubber	50
GPT black (N-299)	45
Aromatic oil	9
VANWAX H Special (Paraffin Wax)	1
ANTOZITE 67P (Antiozonant)	2
WINGSTAY 100	1
Stearic acid	3
Zinc oxide	3
Sulfur	1.6
AMAX (N-oxydiethylene benzothiazo-ε-2-sulfanamide)[a]	0.8
VANAX DPG (N, N-diphenyl guanidine)[b]	0.4
Total	166.8

Conditions

Rheometer at 150°C (300°F)	
ts 1, minutes	4.8
t′c (90), minutes	15.8
MH, Nm (in. lb.)	3.8 (33.7)
ML, Nm (in. lb.)	0.7 (6.0)

Physical properties

Cured 18 min at 150°C (300°F)	
Stress at 300%, MPa (psi)	8.1 (1170)
Tensile strength, MPa (psi)	19.5 (2820)
Elongation at break, %	540
Rebound,* % 22°C	61
100°C	72

*ASTM D1054, cured 28 min at 150°C (300°F)

Dynamic mechanical properties

Vibrotester, 100°C (212°F), ASTM D2231	
Storage modulus, MPa (psi)	5.7 (830)
Resilience, %	38.9
Tan delta	0.139

Source: *The Vanderbilt Rubber Handbook*, R. F. Ohm, Ed., R. T. Vanderbilt Co., Norwalk, CT, 14th ed. 1995 (CD ROM version).
[a] Accelerator, delayed action type.
[b] A secondary accelerator.

While all of these elastomers are immiscible with each other, the various rubber phases are covulcanized together to attain maximum elasticity and endurance (54). The rubber tread is compounded for wear, traction, low rolling resistance, and durability. Passenger tire treads are normally composed of a

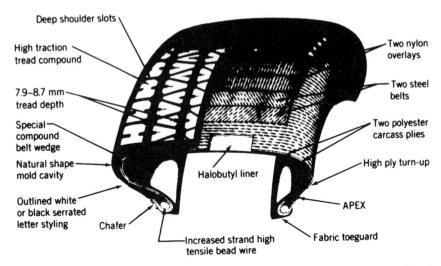

Figure 3.19. Cross section of Goodyear's all season, high-performance passenger tire, the Eagle GT + 4. Note that a modern tire is both a complex blend and composite.

blend of SBR and BR, while truck tires and aircraft tires use NR and BR, because of their greater tearing and shredding resistance. Both NR and BR are capable of crystallization on extension, providing extra tear resistance during abrasion. Under use conditions, tiny shreds of rubber tear loose from the tire tread. Each time the tire goes around and contacts the road, these shreds are stretched. When the NR or BR are stretched, they crystallize, providing a special kind of "smart material" reinforcement. When a particular shred leaves the road, it retracts, returning to its amorphous condition.

The elastomer SBR provides low cost and high extensibility. As a statistical copolymer, it cannot crystallize.

The elastomer EPDM protects the tire against ozone chemical attack. Ozone attacks double bonds, cutting them chemically (55, 56). Even under normal use, tires develop microscopic cracks. Each time the tire goes around, at some point the crack is opened. If ozone is present, it is admitted to the surface. (Ozone has a very low diffusion coefficient in rubber and cannot enter sound rubber rapidly.) Under these conditions, the crack may grow at 15 Å a tire revolution due to chemical degradation of the rubber to an oil. This is particularly true in areas of California, where smog is a particularly bad problem. While 15 Å per revolution does not sound like much, that corresponds to about 10 cm of crack growth over 50,000 miles, or that the tire blew out if not otherwise protected. Since ozone cannot easily cut the saturated EPDM, a separate phase of this elastomer tends to arrest such cracks.

The elastomer IIR, see Table 3.4, is highly impermeable to air gases, and hence helps keep the air in the tires. It is still the main component of bicycle tire tubes, for example.

There are, of course, many other components making up passenger tire treads. As noted in Table 3.5, zinc oxide and many complex organic compounds such as N-cyclohexyl-2-benzothiazolsulfenamide are used to accelerate the vulcanization process. Antioxidants include such compounds as N-(1,3-dimethylbutyl)-N'-phenyl-paraphenylenediamine. Other formulations include a heavy aromatic oil, which serves as a plasticizer for the rubber, and a processing oil (57). Here, plasticizer is used in the sense of softening, rather than reduction of the glass transition temperature.

3.7. SUMMARY

In brief, either of the quantities G_c or K_c, related through equations (3.12) and (3.13), provide an energetic basis for determining the resistance of a material to failure. These methods are better than the Izod or Charpy impact tests, for example, although these still play an important role in providing a simple reference number. The molecular basis of fracture calculations open future routes to fundamental understanding.

Part I of this book has largely been concerned with the fundamentals of multicomponent polymer materials. Many introductory aspects of polymer blends and composites have been treated in an integrated fashion. Thus, the several items of nomenclature, morphology, modulus, dynamic mechanical behavior, modeling, thermodynamics, phase separation, and crack growth and fracture behavior were all briefly examined to introduce the reader to the field. The field is vast, and this whole book constitutes but an introduction. This first part has now prepared the reader for an examination of polymer surfaces and interfaces, the content of Part II. Part III will be concerned with specific systems, emphasizing impact resistant plastics, block copolymers, and interpenetrating polymer networks (IPN)s.

REFERENCES

1. J. G. Williams, *Fracture Mechanics of Polymers*, Ellis Horwood, Chichester, England, 1984, Chapter 2.
2. A. J. Kinloch and R. J. Young, *Fracture Behavior of Polymers*, Applied Science, London, 1983.
3. Y. H. Kim and R. P. Wool, *Macromolecules*, **16**, 1115 (1983).
4. H. Zhang and R. P. Wool, *Macromolecules*, **22**, 3018 (1989).
5. R. P. Wool, *Polymer Interfaces Structure and Strength*, Hanser, Munich, 1995.
6. N. Mohammadi, A. Klein, and L. H. Sperling, *Macromolecules*, **26**, 1019 (1993).
7. M. Sambasivam, A. Klein, and L. H. Sperling, *J. Appl. Polym. Sci.*, **58**, 357 (1995).
8. X. Lu, N. Ishikawa, and N. Brown, *J. Polym. Sci.: Part B: Polym. Phys.*, **34**, 1809 (1996).

9. H. R. Azimi, R. A. Pearson, and R. W. Hertzberg, *J. Appl. Polym. Sci.*, **58**, 449 (1995).

10. D. Haderski, K. Sung, J. Im, A. Hiltner, and E. Baer, *J. Appl. Polym. Sci.*, **52**, 121 (1994). This whole issue *J. Appl. Polym. Sci.*, **52(2)**, is largely concerned with the micromechanical approach to fracture.

11. A. A. Griffith, *Philos. Trans. R. Soc.*, **A221**, 163 (1921).

12. R. S. Rivlin and A. G. Thomas, *J. Polym. Sci.*, **10**, 291 (1953).

13. R. A. Hertzberg, *Deformation and Fracture Mechanics of Engineering Materials*, 3rd ed., Wiley, 1989, Chapter 8.

14. E. Orowan, *Phys. Soc. Rep. Prog. Phys.*, **12**, 186 (1948).

15. G. R. Irwin, in *Fracture, Handbuch der Physik*, Vol. VI, S. Flugge, Ed., Springer, Berlin, 1958, p. 551.

16. L. E. Nielsen and R. F. Landel, *Mechanical Properties of Polymers and Composites*, 2nd ed., revised and expanded, Marcel Dekker, New York, 1994.

17. A. S. Argon, R. E. Cohen, O. S. Gebizlioglu, H. R. Brown, and E. J. Kramer, *Macromolecules*, **23**, 3975 (1990).

18. S. H. Spiegelberg, A. S. Argon, and R. E. Cohen, *J. Appl. Polym. Sci.*, **48**, 85 (1993).

19. F. P. Gerstle, Jr., in *Encyclopedia of Polymer Science and Engineering*, Vol. 3, J. I. Kroschwitz, Ed., Wiley, New York, 1985, p. 776.

20. ASTM D 256–93a, ASTM Standards, 1995. American Society for Testing Materials, Section 8, Annual book of ASTM Standards.

21. D. G. Cook, A. Rudin, and A. Plumtree, *J. Appl. Polym. Sci.*, **48**, 75 (1993).

22. G. C. Adams, R. G. Bender, B. A. Crouch, and J. A. Williams, *Polym. Eng. Sci.*, **30**, 241 (1990).

23. R. H. Zee, C. J. Wang, A. Mount, B. Z. Jang, and C. Y. Hsieh, *Polym. Compos.*, **12**, 196 (1991).

24. E. L. Rodriguez, *Polym. Compos.*, **12**, 75 (1991).

25. P. E. Reed, in *Developments in Polymer Fracture*, E. H. Andrews, Ed., Applied Science, London, 1979 , p. 121.

26. R. W. Hertzberg and J. A. Manson, *Fatigue in Engineering Materials*, Academic, Orlando, FL, 1980.

27. J. Michel, R. W. Hertzberg, and J. A. Manson, *Polymer*, **25**, 1657 (1984).

28. N. Mohammadi, J. N. Yoo, A. Klein, and L. H. Sperling, *J. Polym. Sci., Polym. Phys. Ed.*, **30**, 1311 (1992).

29. G. J. Lake and A. G. Thomas, *Proc. R. Soc. London*, **A300**, 108 (1967).

30. N. Mohammadi, A. Klein, and L. H. Sperling, *Macromolecules*, **26**, 1019 (1993).

31. L. H. Sperling, A. Klein, M. Sambasivam, and K. D. Kim, *Polym. Adv. Technol.*, **5**, 453 (1994).

32. M. Sambasivam, A. Klein, and L. H. Sperling, *Macromolecules*, **28**, 152 (1995).

33. M. Sambasivam, A. Klein, and L. H. Sperling, *J. Appl. Polym. Sci.*, **58**, 357 (1995).

34. N. Mohammadi, R. Bagheri, G. A. Miller, A. Klein, and L. H. Sperling, *Polym. Testing*, **12**, 65 (1993).

35. R. P. Wool, *Polymer Interfaces: Structure and Strength*, Hanser, Munich, 1995, pp. 409–410.

36. E. J. Kramer, in *Advances in Polymer Sciences: Crazing*, H. H. Kausch, Ed., Springer, Berlin, 1983.

37. P. G. Saffman and G. Taylor, *Proc. R. Soc., Ser. A. Math. Phys. Sci.*, **245**, 312 (1958).

38. H. R. Azimi, R. A. Pearson, and R. W. Hertzberg, *J. Appl. Polym. Sci.*, **58**, 449 (1995).

39. R. A. Pearson and Y. F. Yee, *J. Mater. Sci.*, **26**, 3828 (1991).

40. A. F. Yee, *Polym. Eng. Sci.*, **17**, 213 (1977).

41. C. B. Bucknall, *Toughened Plastics*, Applied Science, London, 1977, p. 193.

42. J. F. Hwang, J. A. Manson, R. W. Hertzberg, G. A. Miller, and L. H. Sperling, *Polym. Eng. Sci.*, **29**, 1477 (1989).

43. M. P. Lee, A. Hiltner, and E. Baer, *Polym. Eng. Sci.*, **32**, 909 (1992).

44. J. X. Li, A. Hiltner, and E. Baer, *J. Appl. Polym. Sci.*, **52**, 269 (1994).

45. F. P. Gerstle, Jr., in *Encyclopedia of Polymer Science and Engineering*, 2nd ed., J. I. Kroschwitz, Ed., Wiley, New York, vol. 3, pp. 1085, 1985.

46. M. R. Piggott, in *Composite Applications: The Role of the Matrix, Fiber, and Interface*, T. L. Vigo and B. J. Kinzig, Eds., VCH Publishers, New York, 1992.

47. D. R. St. Cyr, in *Encyclopedia of Polymer Science and Engineering*, 2nd ed., J. I. Kroschwitz, Ed., Wiley, New York, Vol. 14, pp. 687, 1988.

48. R. S. Bhakuni, S. K. Mowdood, W. H. Waddell, I. S. Rai, and D. L. Knight, in *Encyclopedia of Polymer Science and Engineering*, 2nd ed., J. I. Kroschwitz, Ed., Wiley, New York, Vol. 16, p. 834, 1989.

49. W. M. Hess, P. C. Vegvari, and R. A. Swor, *Rubber Chem. Technol.*, **58**, 350 (1985).

50. J. A. C. Harwood, L. Mullins, and A. R. Payne, *J. IRI*, **1**(Jan/Feb), 17 (1967).

50a. L. Mullins and N. R. Tobin, in *3rd Rubber Tech. Conf.*, London, 397, 1954.

50b. F. Bueche, in *Reinforcement of Elastomers*, G. Krause, Ed., Interscience, New York, 1956.

50c. E. M. Dannenberg, *Trans. IRI*, **42**, T.26 (1966).

50d. B. B. Boonstra, in *Reinforcement of Elastomers*, G. Krause, Ed., Interscience, New York, 1956.

51. C. R. Herd, G. C. McDonald, R. E. Smith, and W. M. Hess, *Rubber Chem. Technol.*, **66**, 491 (1993).

52. S. J. Kim and D. H. Reneker, *Rubber Chem. Technol.*, **66**, 559 (1993).

53. W. H. Waddell, R. S. Bhakuni, W. W. Barbin, and P. H. Sandstrom, in *The Vanderbilt Rubber Handbook*, R. F. Ohm, Ed., R. T. Vanderbilt Co., Inc., Norwalk, CT, 14th ed., 1995 (CD-ROM version).

54. W. M. Hess, C. R. Herd, and P. C. Vegvari, *Rubber Chem. Technol.*, **66**, 329 (1993).

55. J. F. O'Mahoney, *Rubber Age*, **102**, March, 47 (1970).

56. G. J. Lake, P. B. Lindley, and A. G. Thomas, in *Proc. 2nd Int. Conf. on Fracture*, Brighton, England, P. L. Pratt, Ed., Chapman & Hall, London, 1969.

57. A. N. Gent and D. I. Livingston, in *Concise Encyclopedia of Composite Materials*, revised ed., A. Kelly, Ed., Pergamon, Elsevier, Tarrytown, NY, 1994.

MONOGRAPHS, EDITED WORKS, AND REVIEWS

Arends, C. B., Ed., *Polymer Toughening*, Marcel Dekker, New York, 1996.

Chandra, C. and Y. Ohama, *Polymers in Concrete*, CRC Press, Boca Raton, FL, 1994.

Jang, B. Z., *Advanced Polymer Composites: Principles and Applications*, ASM International, Materials Park, OH, 1994.

Kelly, A., Ed., *Concise Encyclopedia of Composite Materials*, revised ed., Pergamon, Elsevier, Tarrytown, NY, 1994.

Lipatov, Yu. S., *Polymer Reinforcement*, ChemTec, Toronto–Scarborough, Ontario, 1995.

Mark, J. E., B. Erman, and R. R. Eirich, Eds., *Science and Technology of Rubber*, 2nd ed., Academic, San Diego, 1994.

Nielsen, L. E. and R. F. Landel, *Mechanical Properties of Polymers and Composites*, 2nd ed., revised and expanded, Marcel Dekker, New York, 1994.

Woodward, A. E., *Understanding Polymer Morphology*, Hanser, Munich, 1995.

Wool, R. P., *Polymer Interfaces: Structure and Strength*, Hanser, Munich, 1995.

STUDY PROBLEMS

1. A ternary blend of PP, EPDM, and PE are mixed in 75/15/10 proportions as described in Section 2.3.3.1. The ternary blend is then cooled to 25°C. Young's moduli of the three components are, respectively, 2×10^9 dyn/cm^2, 3×10^7 dyn/cm^2, and 7×10^8 dyn/cm^2. Using the Takayanagi models, what is Young's modulus of the blend?

2. If the viscosity of the polypropylene of problem 1 is 3×10^7 Pa/s, what possible values of the viscosity can the other two components have if the polypropylene is to be the continuous phase, the EPDM discontinuous, with domains of PE only inside of the EPDM?

3. What are the chemical structures of the following?
   a. polyisoprene-*graft*-polystyrene
   b. polybutadiene-*block*-poly(methyl methacrylate)
   c. *net*-polyethylene

4. How does osmium tetroxide act to stain polymer blends for transmission electron microscopy? What kinds of blends can be usefully stained?

5. What applications (actual or imaginary) do the following composites have?
   a. long oriented fiber–plastic
   b. carbon black–rubber
   c. wood–poly(methyl methacrylate)
   d. concrete–poly(methyl methacrylate)

6. A sample of rubber, Young's modulus of 1.5×10^6 Pa at 25°C, is loaded with 20% v/v of glass spheres, Young's modulus of 8×10^{10} Pa. What modulus do you predict for the composite? Use the following:
   a. The appropriate Takayanagi model
   b. The Kerner equation (assume $\nu = 0.40$)
   c. The Guth–Smallwood equation
   d. The Davies equation
   Do you think that all of the above models apply equally well? Explain.

7. The rubber-toughened epoxy in Table 3.1 is formed into a wide, thin sheet. If a stress of 100 MPa is applied, what sized center-notched crack can be tolerated without failure?

8. You are asked to formulate rubber blends suitable for making tires in (a) Los Angeles, (b) Alaska, and (c) Kansas. What compositions do you recommend, and how will they differ?

9. An elastomer of Young's modulus of 2×10^6 Pa at 25°C is blended at a 20% level with a plastic of Young's modulus of 3×10^9 Pa. (a) Using the Takayanagi models, what is Young's modulus of the blend?

10. If the blend in Problem 9 is mixed at 200°C, and the elastomer has a melt viscosity of 1×10^4 Pa/s, and it is required that the plastic phase be continuous, what are the viscosity requirements of the plastic?

PART II

POLYMER SURFACES
AND INTERFACES

4

BASIC PRINCIPLES AND INSTRUMENTS

4.1. CLASSES OF SURFACES AND INTERFACES

There are five basic classes of surfaces and interfaces involving polymers:

1. A surface, which refers to that portion commonly exposed to air, that people can see and touch. Strictly speaking a surface (or *free* surface) refers only to a part of a pure condensed substance exposed to a vacuum. In the ordinary sense, however, the surface may be oxidized, oily, or dirty; see Figure 4.1. However, for purposes of this chapter, the surface will be assumed pure polymer, unless otherwise stated.

2. A dilute polymer solution–colloid interface, where polymer chains in solution are partly adsorbed onto a surface, often colloidal in nature. A single polymer chain may be adsorbed at a number of sites, the remaining mers sticking out into the solution. The bonding and characteristics of such polymer chains have been treated by Fleer et al. (1).

3. A symmetric interface, where two more or less identical polymers are in contact with each other. The healing and fracture of symmetric polymer interfaces has been recently treated by Wool (2). Briefly, healing or welding involves the contact of two melt surfaces not previously in contact. Wool describes the process on a molecular scale via a minor chain reptation model, where chains disengage themselves from their initial tubes, these minor chains then poking themselves through the interface, gradually healing it.

4. An asymmetric polymer interface, involving two different polymers, where the interface may remain indefinitely if the polymers are immiscible. Interpenetration at the interface ranges from a depth of a few to several nanometers, depending on the statistical segment length and χ_1. The term *interphase* is commonly used in such cases, and the interphase region may have physical properties distinctly different from either polymer. The

119

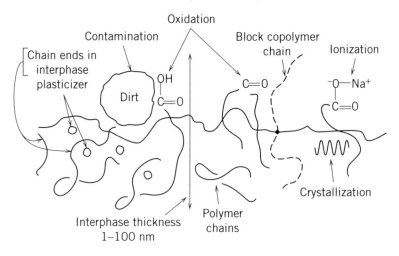

Figure 4.1. Some of the features of polymer surfaces and interfaces.

interphase may contain chemical bonds uniting the two surfaces such as graft or block copolymers, or physical bonds such as hydrogen bonds.

5. A composite interface between a polymer and a nonpolymer solid phase such as glass or boron fibers, steel, or calcium carbonate particles. In such cases, the polymer is usually unable to interdiffuse into the contacted surface, but may adhere to it with a variety of physical or chemical bonding modes.

Since this book centers on polymer blends and composites, the asymmetric and composite interfaces will be the principal focus. However, all of the above have a great deal in common. In addition, one cannot properly understand any kind of polymer interface in detail without some understanding of the properties of each of the other possibilities. Therefore, this chapter will consider important aspects of each of the other classes of surfaces and interfaces.

4.2. SURFACE TENSION, INTERFACIAL TENSION, AND WORK OF ADHESION

All of the mixtures of polymers with reinforcing agents, and all of the blends of two or more polymers except those that are thermodynamically miscible exhibit interfaces. The reinforcing capability of the added phases depends critically on the strength of these interfaces. Usually , the better the bonding, the better the

Table 4.1   Surface tensions of selected polymers

Polymer	Surface Tension, γ (erg/cm^2)		$-(d\gamma/dT)$ (erg/cm^2-deg)
	20°C	140°C	
Polyethylene	35.7	28.8	0.057
Polystyrene	40.7	32.1	0.072
Poly(methyl methacrylate)	41.1	32.0	0.076
Poly(dimethyl siloxane)	19.8	14.0	0.048
Poly(ethylene oxide)	42.9	33.8	0.076
Poly(ethylene terephthalate)	44.6	28.3	0.065
Polycarbonate	49.2	35.1	0.060
Polytetrafluoroethylene	23.9	16.9	0.058
Poly(n-butyl methacrylate)	31.2	24.1	0.059
Poly(vinyl acetate)	36.5	28.6	0.066
Polychloroprene	43.6	33.2	0.086

Source: S. Wu, *Polymer Interface and Adhesion*, Marcel Dekker, Table 3.7, p. 88.

material. However, there may be a maximum in mechanical properties, depending on the bonding energy per unit of interface area. An important measure of surface and interface energy is the surface tension.

Since creation of a new surface requires work, the free energy of its formation is always positive. The work, W, required to create a unit of surface area, dA, is given by (3)

$$W = \gamma dA \qquad (4.1)$$

where the surface tension, γ, has cgs units of ergs/cm^2, or SI units of J/m^2. However, since work can be expressed as force times distance, the units are sometimes presented as dynes per centimeter or Newtons per meter.

Table 4.1 (4) shows the surface tensions for various polymers. It is immediately noted that the surface tension of most polymers is low compared with inorganic materials. For example, the surface tension of mercury is 486.8 ergs/cm^2, and that of water is 72.94 ergs/cm^2.

Surface tension is generally a decreasing function of temperature so that $-(d\gamma/dT)$ in Table 4.1 is positive. This will be interpreted below in terms of disorder and entropy changes at the surface.

While liquid polymer–air surface tensions are of importance in their own right, a quantity of great interest to polymer blend people is the interfacial tension, γ_{ab}, between two polymers, a and b. The interfacial tensions are usually significantly smaller than the polymer–air surface tensions.

Another quantity of interest is the spreading coefficient, $S_{b/a}$, which determines whether liquid b will spread over liquid a, or bead up. The thermodynamic basis is as follows (3).

Consider a drop of liquid b placed on the liquid surface of a. (For the present purposes, both a and b are polymers.) The surface free energy change on increasing the contact area by dA is given by

$$dG = (\partial G/\partial A_a)dA_a + (\partial G/\partial A_b)dA_b + (\partial G/\partial A_{ab})dA_{ab} \qquad (4.2)$$

However, the same areas are involved:

$$dA_b = -dA_a = dA_{ab} \qquad (4.3)$$

and as noted above, $\partial G/\partial A_a = \gamma_a$, and so forth. The coefficient, $-(\partial G/\partial A_b)$ represents the free energy change for the spreading of liquid polymer b over liquid polymer a, and is given the special term *spreading coefficient* of b over a. Thus,

$$-\partial G/\partial A_b = S_{b/a} = \gamma_a - \gamma_b - \gamma_{ab} \qquad (4.4)$$

If $S_{b/a}$ is positive, spreading is accompanied by a decrease in free energy and is a spontaneous process. If

$$\gamma_{ab} > |\gamma_a - \gamma_b| \qquad (4.5)$$

then $S_{b/a}$ must always be negative, and neither polymer liquid spreads on the other. Table 4.2 (4) shows the interfacial tensions of various polymer blends.

Another quantity of interest is the work of adhesion, W_a, defined as the reversible work per unit area required to separate the interface between two bulk phases a and b from their equilibrium positions to infinity:

$$W_a = \gamma_a + \gamma_b - \gamma_{ab} \qquad (4.6)$$

This is a lower bound value.

Table 4.2   Polymer blend interfacial properties

Polymer Pairs, b/a [a]	γ_{ab} (erg/cm², 140°C)	$\lambda_{b/a}$ [a] (erg/cm², 140°C)	$-d\gamma_{ab}/dT$ (erg/cm²-deg)	W_a (erg/cm², 140°C)	Shear Strength (dyn/cm²)
PMMA/PS	1.7	−1.6	0.013	62.4	6.5×10^7
PE/PMMA	9.7	−6.5	0.018	51.1	5.2×10^7
PnBM/ PVA	2.9	1.6	0.010	49.8	9.6×10^7
PDMS/PCP	6.5	12.0	0.0050	40.8	1.3×10^8
PB/PS[b]	3 (100°C)	—	~0	—	—

[a] Abbreviations: PMMA, poly(methyl methacrylate); PDMS, poly(dimethyl siloxane); PS, polystyrene; PCP, polychloroprene; PE, polyethylene; PnBM, poly(n-butyl methacrylate); PVA, poly(vinyl acetate).
[b] From U. Bianchi, E. Pedemonte, and A. Turturro, *Polymer*, **11**, 268 (1970).
Source: From S. Wu, *Polymer Interface and Adhesion*, Marcel Dekker, New York, Table 11.1, p. 362, and Table 3.20, p. 126.

The surface tension, spreading coefficient, and the work of adhesion are all material properties, governed primarily by the chemical structure of the polymers, assuming equilibrium conformations, and so forth. However, the work of adhesion and other quantities such as the shear strength between the surfaces can be significantly improved by higher molecular weights and/or especially by various interfacial additives. Thus, a block or graft copolymer is sometimes used as a compatibilizer, with the effect of bonding the two surfaces together (5,6). Alternate routes involve actual grafting of the two surfaces together with primary chemical bonds, or through the use of secondary bonding such as hydrogen bonding. Each of these items will be major sections in this part of the book.

4.3. EXAMPLE CALCULATION: SPREADING COEFFICIENTS

Will poly(methyl methacrylate) (PMMA) spread over polystyrene (PS) at 140 °C, when both are in the melt state?

From Table 4.1, $\gamma(\text{PMMA}) = 32.0\,\text{ergs/cm}^2$, and $\gamma(\text{PS}) = 32.1\,\text{ergs/cm}^2$, and from Table 4.2, $\gamma(\text{PS/PMMA}) = 1.7\,\text{ergs/cm}^2$. Starting with equation (4.4),

$$S(\text{PS/PMMA}) = (32.1 - 32.0 - 1.7)\,\text{erg/cm}^2 = -1.6\,\text{ergs/cm}^2$$

Since $S(\text{PS/PMMA})$ is negative, poly(methyl methacrylate) will not spread over polystyrene. From equation (4.5),

$$|32.1 - 32.0| < 1.7$$

so that, in fact, neither polymer melt will spread over the other.

4.4. SURFACE TENSION THERMODYNAMICS

The inhomogeneous organization of the atoms at the surface of a condensed phase causes the phenomenon known as surface tension, γ. Again, γ is the reversible work to create a unit surface area in a substance. The surface tension is sometimes called the specific surface energy, the intrinsic surface energy, or the true surface energy. The free energy associated with the surface is called the surface free energy. It is equal to one half of the work of cohesion, W_c, which is the reversible work required to separate two identical surfaces of unit area, see Figure 4.2. The units are energy/area. By contrast, the work of adhesion, W_a, is the reversible work required to separate two dissimilar surfaces, per unit area, see equation (4.6). The simple relationship $W_a = 2\gamma$ forms the lower bound of the adhesion energy. Interdiffusion, chain entanglements, and primary and secondary interfacial bonding may raise W_c by a factor of several hundred.

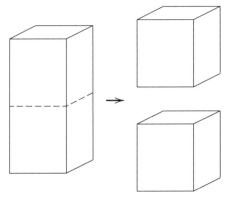

Figure 4.2. Literal illustration of the work of cohesion. $W_{coh} = 2\gamma$ through the creation of *two* surfaces.

Just as bulk materials have thermodynamic properties, so do surfaces and interfaces. The change in the free energy of a surface, dG, on the change in the area of a surface dA, is given by

$$dG = \gamma dA \qquad (4.7)$$

The surface free energy, G^s, is given by

$$G^s = (\partial G/\partial A)_{T,P} = \gamma \qquad (4.8)$$

where the superscript s means a surface property. Following classical thermodynamics,

$$(\partial G^s/\partial T)_P = -S^s \qquad (4.9)$$

and

$$d\gamma/dT = -S^s \qquad (4.10)$$

Thus, the quantities $-d\gamma/dT$ shown in Tables 4.1 and 4.2 can be interpreted in thermodynamic terms. As shall be seen, this information provides evidence for changes in the orientation of polymer chains at surfaces and interfaces.

4.5. SURFACE TENSION EXAMPLE CALCULATIONS

What is the surface entropy of a polymer with a surface tension of 19.8 ergs/cm^2 at 20°C and a surface tension of 14.0 ergs/cm^2 at 140°C?

From equation (4.10),

$$S^s = -(14.0 - 19.8)/(140 - 20) \, \text{erg/cm}^2 \, \text{K} = +0.048 \, \text{erg/cm}^2 \, \text{K}.$$

This is the value calculated from Table 4.1 for poly(dimethyl siloxane).

4.6. CHARACTERIZATION OF POLYMER SURFACES AND INTERFACES

A number of methods have been developed to examine the properties of polymer surfaces and interfaces, see Table 4.3 (7,8). This section will provide a brief summary of the methods and instruments available.

4.6.1. Analysis of Polymer Surfaces

4.6.1.1. *Measurements of Surface Tension*   Surface tensions in polymers can be easily measured in the melt state. Methods such as the pendant drop method or the sessile drop method (3) provide reasonable results. However, measurement of the surface tension of a solid polymer cannot be made directly, as reversible formation of its surface is difficult. A variety of indirect methods have been used successfully, including the liquid homolog molecular weight dependence method, the polymer melt–temperature extrapolation method, and some theoretical approaches (4).

For example, the polymer melt–temperature extrapolation method can be used by measuring the surface tension of a polymer in the melt state over a range of such temperatures, and extrapolating to the temperature of interest. The quantity $-d\gamma/dT$ varies linearly with temperature, with typical values being about 0.05 erg/cm^2 deg. Of course, as shown in equation (4.10), this quantity is also the surface entropy.

The interfacial tension between polymers is often more difficult to determine. One method makes use of a variation of the polymer melt–temperature method mentioned. Measurements of both polymers are made in the melt state, and the values extrapolated to room temperature.

Generally, surfaces can be classified into two types: high-energy surfaces and low-energy surfaces. Materials with high-energy surfaces include metals, metal oxides, silica, and diamond. For example, mercury has a surface tension of 486.8 ergs/cm^2, and water has a value of 72.94 ergs/cm^2. The low-energy surface materials include most organic compounds and polymers. Here, toluene has a surface tension of 28.52 ergs/cm^2, while polystyrene has a value of 40.70 ergs/cm^2. The low-energy materials tend to adsorb strongly onto the high-energy surfaces, greatly decreasing the surface energy of the system. Since fillers and reinforcing agents in polymer composites are usually high-energy materials, it is of direct interest to determine the interfacial surface tensions and contact angles of such systems.

Perhaps the most versatile method for determination of contact angles is known as the Wilhelmy plate method. In this experiment, a thin plate of the high-surface energy material is immersed in a melt of the low-energy surface material. Through a measure of the advancing and receding contact angles, much about the interaction energies between the polymer and the filler can be determined (9). Depending on the experimental requirements, either water or an organic liquid can be used to form the drops on the polymer.

Table 4.3   Methods of characterizing surfaces and interfaces

Technique	Advantages	Disadvantages
Dynamic light scattering	Yields thickness of adsorbed polymer layers on latexes and particles	Limited by the diffusion coefficients of the particles
Force–balance apparatus (surface forces apparatus)	Determines adsorbed layer thickness on solid–liquid surfaces	Electrostatic or other forces must be absent
Electron spectroscopy for chemical analysis (ESCA), also called XPS	Sampling depth is relevant to surface interactions ($\sim$10–100 Å) All elements detected (except H and He) High sensitivity Simple sample preparation Well-established theory Nondestructive Many levels of information can be obtained Hydrated samples can be observed	Expensive Artifact-prone High vacuum required
Surface tension and contact angle methods	Inexpensive Rapid Provides information related to surface energetics Hydrated samples can be observed	Artifact-prone: swelling of substrate, leaching, etc. Liquid purity is of critical importance Only surface energetics information is obtained
Atomic force microscopy	Inexpensive Sample may be insulator	Atomic resolution difficult
Scanning electron microscopy (SEM)	Rapid Visually rich Three-dimensional perspective	Differences in secondary electron emission can be confused with topographic features
Attenuated total reflection (ATR) infrared	Readily available Spectra are rich in chemical information Orientation information can be obtained	Analysis to 1-μm depth studies both the bulk and surface Some samples are not amenable to ATR analysis
Secondary ion mass spectrometry (SIMS)	Looks at a highly surface-localized region ($\sim$10 Å) Rich in chemical information Depth profiling (destructive) is possible	Expensive Artifact-prone Theory, particularly for polymers, is not developed
Evanescent wave method	Provides information on adsorbed layers	Adsorbate must interact with radiation

Table 4.3   Continued

Technique	Advantages	Disadvantages
Auger electron spectroscopy (AES) or scanning Auger microprobe (SAM)	Rapid Good spatial resolution $(2.5 \times 10^{-3} \mu m^2)$ All elements detected	Destructive to organics Limited chemical information Quantitative analysis is difficult
Small-angle neutron scattering (SANS)	Interior interfaces can be characterized	Expensive Deuterated probe molecules often required
Scanning tunneling microscopy	Atomic resolution Relatively inexpensive	Resolution limited with insulating materials
Light scattering	Determines interior surface areas	Secondary scattering must be eliminated

Problems of wettability and contact angle are fundamental to understanding interfaces. If a drop of liquid is placed on a surface, either it wets the surface, and (in principle) spreads to cover the entire surface, or else it beads up. For drops that bead up, the contact angle, θ, provides basic information about the interface forces involved, see Figure 4.3 (4).

There are three vector forces determining θ: γ_{LV}, which wants to bead the drop up, γ_{SV}, which wants to cover the surface over; and γ_{SL}, which wants to clear the surface. The subscripts S, L, and V stand for solid, liquid, and vapor, respectively. It is assumed that the vapor from the liquid phase is saturated, and that the system is at equilibrium.

A simple balance of the three vectors leads to Young's equation, dating from 1805 (10):

$$\gamma_{SV} = \gamma_{SL} + \gamma_{LV} \cos \theta \qquad (4.11)$$

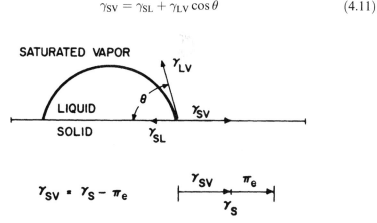

Figure 4.3. Vector forces controlling the contact angle of a liquid drop on a smooth, homogeneous, planar, rigid surface.

Table 4.4   Selected contact angle data

Liquid (γ, ergs/cm^2)	Solid	Advancing θ (deg)	$\dfrac{d\theta}{dT}$	Ref.
Water (72)	Polytetrafluoroethylene	108	—	a
Water (72)	Polyethylene	96	−0.11	b
CH$_2$I$_2$ (67)	Polyethylene	46	—	c
n-Octane (21.6)	Polytetrafluoroethylene	30	−0.12	d

[a] W. A. Zisman, *Adv. Chem. Ser. No.* **43**, 1964.
[b] F. D. Petke and B. R. Ray, *J. Coll. Interf. Sci.*, **31**, 216 (1969).
[c] A. El-Shimi and E. D. Goddard, *J. Coll. Interf. Sci.*, **48**, 242 (1974).
[d] C. L. Sutula, R. Haritala, R. A. Dalla Betta. and L. A. Mitchel, *Abstracts*, 153rd Meeting, American Chemical Society, April, 1967.
Source: Based on A. W. Adamson, Ref. 3.

Young's equation has two measurable quantities, $\cos\theta$ and γ_{LV}. Unfortunately, determining the other two quantities independently still presents a problem.

While Figure 4.3 deals with the equilibrium situation for a drop on an idealized surface, more can be learned by a study of a drop in motion. Consider a drop on an inclined plane. There are two contact angles, the advancing contact angle, θ_a, and the receding contact angle, θ_r. There are three main causes of the hysteresis between the two angles:

1. Contamination of either the solid or liquid surface
2. Rough surfaces
3. Low mobility in the form of either slow desorption or hindered spreading

Selected contact angles are shown in Table 4.4. As might be expected from common experience, water beads up on polytetrafluoroethylene, while *n*-octane, a much less polar material, spreads but does not completely wet polytetrafluoroethylene either. The contact angle also usually decreases as the temperature increases, suggesting increased tendency to wet the surface.

4.6.1.2. *ESCA, AES, STM, TEM, AFM, and Evanescent Wave Methods*   Electron spectroscopy for chemical analysis (ESCA), sometimes known as X-ray photoelectron spectroscopy (XPS), is an experimental method that measures the electron binding energies of the various elements within 5 nm or so from the surface. The binding energies differ from bond type to bond type, and so are sensitive to the chemical environment (11,12). Often, instruments operate with an Al Kα X-ray source at 1486.6 eV.

ESCA is extremely valuable for identifying chemical structures at the surface. For example, Yilgor and McGrath (13) found that in a block copolymer blend of poly(dimethyl siloxane) and polycarbonate, only about 1% of poly(dimethyl siloxane) was sufficient to cover 50–70% of the surface, see Figure 4.4. The

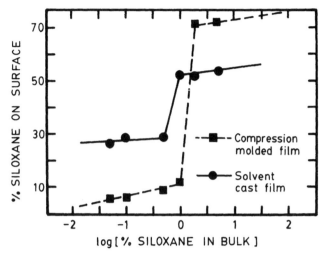

Figure 4.4. Surface segregation in polycarbonate-*blend*-[polycarbonate-*block*-poly(dimethyl siloxane)] copolymer blends. The poly(dimethyl siloxane) portion covers 50–70% of the surface with only 1% of the bulk concentration.

lower surface tension of the poly(dimethyl siloxane) causes it to cover the higher surface tension polycarbonate. LeGrand and Gaines (14) had already found that contact angles of similar blends displayed values very close to that of pure poly(dimethyl siloxane), providing important supporting evidence. In a more recent ESCA study, the surface of a poly(ethylene terephthalate)–perfluoro polyether block copolymer was found to be 10–20 times enriched in the fluoropolymer (15).

An important requirement in these systems is that the polymer to be identified have at least one atom different from the atoms of the other polymer(s). Thus, polymers bearing halogen, oxygen, sulfur, or silicon atoms can be quantitatively identified on surfaces in the presence of polymers that lack these atoms. Different bonding types can also sometimes be noted.

In Auger electron spectroscopy (AES), an atom in an M^+ excited state emits another electron to form an M^{2+} lower lying state. Surface examination is limited to 0.3–0.6 nm. In this process, an electron beam gun fires electrons at a low angle of incidence to the sample, thus increasing the interaction with surface atoms. The energy and number of emitted Auger electrons are detected by an energy analyzer and counter. AES is used for measuring surface oxidation, surface strain, and a variety of lubrication problems.

Scanning tunneling microscopy (STM) was invented in 1982 by Binnig and Rohr (16). This is a powerful tool for the study of conducting surfaces at atomic resolution (17,18). Under good conditions, STM can resolve individual atoms on a surface; see Figure 4.5 (19). The motion of the electrons between the two electrodes is due to a quantum mechanical tunneling effect. To build up a three-dimensional topographic surface map, the probe is scanned in a rasterlike

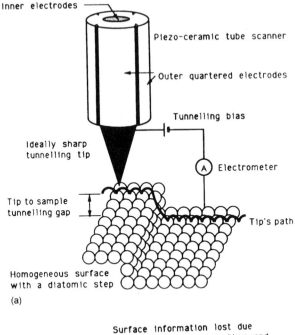

Inner electrodes

Piezo-ceramic tube scanner

Outer quartered electrodes

Tunnelling bias

Ideally sharp
tunnelling tip

(A) Electrometer

Tip to sample
tunnelling gap

Tip's path

Homogeneous surface
with a diatomic step

(a)

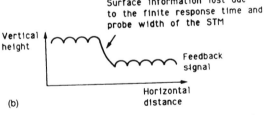

Surface information lost due
to the finite response time and
probe width of the STM

Vertical
height

Feedback
signal

Horizontal
distance

(b)

Figure 4.5. Scanning tunneling microscopy illustrating (a) the scanning head and (b) the feedback signal scan line. Note loss of information due to steep edges.

manner above the surface. The tunneling current is maintained at a constant preset value. A major limiting factor of its use in polymer science is the requirement of electrical conductivity.

An important derivative method to STM is atomic force microscopy (AFM). In this method, an atomically sharp tip, usually made of diamond or silicon nitride, is drawn across a surface and kept in contact by a very soft spring, see Figure 4.6 (19). In a fashion similar to the STM, the top is then raster scanned across the sample. The AFM then records contours of constant force due to the repulsion generated by an overlap between the electron clouds of the tip and those of the surface atoms. The method works well with insulators, and thus can provide useful results with many polymers. Polymeric structures may be resolved to nanometer levels. Figure 4.7 (20) illustrates polyalanine, a polypeptide, proteinlike molecule made up of only one amino acid. A drop of polyalanine solution in 85/15 chloroform/trifluoroacetic acid was placed on a

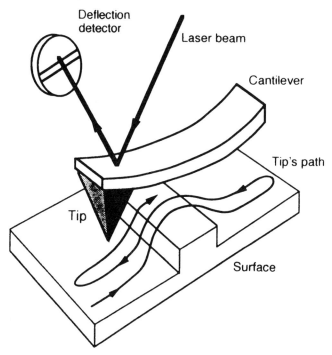

Figure 4.6. Main components of an atomic force microscope. The tip is tracked or rasped across the surface, keeping the contact force constant. The topographic detail is in the subnanometer range.

washed microscope slide, and the excess solution blown off with filtered compressed air. The slide was then washed with filtered deionized water and blown off with clean compressed air. The sample was then put into the AFM and rehydrated with a drop of water. While the polymer chains are clearly visible, the pictures lack atomic resolution. The polymer chains that do not wash off are clearly bonded to the surface by secondary adhesion forces. They are seen to lie in a surprisingly parallel array.

Often, an atomic force microscope of the optical deflection type operates at constant force in the contact (repulsive) mode in which the tip always touches the surface when the feedback loop is on. Such an instrument was used to study the fracture behavior of rubber-toughened epoxies (21). In particular, core–shell latexes, with styrene–butadiene rubber (SBR) cores and poly(methyl meth-acrylate-*stat*-acrylonitrile) shells were used to toughen an epoxy matrix of DGEBA epoxy (DER 331 from Dow Chemical), the latter cured with piperidine. The epoxy contained 10 vol/% of the latex particles. An atomic force microscope figure of the stress-whitened zone is shown in Figure 4.8 (21). This AFM scan strongly suggests cooperative cavitation; note troughs of particles. Also, rims around the cavitated particles can be seen. While Figure 4.8 was taken at a lower magnification than Figure 4.7, the objective was quite different.

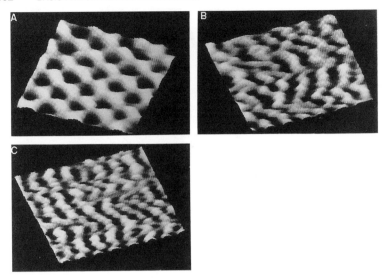

Figure 4.7. Computer-processed atomic force microscopy images. (a) Crystalline mica. Note hexagonal array characteristic of cleaved mica. The image is 26 Å × 26 Å. (b) Polyalanine adsorbed on glass and covered with water. (c) Same as (b), but dry. Note the outline of the polymer chains, which appear to be closer packed without the water.

Optical microscopy has been used to study surfaces of materials for many years, continuing into the present. However, conventional optical microscopy can resolve features only down to about 500 nm. Transmission electron microscopy (TEM), sees wide use for the study of the interior of materials, down to 1–10 nm. TEM works by having electrons pass through the sample. Regions of space that have higher density, and hence more electrons per unit volume, tend to scatter and/or absorb more electrons than less dense regions. Frequently, this effect is enhanced in polymers by the addition of heavy-metal

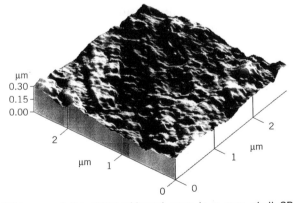

Figure 4.8. AFM scan of the stress-whitened zone in a core–shell SBR/P(MMA-*stat*-AN) toughened epoxy. Note appearance of craze line.

staining agents, osmium tetroxide, for example. In other cases, the natural density differences between polymers may be enhanced by the use of high contrast photography. In any case, for polymer blends, the end result is that the more dense phase appears darker. The morphology of polymer blends and composites is easily studied; see Figure 2.5, for example. However, surface properties as such are often difficult to determine.

While TEM is used to examine thin sections, scanning electron microscopy (SEM), is used to examine surfaces. Resolution down to a few hundred nanometers is often possible. A focused beam of electrons scans the surface. The intensity of the secondary, or scattered electrons, is monitored. The output from the secondary electron detector can be seen on a cathode-ray tube. The strength of the image varies from point to point according to the intensity of the secondary electron production. The images from SEM characteristically have a wide range of contrast, producing great depth of focus, with results that are often exciting to behold; see Figure 2.3.

Light can be totally internally reflected when a ray tries to pass from a medium of higher refractive index to one of lower refractive index. Under conditions of total reflection, a portion of the incident wave does, however, penetrate the medium of the lower refractive index to a depth of up to about 60 nm and may be absorbed or cause fluorescence, thus providing information about adsorbed polymer molecules. This nonclassical penetration by light is called the evanescent wave method. In a typical experiment, the light is passed down a rectangular glass prism, onto which a polymer is adsorbed. The light strikes the upper and lower surfaces of the glass, being slightly attenuated during reflection, but interacting with the polymer adsorbed on the surface, called total internal reflectance fluorescence (TIRF). Often, the experiment is carried out under a solvent or dispersant for the polymer.

4.6.2. Interior Polymer Interfaces

The polymer interface area in the interior of a polymer blend or composite sample can be determined via small-angle neutron scattering (SANS) or small-angle X-ray scattering (SAXS). Debye et al. (22) derived the basic theory for SAXS. Debye and co-workers examined porous particles of interest in catalysis. Such materials yield very ample differences in electron density, required for good scattering intensity.

When used for SANS, one phase of the material is usually deuterated to provide proper contrast (23,24). Debye et al. (22) assumed a randomly structured material in which the probability of striking a surface at an arbitrary distance could be provided by a correlation function in an exponential form:

$$\gamma(r) = \exp(-r/a) \tag{4.12}$$

where $\gamma(r)$ represents the correlation function for a characteristic distance r between scattering centers, and the quantity a represents the correlation distance defining the size of the heterogeneities. The specific interfacial surface area, S_{sp}, is defined as the ratio of interfacial surface area, A, to the mass, m,

$$S_{sp} = A/m = 4\phi(1 - \phi)/ad \qquad (4.13)$$

where ϕ represents the volume fraction of either phase, and d is the density. (Then, of course, $1 - \phi$ represents the volume fraction of the remaining phase.)

The theory calls for a plot of $I^{-1/2}$ vs. K^2, where I is the excess scattered intensity (after removal of background, etc.), and $K = 4\pi\lambda^{-1}\sin(\theta/2)$, where λ represents the wavelength, and θ is the scattering angle. The quantity a is then given by

$$a = (\lambda/2\pi)\,(\text{slope/intercept})^{1/2} \qquad (4.14)$$

where the slope and the intercept refer to the line drawn by plotting $I^{-1/2}$ vs. K^2.

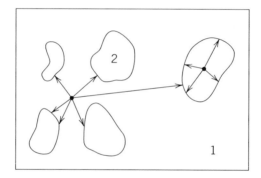

(a) Correlation distance, a

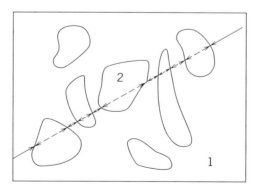

(b) Transverse length, l_1, l_2

Figure 4.9. (a) The correlation distance a is taken from a random spot in a random direction. The distance to an interface, on average, is a. (b) A straight line is drawn through the material. The average distance between two interfaces determines L_1 and L_2.

The same theory can also be interpreted in terms of the transverse lengths across the domains, L_1 and L_2 of phases 1 and 2;

$$L_1 = a/\phi_2, \qquad L_2 = a/\phi_1 \tag{4.15}$$

The correlation distances and the transverse lengths are illustrated in Figure 4.9. Typical results are shown in Figure 4.10 (24) for a *cross*-polybutadiene-*inter*-*cross*-polystyrene interpenetrating polymer network (IPN). A maximum in specific area is shown in the midrange compositions. Specific surface areas were of the order of 50–300 m^2/g. These values are typical for colloidally dispersed materials and finely divided polymer blends, as well as many IPNs. If the SANS or SAXS data exhibit peaks in scattering intensity, a further treatment of such data will be found in Appendix 9.1, including a method of determining the interfacial thickness.

4.6.3. Surface Forces Apparatus

Some of the techniques already described are also useful for adsorbed polymer chains, particularly the evanescent wave method. The surface forces apparatus has been particularly useful for such studies.

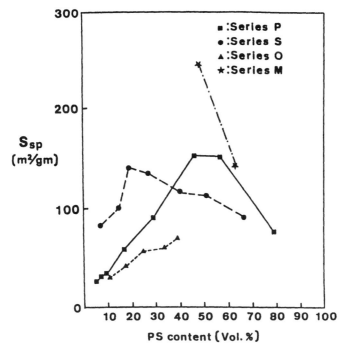

Figure 4.10. Specific interfacial surface area goes though a broad maximum in the midrange compositions in an IPN.

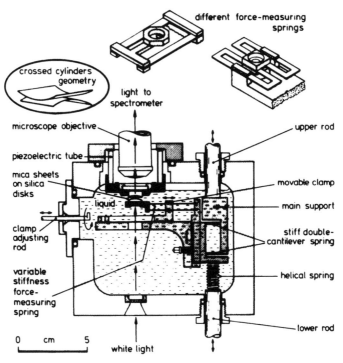

Figure 4.11. Schematic illustration of the surface forces apparatus, used for directly measuring the forces between surfaces in liquids or vapors at the Ångstrom resolution level. Both normal loads and shear stresses may be applied to the crossed cylinders.

Consider a polymer adsorbed on a surface, all immersed in a liquid phase. The polymer chains are, in general, partly dissolved, and those portions of the polymer, composed of loops and tails, may have conformations similar to (but not identical to) that of random coils. When two such adsorbed polymer surfaces approach each other, there will be an entropic repulsive force between them, among other possible forces. Measurement of these forces is the purpose of the surface forces apparatus (SFA) (25,26). Figure 4.11 (27,28) illustrates an SFA with which the force between two surfaces in controlled vapors or immersed in liquids can be directly measured. The resolution in distance is about 0.1 nm. Typically, the SFA contains two curved molecularly smooth surfaces of mica of radius of about 1 cm. The interaction forces are measured using force-measuring springs. A popular configuration has the two surfaces in a crossed cylinder configuration, locally equivalent to a sphere near a flat surface or to two spheres close together. The distance between the two cylinders is measured with an optical technique utilizing interference fringes. The motion of the two surfaces can be toward each other, resulting in the measure of entropic interference between the two layers of chains, or laterally, resulting in a measure of friction between the chains.

The SFA is similar in principle to the AFM. In the SFA, forces are measured between two macroscopic surfaces. In the case of AFM, a fine tip within molecular distance of a surface is employed. Tips approaching one atom in dimension are currently of the most interest.

If the polymer is adsorbed on colloidal particles, the effective diameter of the particles is increased by the presence of the polymer, and the effective diffusion coefficient, D, will be decreased. The changes in average motion can be studied by dynamic light scattering, also known as quasi-elastic light scattering or photon correlation spectroscopy. Dynamic light scattering uses monochromatic laser light to probe the particle dynamics via a Doppler broadening technique. If the particle is moving toward the beam, the wavelength of the scattered light will be shortened, and if it is moving away from the beam, the wavelength will be lengthened. The total effect is to broaden the wavelength of the scattered light, in proportion to the average velocity of the particles. The diffusion coefficient is related to the particle radius by the Stokes–Einstein equation:

$$D = kT/6\pi\eta a \qquad (4.16)$$

where a is the particle radius and η is the viscosity of the medium. Ordinarily, the results are obtained through the use of Fourier transforms and computer analysis.

4.6.4. Neutron Reflection

One of the newest methods of determining surface and interfacial characteristics involves neutron reflection (29). The method depends on utilizing systematic differences in the neutron equivalent for refractive index, which is the scattering length. One of the largest differences in scattering lengths is that between the proton and deuteron isotopes, also one of the easiest combinations to prepare. By selective deuteration, the spatial distribution of the different portions of diblock copolymers can be determined to 1500 Å below the surface (30).

Experimentally, the measurement of reflectivity involves radiation impinging neutrons on a surface at a grazing angle θ and measuring the number of neutrons reflected at the same grazing angle (in the forward direction). Either the angle or the wavelength of the experiment can be varied.

Thus, thin layers of block copolymers (with one block deuterated) on flat surfaces have been shown to develop characteristic microlayers, each either one or two blocks thick, depending on the composition and layer thickness; see Section 7.6.

4.6.5. Adhesion and Surface Energies via Direct Contact Deformation

When a curved elastic material comes into contact with another curved or flat material, the forces of adhesion across the interface tend to deform the materials in such a way as to increase their mutual area of contact. This phenomenon is

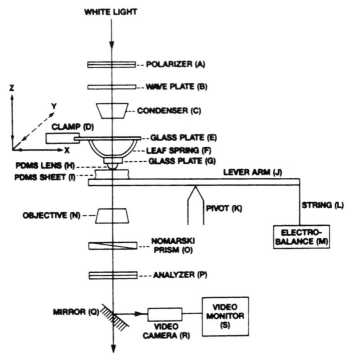

Figure 4.12. Chaudhury-Whitesides apparatus used to measure the contact deformatin between elastic polymers. The polymeric lens (H) should be rubbery and capable of significant deformation. While the video monitor(s) record(s) the contact area, the electrobalance (M) records extra loads.

the basis of the JKR (Johnson, Kendall, and Roberts) theory of contact mechanics (31).

Recently, Chaudhury and Whitesides (32,33) applied the JKR theory to the adhesion energy of chemically modified poly(dimethyl siloxane) surfaces as a function of composition. The instrument they designed used lens-shaped materials that were brought into contact with flat sheets of the same material, as shown in Figure 4.12 (32). Here, the samples are viewed in a video monitor, recording the actual contact deformation. The polymeric lens can be translated up, down, or sideways. Any extra load caused by adhesion is registered on an electrobalance. This experiment can be used for polymer blend and composite contact energies, as well as homopolymers. The samples may be examined dry, or in contact with liquids or special atmospheres that may modify the energies of interaction.

4.6.6. Solid-State NMR

Solid-state nuclear magnetic resonance (NMR) can be used to determine interphase thicknesses and details of morphology in such materials as polymer blends and core–shell latex particles (34,35). The experiments use combinations of ^1H spin-diffusion, two-dimensional (2D) wideline separation (WISE), cross polarization, high-power proton decoupling, and magic angle spinning (CPMAS–NMR), and so on, to elucidate morphologies. Of particular interest, this new field makes use of ordinary diffusion kinetics, where the mean-square distance, $\langle x^2 \rangle$, of the magnetization moves with time, going as $\langle x^2 \rangle = aDt$, where D is the spin-diffusion coefficient. The prefactor a depends on the geometry of the packing. Thus, distances across small domains as well as extent of mixing can be estimated. For morphology studies, the experiments are often done in conjunction with transmission electron microscopy, the NMR providing details at very short distances.

For example, the 2D-WISE experiment allows the combination of structural and dynamic information obtained from the chemical shift in the ^{13}C dimension and the proton line shape in the ^1H dimension, respectively. Such 2D solid-state NMR spectra reveal changes in mobility in different chemical surroundings making use of the ^1H NMR line widths. Polymers below their glass transition temperature tend to present broadened lines in the ^1H dimension due to the large dipolar couplings. Narrow lines may be superimposed on top of these, indicating greater chain mobility. In interphases, the dynamics of both rigid and mobile polymers are different from their respective dynamical behavior in the pure phases, allowing analysis of the interphase composition.

REFERENCES

1. G. J. Fleer, M. A. Cohen Stuart, J. M. H. M. Scheutjens, T. Cosgrove, and B. Vincent, *Polymers at Interfaces*, Chapman & Hall, London: 1993.

2. R. P. Wool, *Polymer Interfaces: Structure and Strength*, Hanser, Munich, 1995.

3. A. W. Adamson, *Physical Chemistry of Surfaces*, 5th ed., Wiley-Interscience, New York, 1990.

4. S. Wu, *Polymer Interface and Adhesion*, Marcel Dekker, New York, 1982.

5. D. J. Lohse, S. Datta, and E. N. Kresge, *Macromolecules*, **24**, 561 (1991).

6. C. C. Chen and J. L. White, *Polym. Eng. Sci.*, **33**, 923 (1993).

7. B. D. Ratner, *Ann. Biomed. Eng.*, **11**, 313 (1983).

8. B. D. Ratner, S. C. Yoon, and N. B. Mateo, in *Polymer Surfaces and Interfaces*, W. J. Feast and H. S. Munro, Eds., Wiley, Chichester, England, 1987.

9. G. S. Ferguson, M. K. Chaudhury, H. A. Biebuyak, and G. M. Whitesides, *Macromolecules*, **26**, 5870 (1993).

10. T. Young, *Phil. Trans. R. Soc. London*, **95**, 65, 84, (1805).

11. G. Beamson and D. Briggs, *High Resolution XPS of Organic Polymers*, Wiley, New York, 1992.

12. T. L. Barr, *Modern ESCA*, CRC, Boca Raton, FL, 1994.

13. I. Yilgor and J. E. McGrath, *Adv. Polym. Sci.*, **86**, 1 (1988).

14. D. G. LeGrand and G. L. Gaines, Jr., *Polym. Prepr.*, **11**, 442 (1970).

140 BASIC PRINCIPLES AND INSTRUMENTS

15. F. Pilali and M. Toselli, *Macromolecules*, **23**, 348 (1990).

16. G. Binnig, H. Rohrer, Ch. Gerber, and E. Weibel, *Phys. Rev. Lett.*, **49**, 57 (1982).

17. V. M. Hallmark, S. Chaing, J. F. Rabolt, J. D. Swalen, and R. J. Wilson, *Phys. Rev. Lett.*, **59**, 2879 (1987).

18. B. Hoffmann-Millack, C. J. Roberts, and W. S. Steer, *J. Appl. Phys.*, **67**, 1749 (1990).

19. M. C. Davies, D. E. Jackson, C. J. Roberts, S. J. B. Tendler, K. M. Kruesel, M. J. Wilkins, and P. M. Williams, in *Polymer Surfaces and Interfaces II*, W. J. Feast, H. S. Munro, and R. W. Richards, Eds., Wiley, New York, 1993, Chapter 9.

20. B. Drake, C. B. Prater, A. L. Weisenhorn, S. A. Gould, T. R. Albrecht, C. F. Quate, D. S. Cannell, H. G. Hansma, and P. K. Hansma, *Science*, **243**, 1586 (1989).

21. O. L. Shaffer, R. Bagheri, J. Y. Qian, V. Dimonie, R. A. Pearson, and M. S. El-Aasser, *J. Appl. Polym. Sci.*, **58**, 465 (1995).

22. P. Debye, H. R. Anderson, Jr., and H. Brumberger, *J. Appl. Phys.*, **28**, *679 (1957)*.

23. J. H. An, A. M. Fernandez, and L. H. Sperling, *Macromolecules*, **20**, 191 (1987).

24. J. H. An and L. H. Sperling, in *Cross-Linked Polymers: Chemistry, Properties, and Applications*, R. A. Dickie, S. S. Labana, and R. S. Bauer, Eds., ACS Symposium Ser. No. 367, American Chemical Society, Washington, DC, 1988.

25. J. Israelachvili, *Intermolecular and Surface Forces*, 2nd ed., Academic, London, 1992.

26. J. N. Israelachvili and D. Tabor, *Proc. R. Soc. Lond.* **A331**, 19 (1972).

27. J. N. Israelachvili and G. E. Adams, *J. Chem. Soc. Faraday Trans.* I **74**, 975 (1978).

28. J. N. Israelachvili, *Accounts Chem. Res.* **20**, 415 (1987).

29. T. P. Russell, *Physica*, **B221**, 267 (1996).

30. A. M. Mayes, R. D. Johnson, T. P. Russell, S. D. Smith, S. K. Satija, and C. F. Maikrzak, *Macromolecules*, **26**, 1047 (1993).

31. K. L. Johnson, K. Kendall, and A. D. Roberts, *Proc. R. Soc. London*, **A324**, 301 (1971).

32. M. K. Chaudhury and G. M. Whitesides, *Langmiur*, **7**, 1013 (1991).

33. M. K. Chaudhury, *Mat. Sci. Eng.*, **R16**, 97 (1996).

34. K. Landfester, C. Boeffel, M. Lambla, and H. W. Spiess, *Macromolecules*, **29**, 5972 (1996).

35. H. W. Spiess, *Ann. Rev. Mater. Sci.*, **21**, 131 (1991).

5

POLYMER CHAINS AT SURFACES AND INTERFACES

Today, polymer chains at surfaces and interfaces are known to exhibit non-Gaussian coil conformations, thus lowering their entropy and actually providing a repelling force in some cases. However, polymer chains at interfaces may be interdiffused, providing strengthening entanglements, and/or bonding to the "other side" by either primary or secondary chemical bonds. Thus, both the entropic and enthalpic aspects of the thermodynamics at the interfaces are important in determining the strength of the resulting blend or composite.

While this book is not primarily concerned with chains adsorbed from dilute solutions, there are two points that make such studies important for polymer blends and composites. First of all, such systems form the basis for many coatings and adhesives. Both in terms of formulation and properties on drying, these materials are well within the range of this book. Second, much more is still known about polymers adsorbing from dilute solutions than is known about bulk systems. An objective of this chapter will be to try to relate the dilute-solution adsorbed chains to their counterpart bulk materials.

5.1. CONFORMATION OF POLYMER CHAINS AT SURFACES AND INTERFACES

The simplest case to consider involves a dilute polymer solution in contact with a solid surface (1). Some of the chains in the solution are adsorbed on the surface. The adsorbed polymer chains have three main parts, see Figure 5.1. There are trains of polymer that have all of their mers in contact with the substrate, loops of polymer that are unbound and connect the trains, and the tails, which are nonadsorbed chain ends. An important parameter in calculations is the train fraction or bound fraction, p.

At very low surface coverage, the chains have a tendency to flatten out, or "pancake," since this lowers the free energy the most. At higher coverage, the chains stick up much more. In particular, the chain ends form a structure

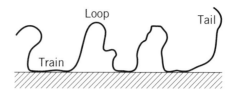

Figure 5.1 Polymer chain partly adsorbed on a surface.

commonly referred to as a "brush;" see Figure 5.2. The surface concentration, Γ, of polymers adsorbed from dilute solution is usually in the range of a few milligrams per meter squared. A third kind of conformation is sometimes called a *mushroom*, resulting from isolated end-bonded chains in solution. Such mushrooms may also be important in bulk composite structures as well.

If the polymer is polydisperse, three concentration regions of adsorption can be distinguished. When the total concentration is low, all the polymer chains find a place on the surface. As the concentration increases, adsorption becomes selective with the high-molecular-weight fraction usually adsorbing in preference to the lower molecular weight fraction. As the surface becomes saturated with the large molecules, the surface concentration becomes constant, even though the bulk solution concentration continues to increase.

5.1.1. Polydispersity Effects

The problem of whether high or low molecular weights adsorb preferentially is complex, however. Simple adsorption, chemically bound chains, differing rates of diffusion, and entropic depletion zones all interact simultaneously. The reader will see that there are many important cases where the short chains are preferentially found at the surface (2,3).

Several events must transpire in order for a polymer chain to adsorb on a surface. First, it must diffuse toward the interface, and then contact an empty site in such a way that it sticks. During this time, it must change its average conformation away from a random coil, especially if it cannot penetrate the

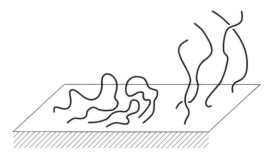

Figure 5.2 Two extreme cases of polymer adsorption on a substrate. Left: pancakes; right, brushes.

surface. However, its total free energy must be minimized. As the process of adsorbing on the surface approaches an equilibrium, some polymer chains will be desorbing while others are adsorbing.

Except for special circumstances, a polymer chain will adsorb on multiple sites; see Figure 5.1. While the bonding strength of each mer to the surface may be small (not more than 5 kcal/mol unless actual covalent bonds are formed), all of the bound mers must simultaneously be desorbed in order for the polymer chain to be released from the surface. This requirement has meant that very slow kinetics of desorption are the rule, for example.

In the presence of flow fields, the loop and tail portions of the chain become oriented in the direction of flow. The effect is slightly similar to the orientation seen in weeds growing in stream beds. Strong enough flow fields may cause the polymer chain to desorb. If the polymer chain is bound to multiple colloidal particles, it may be constantly adsorbing and desorbing depending on binding forces, diffusion coefficients, and the shear forces at the instant.

This is the basis for the so-called non-drip paints. In the absence of shear fields, the polymer chains adsorb on the surfaces of the latex particles and filler and also self-associate. However, the bonds are weak. In the presence of shear fields, the bonds break, and the viscosity of the system falls. Thus, when a paint brush is dipped into the paint, a high shear field is created locally, and the viscosity falls, allowing the paint to flow onto the brush. When the brush is removed from the paint, the shear field drops to zero, allowing the polymer chains to readsorb, causing the viscosity of the system to increase. Again, when the paint-laden brush is applied to walls and ceilings, the shear field causes the bonds to break again, and the viscosity drops. However, if a drop of paint starts to form a drip, it does so in a low shear field, when readsorption of the polymer chains has occurred, with a consequent increase in viscosity. Thus, paint on walls and ceilings see low shear fields, minimizing "sag," running, and dripping.

The preceding discussion has its analogy in bulk materials, where polymer chains adsorb on particles, fibers, and other surfaces to form composites. The following example illustrates the energies involved in a simple adsorption case.

5.1.2. Example Calculation: A Chain Bound to Two Colloid Particles

Consider what might be a portion of latex paint, with latex particles and colloidal filler dispersed in an aqueous medium. An associative poly(ethylene oxide) chain (PEO) dispersed in the aqueous phase becomes bonded to two different particles (could be either latex or filler); see Figure 5.3. The molecular weight of the PEO chain is 1×10^5 g/mol. The bonding energy to each of the two particles is 8 kcal/mol. A shearing force is applied to the system, tending to separate the two bonded particles. The question is: How far must the particles be separated before the PEO debonds from one of them?

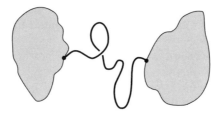

Figure 5.3 Two particles bonded by a single chain.

The PEO chain at rest has a radius of gyration governed by the relation (1):

$$R_{go}(nm) = 343 \times 10^{-4} M^{0.5} \tag{5.1}$$

which yields 10.8 nm for R_{go}. Thus, when the dispersion is at rest, the average distance between the bound surfaces of the particles will be 10.8 nm.

The theory of rubber elasticity for single chains is invoked next. The force, f, necessary to hold a chain at an end-to-end distance of r ($r^2 = 6R_g^2$) is given by (4):

$$f = 3kTr/r_0^2 \tag{5.2}$$

where r_0 is the value of r under Flory θ-temperature conditions, that is, that approximated by the bulk state. Water is very nearly a θ solvent for PEO. The force here is purely entropic in nature, generated by the reduction in the number of chain conformations possible as the end-to-end distance is increased. Noting that work, W, equals force times distance,

$$W = \int_0^r (3kTr/r_0^2) \, dr \tag{5.3}$$

$$W = (\tfrac{3}{2})kT(r^2/r_0^2) = (\tfrac{3}{2})kT\alpha^2 \tag{5.4}$$

where α represents the ratio of the actual end-to-end distance to that value in a θ solvent. Noting that 8 kcal/mol translates into 5.55×10^{-20} J/chain,

$$5.55 \times 10^{-20} \, \text{J/chain} = (\tfrac{3}{2})(1.38 \times 10^{-23} \, \text{J/K}) \times 298 \, K\alpha^2 \tag{5.5}$$

Solving algebraically, $\alpha^2 = 8.99$, or $\alpha \cong 3$. Then, $3R_{go} = 32.5$ nm. Thus, when the chain is extended three times its relaxed radius of gyration, the entropic retractive forces are sufficient to break one of the bonds, and the two particles will be free to separate. (The two corresponding end-to-end distances are obtained by multiplying the two radii of gyration by the square root of 6.)

Of course, the above calculation is only approximate. Usually, the chain ends are buried somewhere within the polymer coil. Being attached to colloidal particles (or any surface) forces the chain ends outward. There are two limiting cases:

1. The colloidal particles are large with respect to the radius of gyration of the polymer chain, and the surface is smooth. Then, the particle can be approximated by a flat plane.
2. If the particles are significantly smaller than the radius of gyration of the chain, a point approximation may be valid, and the conformation of the random coil may be substantially unperturbed. This latter case may approximate reinforcing carbon blacks, which are often *smaller* than the rubber chains they are there to reinforce.

In reality, most colloid and reinforcing particle sizes lie between these two extremes. The conformations of polymer chains attached to an interface have been discussed by de Gennes (5). In many cases, the polymer tries to avoid the wall, even though attached.

5.2. NONBONDING AND DEPLETION ZONES

When a polymer chain approaches a surface, the required conformational changes causes a loss of entropy. If the surface and the particle do not attract one another, and the polymer chains do not adsorb, there may be an actual zone of depletion. Thus, the free (air) surface of a dilute polymer solution may be relatively rich in solvent because the polymer, effectively, is forbidden by loss of free energy to have significant numbers of loops or ends sticking in the air. This may affect vapor pressure or surface tension, for example (6). This depletion zone also affects the bonding of polymers to surfaces, causing a positive free energy effect. This is true both in polymers adsorbing from dilute solutions and polymers adsorbing in bulk cases.

Another example of the depletion zone is the core–shell effects in the formation of latexes (7). For partly polymerized systems, whether the remaining monomer is the same or different from the already formed polymer, many systems have most of the polymer in the interior, and most of the monomer near the surface. Thus, when polymerizing a monomer II in the latex in the presence of polymer I, even at equilibrium, a core–shell structure may develop. How accentuated the core–shell structure becomes, however, depends on the size of the latex particles relative to the size of the molecules, among other features. The effect reaches a maximum when the radius of the particles is about the same as the radius of gyration of the polymer chains.

5.3. ADHESION IN COMPOSITE MATERIALS

While the colloid chemists talk about adsorption of polymer chains on a surface, their counterpart composite people talk about adhesion. According to Sharpe (8), there are two definitions of adhesion: In physical chemistry, people talk about the atomic or molecular attraction between a solid and a second, usually liquid, phase. The binding energies are determined by contact angles and so forth. The technological definition refers to a mechanical resistance to separation of a system of joined materials, involving such features as breaking stress, fracture energy, and the like. Both definitions, of course, express the same features, but with a different approach, and for different situations.

Also according to Sharpe (8), an interphase is defined as a region intermediate to two, usually solid, phases in contact, the composition, structure, and properties of which vary across the region, and that, in general, differ from either of the two contacting phases. Some examples of interphases in composites include polymers solidified from the melt or polymerized from monomers or prepolymers in contact with a solid. The chains of these polymers usually assume conformations at the surface of the solid that are different from those in the bulk polymer. Such interphases may extend from 1 to 100 nm into the bulk, and are, in general, preserved upon solidification.

The following sections will build on Section 4.2, which introduced the concepts of adhesion, especially in relation to surface tension and related quantities.

5.3.1. Theories of Adhesion

Polymer blend and composite workers want strong bonding between components of materials to produce high-strength materials. In the failure of adhesive bonding, one must distinguish between *adhesive* failure, which talks about debonding at either a sharply defined interface or between an adhesive and one of the components being bonded, and a *cohesive* failure, where failure takes place within one of the materials being bonded. In real cases, both adhesive and cohesive failure mechanisms may be taking place, of course. While the fracture processes themselves will be discussed further in Chapter 6, the theories of adhesion will be developed here.

Sharpe (8) summarized the principal theories of adhesion. In the adsorption theory, materials adhere by physi- or chemisorption. Strong joints result from primary valence forces, or "polar groups," interacting. A second popular theory is the mechanical, or "hooking," theory. The central point in the mechanical theory is that surface roughness causes "puzzle"-like interlocking of the polymer and the substrate. This theory is important in adhesion of polymers to porous materials such as paper, cloth, wood, or metallized plastics, and is sometimes important in polymer blends cooled from the melt, especially if one or both phases crystallize.

There are many other theories of adhesion, each contributing important elements of thought. Deryagin and Silga (9) treated the joint as though it were a capacitor, charged due to the contact of the two materials making up the joint. The strength of the joint depended on the existence of an electrical double layer at the interface. The diffusion theory of Voyutskii (10) maintained that the extent of interdiffusion across the interphase determined the joint strength. This theory is particularly useful for polymer blends or the healing of an interface of two identical polymers. It follows from observations that the breaking strengths of adhesive joints increase as the time of contact increases, and with increasing molecular weight of the interdiffusing polymers.

In earlier times, Bikerman (11) developed a rheological theory of adhesive joints. He stated the strength of an adhesive joint is determined by the mechanical properties of the materials comprising the joint, together with the local stresses in the joint. The interfacial forces are not necessarily the critical elements, especially if the interfacial bonding is above a certain minimum level, because clean failure in adhesion rarely occurs. Bikerman pointed out that failure is almost always cohesive, in the adherents and/or in the adhesive, or in some boundary layer.

5.3.2. Effect of Interface Type

In bulk polymeric materials, the interfaces between two polymers or between a polymer and a nonpolymer are the more important. Miles (12) pointed out that among the most important composites are the fiber-reinforced systems. The fibers can be made of glass, carbon, metal, alumina, or polyaramid. Technically, of course, the last is also a polymer. Other important kinds of composites include the particulate-filled polymers and polymers facing macroscopic surfaces.

The bonding between the polymer and the filler, be it a fiber or particulate, and the polymeric matrix is of critical importance in determining the mechanical behavior of the composite. Miles (12) also pointed out that a moderate density of high-molecular-weight grafts yields better coupling than either a high density of low-molecular-weight grafts or a low density of grafts of any molecular weight; see Figure 5.4.

If the polymer has a broad distribution of molecular weights, the most common case for engineering thermoplastics, a question arises as to whether there is any preferential motion of one molecular species or another to the surface. While this question appears open to speculation in the bulk polymer–composite literature, the dilute solution dispersion literature provides significant hints. Fleer et al. (1) examined two cases: a 1000-mer polymer dissolved in its own monomer and a 1000-mer polymer dissolved in a 100-mer polymer, see Figure 5.5. In both cases, the long chains try to avoid the surface region because they lose conformational entropy. Hence, the high-molecular-weight polymer is depleted from the surface at equilibrium. The thickness of the depletion zone is proportional to $M^{1/2}$. However, the depletion zone is far less severe for the mixture of the two polymers than for the mixture of polymer and monomer. Of course, as the fractions approach each other in molecular weight, the effect should disappear.

These results are important in two ways: If a polymerization is carried out in the presence of a nonbonding filler, the first formed polymer may actually form a depletion zone, avoiding the filler in favor of the remaining monomer. Second,

(a)   **(b)**   **(c)**

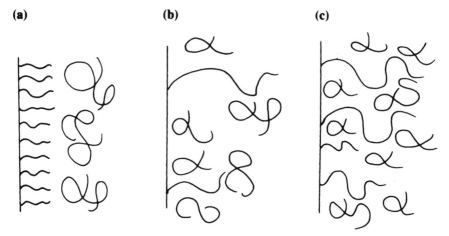

Figure 5.4 Interactions of grafted chains on a filler surface with free chains. (a) High density of low molecular weight grafts results in poor coupling. (b) Low density of grafts also results in poor coupling. (c) Moderate density of high-molecular-weight grafts results in good coupling. In reality, the free chains must have a high enough molecular weight to form good chain entanglements.

if the molecular weight distribution is broad, the shortest chains will tend to be in contact with the filler.

Of course, these effects can be negated significantly by providing actual bonding sites between the polymer and the filler. An important series of cases

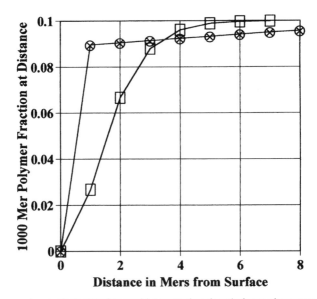

Figure 5.5 Long polymer chains tend to avoid a neutral surface in favor of monomer, solvent, or shorter polymer chains. Ten percent of 1000-mer polymer dissolved in: Squares, its monomer; circles, in 100-mer polymer. (Source: Drawn from table information in Ref. 1, pp. 160–161.)

involves block copolymers bonded onto filler particles from dilute solutions. In an adsorbed block copolymer, usually one block adsorbs preferentially to the other. The adsorbed block is called the anchor. The other block is denoted as the buoy block. Qualitatively, there are two regimes, the anchor regime and the buoy regime, determined by the relative lengths of the two blocks. In the buoy regime, the repulsion between the buoy blocks dominates the extent of bonding and determines the number of chains on the surface. Under these conditions, the buoy may form a brush structure, with the polymer chains sticking up into the solution. The surface may actually remain undersaturated with anchor blocks.

5.3.3. Molecular Weight Dependence of Adsorption

In a model study, Cosgrove et al. (13) examined the data of Wu et al. (14) on diblock copolymers made of poly(dimethyl aminoethyl methacrylate) (DM) and poly(n-butyl methacrylate) (PnBMA). Chains of 200 and 700 mer lengths were studied, as functions of the fraction of DM mers. For low DM fractions, a peak surface concentration of $15\,mg/m^2$ for the 700-mer polymers was attained versus a $10\,mg/m^2$ surface concentration for the 200-mer polymer. However, the concentration of polymer at the surface goes through a maximum at 0.1–0.2 fraction of DM mers.

Recently, the kinetics of exchange of short and long chains of water-soluble associative polymers was examined by Ou-Yang and co-workers by dynamic light scattering (15,16). Poly(ethylene oxide) chains were capped symmetrically on both ends with $C_{16}H_{33}$ or $C_{20}H_{41}$ aliphatic hydrophobes and placed in the presence of uniform polystyrene latexes. This class of material is known as an associative polymer. In one experiment [see Figure 5.6 (17)], 17,000 or 100,000 molecular weight end capped poly(ethylene oxide) was placed in a dispersion of polystyrene latexes containing the other component, and the rates of exchange were observed. The dynamic light scattering measures the effective layer thickness by measuring the changes in the colloidal particle diffusion coefficient; see equation (4.14). Equilibrium was attained after about 8 h, under the conditions used. Most importantly, the authors concluded that the long chains replaced the short chains more effectively than the reverse case.

In the above experiment, the ratio of molecular weights of the long and short chains was roughly a factor of 6. Klein et al. (18) studied polystyrene samples differing by a factor of 15 in molecular weight. The polystyrenes were terminated by the zwitterionic group $N^+(CH_3)_2(CH_2)_2SO_3^-$. The samples were allowed to adsorb on mica surfaces, and the increase in thickness measured in toluene via the surface forces apparatus; see Figure 4.11. Within the range of their experimental parameters, the shorter chains displaced the longer ones completely. Again, this reflects the lower conformational entropy loss when shorter chains approach surfaces, rather than longer chains.

Jones et al. (19) also point out some of the differences between dilute solutions and bulk polymers in their interaction with solid surfaces. In a good

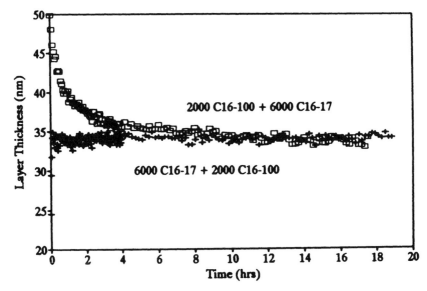

Figure 5.6 Poly(ethylene oxides) containing C16 hydrocarbon tails on both ends, adhering to polystyrene latex particles. Samples of 100,000 g/mol (C16–100) and 17,000 g/mol (C16–17) are used to demonstrate both equilibrium and reversibility by putting first one and then the other on the latex via diffusion and substitution.

solvent, neighboring chains start to interact unfavorably and become significantly stretched as soon as the chains physically overlap, creating the pancake-brush transition. In bulk, the effective excluded volume parameter scales as the degree of polymerization of the "solvent" chains, which in this case are the nonbound chains. They point out that as the degree of polymerization of the matrix (unbound) chains is increased above the degree of polymerization of the brush (bound) chains, one expects the matrix chains to be progressively expelled from the brush, leading to what they call the *dry-brush* regime.

The reader will have noted from the preceding discussion that sometimes long chains stick on surfaces relative to short chains, and sometimes the reverse. The reason is that several competing factors come into play in each case; see Table 5.1. Apparently, sometimes the short chains adhere first, later to be replaced by long chains.

More generally, there is a competition between kinetic effects and thermodynamic effects, see Table 5.2 (20). An important kinetic effect not discussed up to now involves lateral diffusion, which speaks of chains rearranging their position on the surface to provide a lower free energy for the system.

While Tables 5.1 and 5.2 were based primarily on ideas generated from the dilute solution/dispersion case, there is a strong carryover for bulk composite systems. For example, consider a filler such as titanium dioxide being mixed into a polymer melt under high shear conditions. Might not the short chains adhere to the TiO_2 surfaces first (more rapid kinetics, lower shear sensitivity), to be

Table 5.1   Factors affecting the molecular weight dependence of adsorption

Favoring High Molecular Weight	Favoring Low Molecular Weight
More adsorption sites per molecule: Harder to *unstick* all bound sites simultaneously	Smaller conformational entropy losses on approaching the surface
May bind to multiple particles or surfaces, becoming protected from the exterior	Faster diffusion to the surface
	Less subject to shearing effects such as chain extension and degradation
Chemistry/charge distribution/side chains/polarity often are more complex	Fits into smaller pores.

replaced in part by higher molecular weights, perhaps depending on the cooling rate (if the material is in a mold, e.g.)?

5.4. POLYMER BLEND AND COMPOSITE INTERFACES

5.4.1. Rubber–Carbon Black Interfaces

The binding of polymer chains to a substrate suggests that the free volume of the interphase region may be reduced. If this is so, then the glass transition temperature would be expected to increase or at least the glass transition region would broaden on the high side. This was the reasoning of some of the earliest workers on polymer interface problems.

Data by Nielsen et al., Figure 5.7 (21), show that when polystyrene is filled with such materials as mica, calcium carbonate, or asbestos, two effects are noted:

1. The modulus of the material in the glassy state is increased significantly. This can be explained by the use of the Kerner equation (22); (see Section 1.6.

2. The glass transition temperature is seen to rise about 10°C from the neat material to the most highly reinforced material.

Table 5.2. Factors controlling polymers at interfaces

Thermodynamic	Kinetic
Equilibria	Diffusion of single chains
Competitive adsorption	Evolution of structural rearrangements
Extent of spatial entanglement	Segment–surface contact life
Number of bound segments	Lateral diffusion
Adsorbed layer thickness	Replacement of one species by another
Patchiness and bare spots	Effect of dynamic entanglement

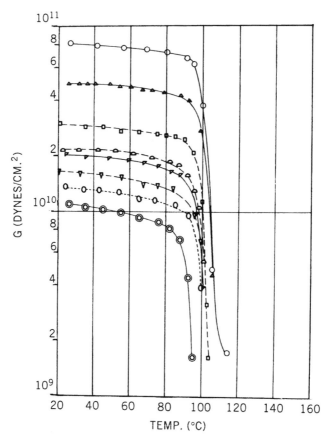

Figure 5.7 Log shear modulus vs. temperature for a series of filled polystyrenes, showing the increase in the glassy modulus, and apparent increase in the glass transition temperature. (⊚) control, (∇) 20% mica, (△) 40% mica, (0) 20% calcium carbonate, (◠) 40% asbestos, (o) 60% mica, (▷) 20% asbestos, (□) 60% asbestos.

A more quantitative approach was taken by Kraus and Gruver (23), who examined the thermal expansion, free volume, and molecular mobility of HAF carbon-black-filled styrene–butadiene rubber (SBR). Continuing the Nielsen et al. (21) experimental approach, Krause and Gruver chose dilatometry as their main instrument. The value of dilatometry for this type of investigation is that the changes in the glass transition temperature are not highly affected by modulus increases.

SBR elastomer is thought to bond to carbon black through a variety of secondary bonding modes; see Section 1.12. Thus, the train portion of the polymer chains near the carbon black is thought to be relatively large and held to the surface rather firmly. This, of course, reduces the free volume of the region.

The results of Kraus and Gruver (22) are summarized in Figure 5.8 and Table 5.3. The width of the glass transition region increased from 9 to 20°C for the

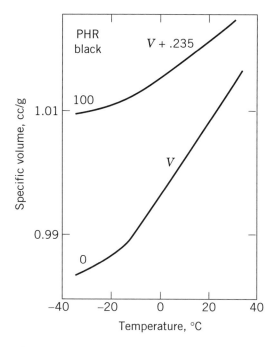

Figure 5.8 Thermal expansion of HAF carbon black reinforced SBR.

bulk material, principally on the upper side of the transition. The glass transition in the restricted region was thus increased some 10°C. The calculated width of the restricted segmental motion region was 30 Å, based on the known HAF surface area and the volume of rubber. Today, that would be called the

Table 5.3   Properties of SBR at carbon black interfaces

Property	Neat	100 phr HAF
A. Basic Properties		
T_g, °C	−11.4	−9.3
$\beta_g \times 10^4$ cm^3/g °C	2.10	1.35
$\beta_r \times 10^4$ cm^3/g °C	5.70	5.56
Surface area, m^2/g	—	80
Width of T_g, °C	9	20
B. Derived Properties		
Thickness of restricted segmental motion, Å	—	30
Increase in T_g in restricted region, °C	—	10
Fraction of polymer restricted	—	0.25

Source: Based on G. Kraus and J. T. Gruver, *J. Polym. Sci., A-2*, **8**, 571 (1970).

interphase thickness. For 100 phr carbon black, the fraction of polymer with restricted motion was about one fourth. Thus, the interphase forms a major component of the whole material.

5.4.2. Polymer–Polymer Interfaces

There are several kinds of polymer–polymer interfaces, including polymer blends, grafts, blocks, or interpenetrating polymer networks (IPNs). Of great importance are those involving rubber-toughened plastics and copolymers of variable composition. The latter form microheterogeneous morphologies, with $200 \, \mathrm{m^2/gm}$ or more of interfacial area. Polymer pairs may be amorphous and crystalline, glassy and rubber, or linear and crosslinked, to name a few nonexclusive possibilities.

5.4.2.1. Composition of the Interphase
The modern story begins with the works of Helfand and Tagami (24,25), who developed a mean-field theory of polymer interfaces, or interphases, as they are now called. They were particularly interested in the composition of the interphase as a function of distance across the interphase, and the calculation of the interfacial tension. From a free energy point of view, the results had to have a balance of energetic and entropic forces, such that the system would be in equilibrium.

They assumed first that the degree of polymerization of the system (DP $= Z$) goes eventually to infinity. The two polymers were assumed symmetrical, especially in regards to density, statistical segment (mer) length, and compressibility. The effective length of a mer, b, was chosen such that Zb^2 equals the mean-square end-to-end distance, $(6^{1/2}R_g)^2$. The density of the polymer in the bulk was ρ_0. In the region of the interphase, the density was the sum of the two polymer densities, divided by the bulk density.

$$\frac{\rho_A(r) + \rho_B(r)}{\rho_0} = \rho \tag{5.6}$$

where r represents a position vector. In equation (5.6), the density contribution of each polymer is variable. If the heat of mixing is positive and the free energy of mixing is positive, then the total density of the system may go through a minimum at the interphase, as shown in Figure 5.9 (25). This is because the A and B polymer chains actually repel each other, relative to the attractive forces for themselves.

5.4.2.2. Interphase Thickness
The interphase surface thickness, S_{th}, is given by $2b/(6\chi)^{1/2}$, and the interfacial tension is given by $(\chi/6)^{1/2}bkT$ (25). These quantities are summarized in Tables 5.4 (25) and 5.5. In Table 5.4, the interfacial tension is calculated as a function of χ and b. The values are compared with the experimental results of Wu (26). The values of the interfacial surface tension of 1–$2 \, \mathrm{erg/cm^2}$ are much smaller than for the free surfaces, already described in Table 4.1. The interfacial thicknesses are in the range of 6–8 mer lengths. The

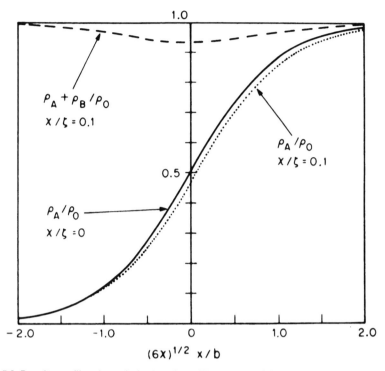

Figure 5.9 Density profiles through the interface. The quantity ζ is a function of the compressibility.

minimum interfacial thickness can also be estimated through dimensions of the de Gennes reptation tube. The tube diameter is calculated from the average distances between entangling chains surrounding the chain in the tube. Frequently, the diameter of the tube is taken as 4–5 nm (27). This is the range across which a polymer chain may interdiffuse without pushing another chain significantly out of the way, because that action costs too much energy. This calculation leads to the conclusion that there is only very limited mixing of the two chain types. From an entropic point of view, however, there is always some mixing.

Table 5.4   Comparison of calculated and measured interfacial tensions

Polymer Pair	Interfacial tension, erg/cm^2		χ	b (Å) (geom. mean)	Spec. vol. (cm^3/mol) (geom. mean)
	Calc.	Exp.			
PS/PMMA	1.0	1.5	0.01	6.5	96
PMMA/PnBMA	2.0	1.8	0.07	6.1	114
PnBMA/PVA	1.9	1.9	0.05	6.3	107

Table 5.5   Interfacial surface thicknesses

Polymer Pair	S_{th} (Å)	χ_{AB}
PS/PB	30	0.03
PS/PMMA	50	0.01

Two other conclusions of the Helfand and Tagami theory (24,25) are very important:

1. The chain ends tend to be slightly more concentrated in the interphase than are segments in the middle of the chain. Thus, there is a slight orientation of the chain end portions perpendicular to the plane of the interphase; see Figure 5.10. In this corner of polymer science, theory is still ahead of experiment, defining experiments still lacking.

2. Small third-molecule components are preferentially adsorbed in the interphase. These usually increase the degree of interpenetration of the two polymers. Here, two forces operate antagonistically. The presence of small molecules dilutes the polymers, tending to weaken the interphase mechanically. The additional interpenetration, however, leads to more entanglements, tending to increase the strength of the interphase region.

In a later study, Helfand and Sapse (28) removed the requirement for symmetry of the two polymers (density, etc.). Agreement with experiment in both cases was described as fair. Of course, the latter case is much more complex.

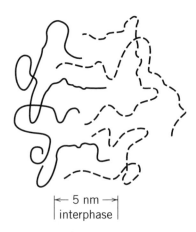

← 5 nm →
interphase

Figure 5.10 Interphase region in polymer blends. Note that the central portions of the polymer chains have a slight tendency to lie parallel to the nominal interface, while the chain ends tend to lie perpendicular, as proposed in modern interphase theories.

5.4.2.3. *Molecular Weight Fractionation at the Interphase*

More recently, Reiter et al. (29) carried out Monte Carlo studies of the interphase existing between two polymer melts. If the polymers have a broad distribution of molecular weights, a certain "fractionation" occurs. The fraction of small chains is larger in the interphase and also in that region where the other polymer constitutes the majority component, compared to the phase where the "shorter" chains form the majority component. This result is similar to that taking place at composite interfaces; see Section 5.3.2. The reason that the shorter chains migrate preferentially to the interphase has to do with a lower loss in entropy. Short chains are more miscible in the other phase because the gain in entropy is larger relative to the enthalpy loss.

5.4.2.4. *Chain Orientation at the Interphase and in Composites*

It was noted earlier that an increase in the density of chain ends occurs near the interphase; see Figure 5.10. Necessarily, the opposite behavior is exhibited by the density of the center of mass of the chains. Because of their larger number of degrees of freedom, the chain ends invade the other phase more readily than centrally placed segments. However, this enhancement is combined with a depletion layer of chain ends just below the enriched layer because the number of chain ends must be conserved within the dimension of a single chain (30). Interestingly, while the chain ends are oriented perpendicular to the surface, the shape of the whole chain is flattened out to a distance of several mers from the nominal interface.

This is particularly true at wall interfaces (30). Figure 5.11 (30) was derived specifically for free surfaces and walls, composite materials. It qualitatively describes the polymer chain parameters expected for immiscible polymer blends. In this area, again, theory is leading experiment.

5.5. SURFACE DENSITIES AND DIFFUSION COEFFICIENTS

The densities of polymers may go up or down at interfaces, depending on interactions. While chemical binding and strong attractive forces (as in many composite surfaces) may yield an increase in the density of a polymer at surfaces and interfaces, lack of such forces is anticipated to yield a decrease in density. A great deal of what we know about polymer blend interphase densities is based on theory; see Figure 5.9. While some experimental data are already available for composites (see Section 5.4.1), less is known experimentally about polymer blend interface densities. However, some progress has been made on homopolymer surfaces. The reader can extrapolate the information in this section to the interface case and at least obtain some additional understanding.

5.5.1. Diffusion in a Latex Film Surface

This section is concerned with the increases in molecular motion caused by decreases in density at a polymer surface. Using atomic force microscopy, Goh

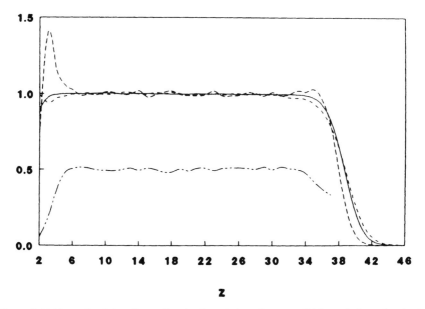

Figure 5.11 Normalized density profiles for the whole polymer, solid line; chain ends; shorter dashed line; and center of mass, longer dashed line. Also shown is the chain deformation factor $R_z^2/(R_x^2 + R_y^2)$, dash-dot line.

et al. (31) studied film formation from poly(butyl methacrylate) latexes. At low ionic strengths and in the absence of surfactant, the latex particles form a face-centered cubic lattice when forming a film above T_g. This cubic lattice persists as the film dries (32,33). In the early stages of film formation, the latex particles on the surface each have a protruding structure resembling a basket of eggs. With time, the surface smoothes via lateral diffusion, the driving force deriving from the excess surface energy. [At the same time, of course, interdiffusion is in progress in the bulk of the film, healing all of the interfaces (34,35).]

The surprising result of Goh et al. (31) was that the diffusion coefficient for the poly(butyl methacrylate) film surface was calculated to be some 10^4 larger than for the bulk polymer homodiffusion coefficient. A possible explanation of this result lies in a reduced density at the surface (36). On a molecular scale, a free surface need not be entirely smooth, but may have a fractal-like nature. Thermally, a vacuum (or air) has a very positive heat of mixing with a polymer, strongly repelling it. Thus, a kind of depletion may be expected to develop. The end result is that a free surface, like some interior composite or blend surface, may have a higher than expected free volume, leading to increased self-diffusion coefficients for the polymers in the interface. This leads to the following example calculations.

5.5.2. Example Calculation: Free Volume at Surfaces and Interfaces

The glass transition of poly(butyl methacrylate) (as above) is 293 K, its density at 25°C is 1.053 g/cm^3 (37), and the volumetric expansion coefficient above T_g,

α_R, is approximately $6 \times 10^{-4}/°C$.

(a) What is the T_g of the surface?
(b) The above surface diffusion experiment was carried out at 70°C. What is the fractional free volume at the surface and in the interior at that temperature? Assume that the driving force from the extra area itself is small, leading to mostly random diffusion from that component.
(c) What is the difference in densities between the surface and the interior of the polymer?

The basis for the calculation is the Williams–Landel–Ferry (WLF) equation (38). This equation has two forms. The theoretical form may be written:

$$\log A_T = \frac{-B/f_0(T - T_0)}{2.303\,\dfrac{f_0}{\alpha_f} + (T - T_0)} \tag{5.7}$$

where f_0 represents the fractional free volume at T_g (or related temperature of interest), α_f is the expansion coefficient of the free volume, T is the temperature, and T_0 a generalized transition temperature very often taken as T_g, A_T represents the reduced variables shift factor, and B is a constant with a value of approximately unity. If the free volume of the polymer is assumed constant below T_g, then $\alpha_R - \alpha_G \simeq \alpha_f$, where the subscript G stands for glassy conditions.

The most widely used experimental form of the WLF equation is

$$\log A_T = -17.44(T - T_g)/[51.6 + (T - T_g)] \tag{5.8}$$

where T_0 has been set as T_g. Equation (5.8) is useful for a wide range of time–temperature related problems from T_g to approximately $T_g + 100\,K$ for amorphous polymers. Comparison of equations (5.7) and (5.8) yields $f_0 = 0.025$ and $\alpha_f = 4.8 \times 10^{-4}\,deg^{-1}$.

First, the T_g of the surface can be calculated. The interior T_g is known to be 293 K. From the WLF equation, A_T is the ratio of two times or two time-related variables, one of which is at T_g. As noted, the diffusion coefficient is 10^4 larger at the surface than in the interior, providing the necessary ratio. Plugging into equation (5.8) yields $T_g = 283.4\,K$, or about 10 K lower than in the interior.

The free volume of the poly(butyl methacrylate) at the surface may be obtained from the relation

$$f = f_0 + \alpha_R\,\Delta T \tag{5.9}$$

where ΔT represents the number of degrees above T_g, $[(273 - 283.4) + 70]°C = 60°C$. Then,

$$f = 0.025 + (6 \times 10^{-4}/°C) \times 60°C \tag{5.10}$$

yielding $f = 0.061$. A similar calculation for the interior yields $f = 0.055$. Thus, the fractional free volume of the surface is increased about 0.006.

The density of the polymer can be calculated from

$$1/\rho_T = 1/\rho_{25} + (\alpha_R/\rho_{25})\,\Delta T \qquad (5.11)$$

where ρ_T is the density at temperature T. The density of the polymer at 25°C is $1.053\,\mathrm{g/cm^3}$. At the temperature of the experiment, 70°C, the density is $1.025\,\mathrm{g/cm^3}$. This is the bulk density, including its free volume. The surface has a fractional free volume excess of 0.006, leading to a surface density of $1.024\,\mathrm{g/cm^3}$ at 70°C. While the difference in densities is small, the difference in diffusion rates is large. While clearly the above calculation is only approximate, it serves to show the relation among some important interfacial quantities. To a significant extent, these concepts must carry over to polymer blend and composite interfaces.

A similar phenomenon is physical aging in the glassy state (39). Here, creep rates decline several orders of magnitude as the polymer ages after being cooled from the melt. The density is known to increase during this time by a small fraction of one percent.

Incidently, since the ratio of the diffusional coefficients for poly(butyl methacrylate) were given, it is also of interest to calculate the absolute values. For a molecular weight of $2.2 \times 10^5\,\mathrm{g/mol}$ (this sample), at 70°C the bulk diffusional coefficient can be estimated to be $7 \times 10^{-17}\,\mathrm{cm^2/s}$ (34). Then, the surface diffusional coefficient is approximately $7 \times 10^{-13}\,\mathrm{cm^2/s}$.

For the three kinds of nonattracting polymer interfaces, free surface (i.e., air), solid surface (composites), and polymer–polymer surfaces (blends), with the latter two assuming no bonding, there seems to be several common features coming to the fore: Chain ends tend to be at the interface, small molecules or lower molecular weight polymer seems to migrate to the interface, the orientation of the chain end portions is perpendicular to the interface, while the central portions of the polymer tend to be oriented parallel to the surface. Lastly, the free volume is larger at nonattracting interfaces, leading to increases in self-diffusion rates. Polymers attracted to surfaces from solutions also tend to have the central portions of their chains parallel to the interface. However, the chain ends are pointed away from the interface, rather than toward it.

REFERENCES

1. G. J. Fleer, M. A. Cohen Stuart, J. M. H. M. Scheutjens, T. Cosgrove, and B. Vincent, *Polymers at Interfaces*, Chapman & Hall, London, 1993.
2. T. Cosgrove, N. Finch, B. Vincent, and J. Webster, *Coll. Surf.*, **31**, 33 (1988).
3. R. A. Croot, A. R. Goodhall, and S. D. Lubetkin, *Coll. Surf.*, **49**, 351 (1990).
4. L. H. Sperling, *Introduction to Physical Polymer Science*, 2nd ed., Wiley-Interscience, New York, 1992.

5. P. G. de Gennes, *Macromolecules*, **13**, 1075 (1980).

6. D. Ausserre, H. Hervet, and F. Rondelez, *Phys. Rev. Lett.*, **54**, 1948 (1985).

7. Se-In Yang, A. Klein, L. H. Sperling, and E. F. Casassa, *Macromolecules*, **23**, 4852 (1990).

8. L. H. Sharpe, in *The Interfacial Interactions in Polymeric Composites*, G. Akovali, Ed., Kluwer, Dordrecht, The Netherlands, 1993.

9. B. V. Deryagin and V. P. Silga, *Proc. 3rd Intern. Cong. Surface Activity*, Cologne, Universitatsdruckerei, Mainz, Vol. 2, p. 349 1960.

10. S. S. Voyutskii, *Autohesion and Adhesion of High Polymers, Polymer Reviews*, Vol. 4, Interscience, New York, 1963.

11. J. J. Bikerman, *The Science of Adhesive Joints*, 2nd ed., Academic, New York, 1968.

12. I. Miles, in *Multicomponent Polymer Materials*, I. S. Miles and S. Rostami, Eds., Longman Sci. & Tech., Bath Press, Avon, UK, 1992.

13. T. Cosgrove, D. A. Guzonas, and M. L. Hair, *Macromolecules*, **25**, 2777 (1992).

14. D. T. Wu, A. Yokohama, and R. L. Setterquist, *Polym. J.*, **23**, 711 (1991).

15. H. D. Ou-Yang, Z. Gao, and L. Dewalt, *Polym. Mater. Sci. Eng. (Prepr.)*, **68**, 55 (1993).

16. Z. Gao and H. D. Ou-Yang, in *Colloid-Polymer Interactions: Particulate, Amphophilic, and Biological Surfaces*, P. L. Dubin and P. Tong, Eds., ACS Symp. Ser. No. 532, American Chemical Society, Washington, DC, 1993.

17. Z. Gao and H. D. Ou-Yang, in *Complex Fluids*, E. B. Sirota, D. Weitz, T. Witten, and J. Israelachvili, Eds., MRS Symposium Proceedings, Vol. 248, MRS, Pittsburgh, 1992, p. 425.

18. J. Klein, Y. Kamiyama, H. Yoshizawa, J. N. Israelachvili, L. H. Fetters, and P. Pincus, *Macromolecules*, **25**, 2062 (1992).

19. R. A. L. Jones, L. J. Norton, K. R. Shull, E. J. Kramer, G. P. Felcher,A. Karim, and L. J. Fetters, *Macromolecules*, **25**, 2359 (1992).

20. Based on M. Santore seminar, Lehigh University, April, 1991.

21. L. E. Nielsen, R. A. Wall, and P. G. Richmond, *Soc. Plastics Eng. J.*, **11** (Sept.), 22 (1955).

22. E. H. Kerner, *Proc. Phys. Soc. London*, **69B**, 808 (1956).

23. G. Kraus and J. T. Gruver, *J. Polym. Sci.*, **A-2 8**, 571 (1970).

24. E. Helfand and Y. Tagami, *J. Polym. Sci.*, **B, 9**, 741 (1971).

25. E. Helfand and Y. Tagami, *J. Chem. Phys.*, **56**, 3592 (1972).

26. S. Wu, *J. Phys. Chem.*, **74**, 632 (1970).

27. P. G. de Gennes, *Scaling Concepts in Polymer Physics*, Cornell University Pres, Ithaca, NY, 1979.

28. E. Helfand and A. M. Sapse, *J. Chem. Phys.*, **62**, 1327 (1975).

29. J. Reiter, G. Zifferer, and O. F. Olaj, *Macromolecules*, **23**, 224 (1990).

30. P. Cifra, E. Nies, and F. E. Karasz, *Macromolecules*, **27**, 1166 (1994).

31. M. C. Goh, D. Juhue, O. M. Leung, Y. Wang, and M. A. Winnik, *Langmiur*, **9**, 1319 (1993).

32. M. Joanicot, K. Wong, J. Maquet, C. Graillat, Pl. Lindner, L. Rios, and B. Cabane, *Prog. Colloid Polym. Sci.*, **81**, 175 (1990).

33. Y. Chevalier, C. Pichot, M. Graillat, M. Joanciot, K. Wong, P. Lindner, and B. Cabane, *Colloid Polym. Sci.*, **270**, 806 (1992).

34. Y. Wang and M. A. Winnik, *J. Phys. Chem.*, **97**, 2507 (1993).

35. K. D. Kim, L. H. Sperling, A. Klein, and G. D. Wignall, *Macromolecules*, **26**, 4624 (1993).

36. L. H. Sperling, unpublished.

37. D. W. van Krevelen, *Properties of Polymers*, 3rd ed., Elsevier, New York, 1990, p. 79.

38. M. L. Williams, R. F. Landel, and J. D. Ferry, *J. Am. Chem. Soc.*, **77**, 3701 (1955).

39. L. C. E. Struik, *Physical Aging in Amorphous Polymers and Other Materials*, Elsevier, New York, 1978.

6

STRENGTHENING THE INTERFACE

Because of the lower density and lack of chemical or physical bonding in the interphase in polymer blends or between the polymer and nonpolymer in composites, these interfaces are frequently weak. Such interfaces constitute a source of mechanical failure under stress. This chapter addresses the methods by which interfacial bonding and hence strengthening may be achieved.

6.1. COUPLING AGENTS: SILANES

In Section 5.3, the discussion centered around the use of stickers, or anchors, on the ends of polymer chains, and the use of end-functional groups for the chemical bonding of polymers to solid surfaces. This section describes the use and properties of coupling agents, defined as molecules that chemically bind to solid surfaces such as silica or glass at one end of the molecule, and which provide bonding sites for the polymer at the other end. Thus, the end result is a polymer material covalently bound to the surface in question. Applications of coupling agents for surface modifications of fillers and reinforcing agents in plastics have generally been directed toward improved mechanical strength and chemical resistance of the resulting composites.

The silanes constitute the most important class of coupling agents. The structure of a silane has three parts:

(a) An Si atom in the middle
(b) Three alkyl ethers attached to the Si atom, each of which ultimately bond to the solid surface
(c) A chemical group, usually attached to the other side of the Si atom through a spacer of one to five $-CH_2-$ groups. The chemical group reacts with the monomer or polymer in contact with it.

Selected silane and other coupling agents are shown in Table 6.1 (1).

Table 6.1 Selected commercial coupling agents

Functional group	Chemical Structure	Name	Application
Vinyl	$CH_2=CHSi(OCH_3)_3$	Vinyltrimethoxysilane	EPDM, PE
Chloropropyl	$Cl(CH_2)_3Si(OCH_3)_3$	Chloropropyltrimethoxysilane	Free radical
Epoxy	$\underset{\displaystyle CH_2CHCH_2O-(CH_2)_3Si(OCH_3)_3}{\overset{O}{\triangle}}$	3-Glycidoxypropyl trimethoxysilane	Epoxy, PBT, acrylic
Methacrylate	$CH_2=C(CH_3)-COO-(CH_2)_3Si(OCH_3)_3$	3-Methacryloxypropyl trimethoxysilane	Unsaturated polyester
Primary amine	$H_2N(CH_2)_3-Si(OC_2H_5)_3$	3-Aminopropyltriethoxy silane	Nylon, urethane
Mercapto	$HS(CH_2)_3-Si(OCH_3)_3$	3-Mercaptopropyl trimethyoxysilane	Free radical
Chrome complex		Volan	High wet strength
Titanate	$[CH_2=C(CH_3)-COO-]_3TiOCH(CH_3)_2$	Tris(methacryl)isopropyl titanate	Inorganics
Phenyl	$C_6H_5Si(OCH_3)_3$	Phenyltrimethoxysilane	Hydrophobing agent

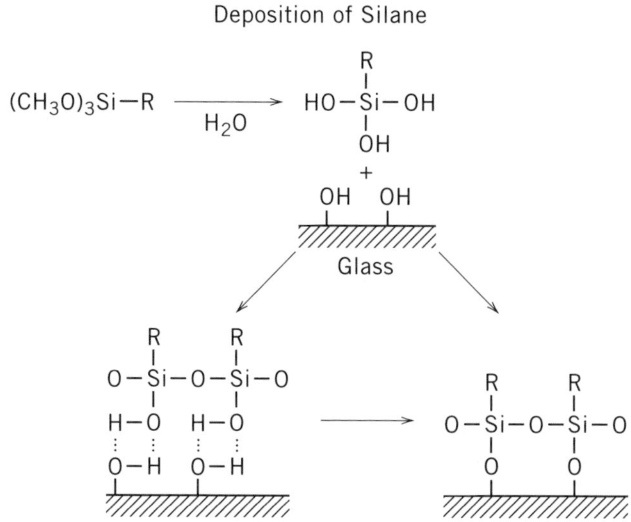

Deposition of Silane

Figure 6.1 Reactions by which silane coupling agents bond to silica surfaces. Note hydrolysis of the methyl ether, followed by hydrogen bonding to the surface hydroxyls, followed by a polymerization of the silane residues.

The silanes react with glass, silica, and other surfaces via an hydrolysis reaction; see Figure 6.1 (2). The R group in Figure 6.1 contains the polymer reactive group. As an illustration, the silane γ-aminopropyltriethoxysilane,

$$NH_2-(CH_2-)_3Si-(OC_2H_5)_3 \tag{6.1}$$

reacts with the oxirane groups of the epoxies through the amino group,

$$R-\overset{O}{\overset{/\backslash}{CH-CH_2}} + H_2N-R' \rightarrow R-\overset{OH}{\overset{|}{CH}}-CH_2-NH-R' \tag{6.2}$$

The bonding between the polymer and the filler increases the mechanical strength of the composite. While this is true for many use conditions, it is especially true for wet, rather than dry, materials; see Figure 6.2 (1). Silanes were introduced in 1948. Figure 6.2 shows the improvements available as a function of silane type and date of introduction.

Although adhesion plays a central role in any coupling mechanism, other factors are important. The treated filler tends to wet out and disperse more readily in the plastic. The treatment also protects the filler against abrasion and cleavage during the mixing process and in the final composite. The same silane coupling agents that are most commonly used on glass are also effective to varying degrees on a number of particulate mineral fillers, such as silica, talc, mica, wollastonite, and clay (3), as well as for primers for metal or ceramic surfaces to promote adhesion of sealants, paints, and adhesives. Coating levels

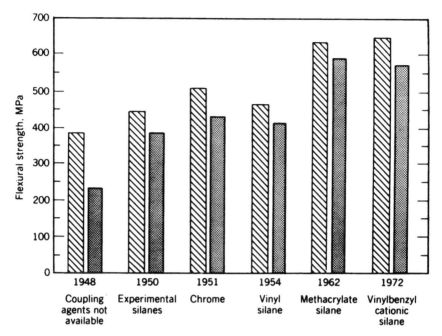

Figure 6.2 In 25 years, the strength of fiber glass–polyester composites has doubled. Diagonal lines, dry; dotted, wet.

vary from 0.2 to 1% silane. In some cases, silane coupling agents have been used to promote adhesion of one organic polymer to another.

6.2. SPECIFIC INTERACTIONS: HYDROGEN BONDING

In Section 2.4, the thermodynamic relations controlling the miscibility of polymer blends were introduced. For high-molecular-weight polymers, the entropy of mixing is small, and the miscibility or nonmiscibility of the blend is controlled primarily by the heat of the mixing term. The chemical nature of the specific interactions between polymer molecules that control miscibility have been extensively studied by Coleman et al. (4). For systems that contain specific chemical interactions between the different kinds of polymer molecules making up a blend, the miscibility of the polymer pair, or the width of the interphase, clearly depend on the number and strength of the bonds capable of being formed per unit area.

Table 6.2 (4) delineates the most frequent types of forces encountered. Of these, perhaps the most important are the interactions developed between polar or hydrogen bonding types of polymers. Of course, in these cases, the chemistry of the individual mers comes to the fore.

Table 6.3 illustrates both weak and strong kinds of association. Weak hydrogen bonds are developed between poly(vinyl chloride) and polyesters

Table 6.2   Frequently encountered interactions, increasing strength

Type of Molecule	Type of Interaction	Interaction Strength
Nonpolar "small" molecules or polymers	Physical	Weak
"Weakly" polar molecules or polymers	Physical	Weak
"Strongly" polar molecules or polymers	Physical	Intermediate
Hydrogen bonded molecules or polymers	Chemical	Intermediate
Molecules or polymers that interact by the formation of charge transfer complexes	Chemical	Strong
Ionomers (hydrocarbon polymers containing ionic groups)	Chemical	Strong

and between polyacrylonitrile and polyesters, while a more strongly associated system involves the formation of carboxylic acid dimers.

For relatively weak favorable intermolecular interactions, dissociation energies of between 1 and 3 kcal/mol are involved. The presence of these bonds encourages mixing significantly, especially when compared with similar mixtures of nonpolar polymer having only dispersive forces between the segments. These favorable interactions tend to counteract the unfavorable contribution to the free energy of mixing expressed in the usually positive χ value.

For relatively strong, favorable, specific intermolecular interactions, dissociation energies of 3–7 kcal/mol are involved (4). Because of the increased strength of these bonds, significantly larger differences in the solubility

Table 6.3   Examples of favorable intermolecular interactions

A. Weak hydrogen bonds

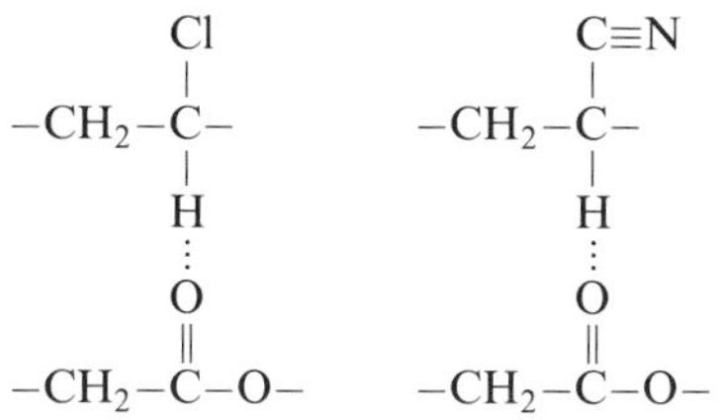

$$
\begin{array}{cc}
\text{Cl} & \text{C}\equiv\text{N} \\
| & | \\
-\text{CH}_2-\text{C}- & -\text{CH}_2-\text{C}- \\
| & | \\
\text{H} & \text{H} \\
\vdots & \vdots \\
\text{O} & \text{O} \\
\| & \| \\
-\text{CH}_2-\text{C}-\text{O}- & -\text{CH}_2-\text{C}-\text{O}-
\end{array}
$$

B. Strongly associated

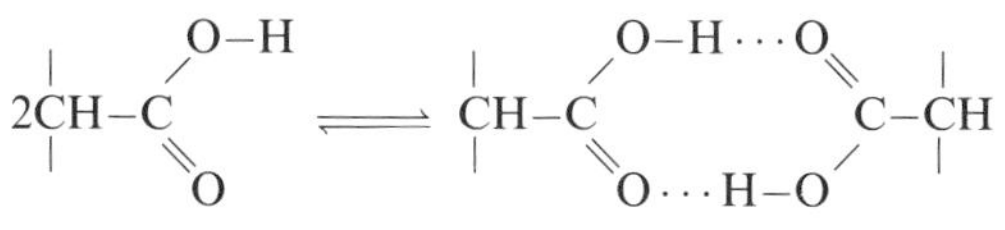

parameters of the two polymers can be overcome, resulting in industrially *compatible* polymers.

Hydrogen bonds exhibit ordinary chemical equilibrium constants that depend on the thermodynamic energy of the bond. Stoichiometry is also critically important. Thus, classical chemistry can be brought to bear on the subject:

$$A-H + B \rightleftharpoons A-H \cdots B \qquad (6.3)$$

$$K = [A-H \cdots B]/[A-H][B] \qquad (6.4)$$

where K represents the equilibrium constant of the hydrogen bond, $H \cdots B$.

Infrared spectrometry provides an excellent way to obtain information about K. Basically, each bonded and unbonded moiety absorbs at its own characteristic frequency; see Figure 6.3 (5). Here, ethylene–methacrylic acid (EMAA)

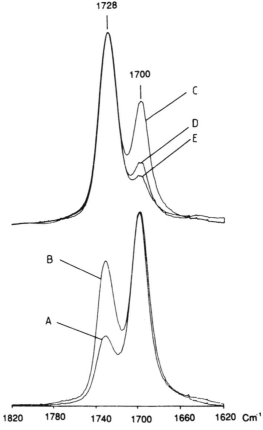

Figure 6.3 Carbonyl stretching region of EMAA(44)/PVME blend samples: (A) 80/20, (B) 60/40, (C) 40/60, (D) 20/80; and (E) 10/90 wt%.

Table 6.4 Polymer blend interactions

Interaction	Example
London dispersive forces (van der Waals)	Polybutadiene–polyethylene
Hydrogen bonding	PVC–aliphatic polyesters
Dipole–dipole	PVF_2–poly(methacrylates)
n-πcomplexes	Polycarbonate–polyesters
Specific rejection	A + BC
Bond interchange	Transesterification
Acid–base	Polyester–poly(vinylidene chloride)

copolymers are blended with poly(vinyl methyl ether) (PVME). The absorption at $1700\,cm^{-1}$ corresponds to the carboxylic acid dimer; see Table 6.3. The absorption at $1728\,cm^{-1}$ corresponds to the carboxylic acid–ether hydrogen bond, and the almost vanishingly small absorption at $1750\,cm^{-1}$ corresponds to the free acid. For a 50/50 blend, the result is about 30% free C=O bonds and about 70% hydrogen bonded C=O bonds. While this particular mixture is made miscible in the thermodynamic sense by the appearance of these bonds, many other such copolymer mixtures are only partly miscible, exhibiting variable interphase properties. This subject is still under intensive investigation.

Plots of $\ln K$ vs. $1/T$ yield both the enthalpic and entropic contributions to the bond strength through the relation

$$\ln K = -\Delta H/RT + \Delta S/R \qquad (6.5)$$

For example, the carboxylic acid association has values of -6.95 kcal/mol and -13.1 eu/mol for the enthalpic and entropic contributions, respectively (5).

The more important favorable polymer blend interactions are summarized in Table 6.4. Bond interchange produces block copolymers, which tend to strengthen the interphase via primary bonds; see Section 6.4.2.

6.3. ACID–BASE INTERACTIONS

Section 6.2 considered hydrogen bonding and other types of chemical interactions at polymer–polymer interphases. According to Fowkes (6), hydrogen bonds are a special case of acid–base bonding, for the heats of hydrogen bonding depend on the acid strength of the hydrogen donor and the base strength of the hydrogen acceptor. Thus, the general cases of Fowkes involves acid–base interaction, with hydrogen bonding being a specific, if very important case. In an acid–base interaction, one of the surfaces is either naturally acid or made to be so through chemical reactions. The other surface, similarly, is either naturally basic, or made to be so. For the most part, this discussion utilizes the Lewis acid–base concept of an electron donor–acceptor interaction. In Section

4.2, the work of adhesion, W_a, was defined in terms of surface and interfacial tensions, $W_a = \gamma_a + \gamma_b - \gamma_{ab}$ for surfaces a and b. The quantity W_a can also be defined in terms of the contributions from the London–Lifshitz dispersion force interactions, W^d, and the Lewis acid–base interactions, W^{ab}, leading to the relation

$$W_a = W^d + W^{ab} \tag{6.6}$$

Dispersion forces are present in all materials, and their contribution to the work of adhesion is assumed to be relatively constant here. The W^{ab} forces are usually much larger.

Fowkes and Mostafa (7) proposed that the acid–base contribution to the work of adhesion be separated into the surface population of acidic or basic sites, n^{ab}, and the enthalpy of acid–base interactions, ΔH^{ab},

$$W^{ab} = -fn^{ab}\Delta H^{ab} \tag{6.7}$$

where f represents a constant for converting enthalpies of interfacial acid–base interactions into free energies of interactions.

6.3.1. Drago Constants

Drago et al. (8) developed a method for predicting the strength of the interaction of two solid materials from constants that characterize each material. These constants were associated with the electrostatic, E, and covalent, C, nature of the specific acids and bases:

$$-\Delta H^{ab} = E_A E_B + C_A C_B \tag{6.8}$$

where ΔH^{ab} represents the enthalpy of acid–base bonding. The parameters E and C for a material can be determined graphically using experimentally determined heats of acid–base interaction with several acids or bases with known E and C parameters. Most of the Drago constants that have been determined are for simple organics in solution (8). A few values of interest include, in units of $(\text{kcal/mol})^{1/2}$ (9), some common chemicals and polymers:

Composition	C_B	E_B
Ethyl acetate	1.73	1.03
Poly(methyl methacrylate)	0.96	0.68

These values suggest that poly(methyl methacrylate) is a weaker base than ethyl acetate. The quantity ΔH^{ab} was determined through measurements of the shift in the infrared spectra of the carbonyl groups.

Table 6.5   **Proton donating and accepting groups**

Proton Donors	Proton Acceptors
Hydroxyl	Ether
Phenol	Nitrile
Sulfonic acid	Carbonyl
Carboxyl	Pyrrolidone

6.3.2. Polymer–Polymer Acid–Base Interactions

The acid–base type of hydrogen bonding in polymer blends was investigated by Kwei et al. (10). These workers pointed out that the London dispersion force interaction between molecules (Table 6.4) leads to a positive contribution to the heat of mixing, being very weak, while specific interactions such as hydrogen bonding or acid–base interactions, leads to a negative contribution to the enthalpy of mixing, that is, aids the mixing process. Kwei et al. (10) describe the acid–base interaction, in part, in terms of the ordinary concepts of proton donors and acceptors and not in terms of the Lewis electron donor–acceptor concepts. Table 6.5 summarizes a number of such groups. On considering the mechanism of bonding of such acid–base pairs, the relationship between what is classically considered hydrogen bonding and acid–base relationships becomes clearer.

6.3.3. Polymer–Filler Acid–Base Interactions

The literature contains a number of reports showing that basic polymers can be easily peeled off glass, but that acidic polymers have a much higher adhesion. Such glasses have a surface rich in silicate groups, which are strongly basic (9). By acid rinsing of the glass surface, the silicates were converted to acidic surface silanols, temporarily reversing the bonding strength order. The finite lifetime of the effectiveness of the acid rinse, presumably, arose from the mobility of the sodium ions deeper in the glass, gradually exchanging with the surface hydrogens.

In a recent set of experiments, Kaczinski and Dwight (11) used a Teflon FEP material to prepare model surfaces. The overall strategy was to vary the surface acid–base character by placing functional groups at the surface. The surface was first defluorinated by reacting with sodium naphthalide radical anion in diglyme. This reagent removes the surface fluorine and leaves free radicals along the polymer backbone instead. These free radicals subsequently react with washing solvents and/or the atmosphere to form a variety of new surface groups; see Figure 6.4 (11). These surfaces were then coated with chlorinated poly(vinyl chloride) (CPVC), a model acidic adhesive and, of course, an important commercial material in its own right. The peel strengths of the surface-modified Teflon FEP surfaces with CPVC are shown in Table 6.6 (11). The acidic Teflon FEP surfaces gave lower peel strengths and an interfacial failure mode. On the basic surfaces, by contrast, the Teflon FEP surface deformed, indicating a stronger type of bonding. According to X-ray photoelectron spectroscopy

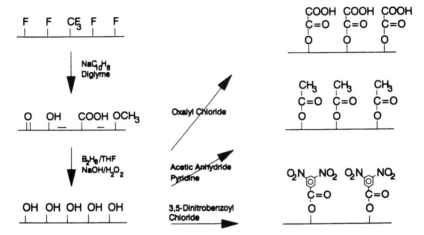

Figure 6.4 Rather unreactive surfaces of perfluorinated polymers can be made to bear special reactive groups as indicated.

(XPS) studies, there was less fluorine on the CPVC side for the acidic surface, and less chlorine on the Teflon FEP side, than when basic surfaces structures were used.

6.4. INTERFACIAL ADHESION AND COMPATIBILIZATION

6.4.1. Definitions and Materials

Sections 6.2 and 6.3 discussed hydrogen bonding and acid–base interactions between phases, respectively. Both of these are weak chemical reactions,

Table 6.6 Adhesion characteristics between surface-modified Teflon FEP films and chlorinated poly(vinyl chloride)

| | Surface Characteristics | | | |
| | Acidic | | Basic | |
Modified FEP Film	OH	COOH	Acetone	Dinitrobenzene
W^{ab} with phenol probe (mJ/m^2)	0	0	30	30
Peel strength with CPVC (g/cm)	10	10	30	50
XPS F/Cl at Teflon side	1	5	19	19
XPS F/Cl at CPVC side	0.03	0.03	1.5	0.7
Failure locus, SEM	Interfacial	Interfacial	Teflon deformation	Teflon deformation

especially compared with ordinary covalent bonding. This chapter is taking the reader to progressively stronger interfacial bonding systems. This section describes the use of block and graft copolymers to improve interfacial adhesion between polymer phases. In Section 1.5, a compatible blend was defined as one that satisfies certain industrial criteria for usefulness (often good mechanical properties), while an incompatible polymer fails to satisfy these criteria and is not useful for the given purpose. (Again, compatibility and incompatibility do not translate into miscibility and immiscibility, the latter terms describing the number of thermodynamically stable phases in the system. Indeed, most people interested in compatibilizing a polymer blend would consider that achieving total miscibility would be a disaster, the multiphase synergisms lost.)

Many polymer blends are incompatible because the interface between the phases is weak, sometimes to the point of giving the blends a cheesy behavior with very poor mechanical strength. A cheesy or chalky optical appearance may be due to a very large interfacial surface tension, which causes large phase domains and hence large amounts of light scattering. This results in a chalk white sample. If the polymer is colored light yellow, it may look like cheese. The poor mechanical properties are caused by the poor interfacial bonding of the material, very few chains traversing the interface.

6.4.2. Graft and Block Copolymers as Compatibilizers

An important industrial cure for the problem utilizes block or graft copolymers. Often, the two polymers comprising the block or graft copolymer are the same two polymers that make up the blend. Then, the two polymers anchor in their respective homopolymer phases, providing a chemical bond between the two phases. Sometimes, however, one or both of the two block or grafts copolymer components are different than the homopolymers, but are miscible or nearly miscible with them. The block or graft copolymer chains must be long enough to form entanglements on both sides of the interface, providing proper holding power. The desired effect is simple: bond the two homopolymers together, thus compatibilizing the material. The block or graft copolymer utilized as a compatibilizer is usually present in the range of 1–5%.

Compatibilizers are also known as modifiers, coupling agents, adhesion promoters, and interfacial agents (12). Utracki (13) defines a polymer alloy as an immiscible polymer blend having a modified interface and/or morphology. Utracki points out that virtually all high-performance engineering blends are actually alloys. Lyatskaya et al. (14,15) point out that the block or graft copolymer must self-assemble at the blend interfaces to provide an effective reduction in the interfacial surface tension. The most important reason for compatibilizing is that it leads to a very fine dispersion of the phases, usually submicron. This arises because of the lowered interfacial surface tension.

Table 6.7 presents selected compatibilized systems. While most of the materials listed utilize ordinary block or graft copolymers of the same chemical nature as the two components of the blend, polycaprolactone, miscible with

Table 6.7 Selected compatibilized systems[a]

No.	Polymer I	Polymer II	Compatibilizer	Reference
1	LDPE	PS	PS-*block*-(hydrogenated)PB	*a*
2	PC	PDMS	(PC-*block*-PDMS)$_n$	*b*
3	HDPE	iPP	EPM	*c*
4	PS	PA-6	PS-*block*-PA-6	*d*
5	PVC	PS	PCL-*block*-PS	*d*
6	iPP	EPM	EP-*graft*-iPP	*e*
7	PS	PI	PS-*block*-PI	*f*
8	PA-66	PET	(PA-66-*block*-PET)$_n$[h]	*i*
9	PDMS	EPDM	Poly(ethylene-*co*-methylacrylate)	*j*
10	PI	BR	PI-*block*-BR	*k*

[a] Abbreviations: BR, polybutadiene; LDPE, low density polyethylene; PS, polystyrene; PC, polycarbonate; PDMS, poly(dimethyl siloxane); HDPE, high density polyethylene; iPP, isotactic polypropylene; PA-6, polyamide (nylon-6); PVC, poly(vinyl chloride); EPM, poly(ethylene-*stat*-propylene); PI, polyisoprene; PA-66, polyamide(nylon)-66; PET, poly(ethylene terephthalate).
[b] R. Fayt, R. Jerome, and P. Teyssie, *J. Polym. Sci., Polym. Lett. Ed.*, **19**, 79 (1981).
[c] E. B. Trostyanskaya, M. B. Zemskov, and O. Ya. *Mikhasenok, Plast. Massy*, **11**, 28 (1983).
[d] L. D'Orazio, R. Greco, E. Martuscelli, and G. Ragosta, *Polym. Eng. Sci.*, **23**, 489 (1983).
[e] R. Fayt, R. Jerome, and P. Teyssie, *Polym. Eng. Sci.*, **27**, 328 (1987).
[f] D. J. Lohse, S. Datta, and E. N. Kresge, *Macromolecules*, **24**, 561 (1991).
[g] W. F. Reichert and H. R. Brown, *Polymer*, **34**, 2289 (1993).
[h] Via amide–ester interchange.
[i] L. Z. Pillon and L. A. Utracki, *Polym. Eng. Sci.*, **24**, 1300 (1984).
[j] S. Kole, R. Santra, and A. K. Bhowmick, *Rubber Chem. Tech.*, **67**, 118 (1994).
[k] D. J. Danzig, F. L. Magnus, W. L. Hsu, A. F. Halasa, and M. E. Testa, *Rubber Chem. Tech.*, **66**, 538 (1993).

poly(vinyl chloride) is used in number 5. A statistical copolymer, EPM, is shown compatibilizing the two respective homopolymers in number 3.

The nylon–polyester block copolymer formed via an exchange reaction was the basis of the commercial fiber Source of Allied Chemical Corp. It was prepared by first blending the molten polymers together until the minor component, the polyester, was dispersed in the form of droplets that were small in comparison to the fiber diameter desired (16,17). Under these melt conditions, sufficient bond interchange took place to allow acceptable adhesion between the phases. Then, the material was melt-spun and drawn in order to orient both constituents and cause the dispersed phase to form elongated cylinders or fibrils. For satisfactory dispersion, the viscosities of both components must be comparable, with the polyester preferably having a slightly higher viscosity; see Section 2.3. This was one of the first commercial compatibilized systems.

The purpose of the new fiber was to prevent flat-spotting in automotive tires. At that time, the new nylon tire cord fibers had too low a glass transition temperature. If a tire was run hot, the tire cord went through its T_g. When it was allowed to cool overnight, the fibers went through the glass transition temperature the other way, forming a flat spot on the bottom of the tire that remained for the first block or two that the car might be driven. This resulted in a rough and noisy ride, until the tires warmed and rounded out. The polyester had a higher glass transition temperature. As dispersed fibrils, the polyester helped keep the tire round as it cooled, thus reducing flat-spotting. The phase-separated nature of the material also caused it to be bright white via scattered light, making it useful for quality carpets.

Number 9 in Table 6.7 is an elastomer blend. The poly(ethylene-*co*-methyl-acrylate) is evidently blocky, with a melting temperature of 81°C. Number 10 is another elastomer–elastomer blend, frequently used for truck tire treads.

6.4.3. Fracture Mechanism Map

When block copolymers are used to toughen an interface, and the material subsequently stressed, it may fail in a variety of modes (all at higher stress levels than the original blend), depending on both the concentration and molecular weight of the block copolymer used. Xu et al. (18) and Washiyama et al. (19) developed a fracture mechanism map; see Figure 6.5 (18). The basic model starts with the behavior of homopolymers at a growing crack, subject to a tensile stress. The chains at the crack tip may either pull-out, scission, or be involved in crazes, depending on the molecular weight of the polymer in the path of the crack, the velocity of the crack, and the temperature of the material relative to the glass transition temperature of the polymer. If the chain density crossing the

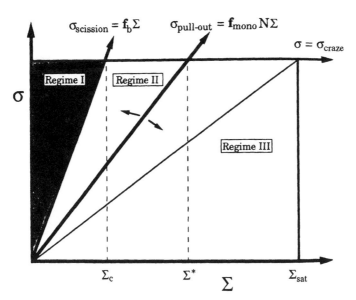

Figure 6.5 Fracture mechanism map. Regime I, chain scission; regime II, pull-out; regime III, elastic deformation and/or crazing.

crack tip is high, and the chains long enough, they may survive the simple pull-out and scission stresses, in which case the polymer crazes.

For polystyrene at room temperature, for example, there is about 50% chain scission and 50% chain pull-out at room temperature for a molecular weight of about 150,000 g/mol (20–22). At very high molecular weights, about 4×10^5 g/mol, there is 90% or more chain scission. At about 30,000 g/mol, the molecular weight between physical crosslinks (entanglements), there is substantially 100% pull-out.

These kinds of results are extended to having block copolymers at a polymer blend interface, where the block copolymers are substantially the only chains traversing the interface. In Figure 6.5, Σ represents the areal density of the block (or graft) copolymer chains at the interface, N is the degree of polymerization of either block, and σ represents the maximum tensile stress ahead of the crack tip and normal to the interface. There are three proposed fracture mechanisms. Regime I involves chain scission, at the lowest areal density of block copolymer chains. Then, there is a transition from chain scission to crazing, where $N \gg N_e$, where the latter denotes the degree of polymerization between entanglements of its respective homopolymer. Regime II represents chain pull-out, bordered by the transition from chain pull-out to crazing, where $N < N_e$ for one block, if a high enough Σ can be achieved before the interface is saturated with the block copolymer, Σ_{sat}. Regime III corresponds to the case of crazing at the interface.

The results of the Washiyama et al. (19) and other studies of block copolymers at interfaces is compared with the research of Sambasivam et al. (21,22) on homopolymer polystyrenes in Table 6.8. It must be noted that the understanding of the molecular mechanisms of fracture depend on the assumption as to how many bonds are stressed in order to achieve chain scission. While

Table 6.8 Molecular basis of fracture at block copolymer and homopolymer interfaces

Property	Block Copolymer	Ref.	Homopolymer	Ref.
Force to break a single chain	2×10^{-9} N/bond	a	5.7×10^{-9} N/bond	d
Static friction per mer	6.3×10^{-12} N/mer	b	1.2×10^{-10} N/mer	d
Fracture toughness, G_c	15 J/m^2	b	3×10^2 J/m^2	e
Area/chain density at saturation	0.11 chains/nm	b	1.93 chains/nm^2	d
Interfacial tension	3.2 ergs/cm^2 (PS/PMMA)	c	0	—
Energy per chain	1.36×10^{-16} J	—	1.55×10^{-16} J	—

[a] C. Creton, E. J. Kramer, C. Y. Hui, and H. R. Brown, *Macromolecules*, **25**, 3075 (1992).
[b] J. Washiyama, E. J. Kramer, and C. Y. Hui, *Macromolecules*, **26**, 2928 (1993).
[c] S. Wu, *J. Phys. Chem.*, **74**, 632 (1970).
[d] M. Sambasivam, A. Klein, and L. H. Sperling, *Macromolecules*, **28**, 152 (1995).
[e] N. Mohammadi, A. Klein, and L. H. Sperling, *Macromolecules*, **26**, 1019 (1993).

some workers have assumed that only one bond is excited, and finally broken, the Sambasivam and co-workers (21,22) research was based on the assumption that all of the bonds between physical entanglements, about 300 for polystyrene, are excited, with one of the bonds finally failing. The energy required to excite all of the other bonds is degraded into heat. The conclusion is that much more work is required to chain scission, which happens at high molecular weight, than chain pull-out, which happens at lower molecular weight.

6.4.4. Calculations of the Numbers of Chains at an Interface

The values shown in Table 6.8 will now be explained through the following calculations. First, assume homopolymer polystyrene, with a density of $1.05\,g/cm^3$ and a mer molecular weight of $104\,g/mol$. The volume per mer is $104\,g/mol / 1.05\,g/cm^3 = 99\,cm^3/mol$. The volume of one mer is given by $99\,cm^3/mol / 6.02 \times 10^{23}\,mers/mol = 1.6 \times 10^{-22}\,cm^3/mer$.

Assuming a cubic model for the shape of a mer (lattice model), with the length given by the cube root of the volume, the length of one side of the cube is $5.48 \times 10^{-8}\,cm/mer$. The area of one face of the cube is given by the length squared, or $3.0 \times 10^{-15}\,cm^2/mer$. The inverse gives the concentration per unit area, or $3.33 \times 10^{14}\,mers/cm^2$, or $3.33\,mers/nm^2$. This assumes a vertical orientation of the chains. If the chains are random in orientation, and the average angle of orientation is $54°$ from the vertical, then the average number of mers is given by $3.33\,mers/nm^2 \times \cos 54° = 1.93\,mers/nm^2$.

This is the value for homopolymer polystyrene. It is the maximum number of mers at the plane of a crack for a styrenic polymer. The value for the block copolymer interface at saturation is $0.11\,chains/nm^2$ (Table 6.8), about 5% of the homopolymer value.

For the case of the block copolymer, Washiyama et al. (19) had a block copolymer of a degree of polymerization of the polystyrene of 580 and that of poly(2-vinyl pyridine) of 220, the basis of the present calculation. As a simplification, the poly(2-vinyl pyridine) will be assumed to have a density and mer molecular weight essentially the same as that of polystyrene. The whole chain then has a molecular weight of $800 \times 104\,g/mol = 83,200\,g/mol$.

Using the same type of mathematical approach as above, the area of one face of a cube of one whole polymer molecule is approximately $2.58 \times 10^{-13}\,cm^2/molecule$. The surface concentration is the inverse, $3.86 \times 10^{12}\,molecules/cm^2$. After correcting for the average angle of $54°$, this value is $0.022\,mers$ of block copolymer junction per nanometer squared. This value is about one fifth of the value shown in Table 6.8, arrived at by more sophisticated models, and about 1% of the value for a homopolymer polystyrene, above. Of course, in the real case, the block copolymer chains at the interface must be diluted somewhat by the pre-existing homopolymer.

If an average polymer blend has an interfacial area of about $100\,m^2/g$, this corresponds to a concentration of 0.5% of block copolymer needed to saturate the surface, Σ_{sat}. Since at least 1% of polymer is used to toughen interfaces in most actual blends, it has to be that some of the block copolymer remains in the interior of one phase or the other, being ineffective.

The force to break a single chain, in the case of the homopolymer calculation, was based on a simple model where 83 kcal of work are required to stretch a carbon–carbon bond from 1.54 to 2.54 Å, presumably arriving at the top of the energy well. Simple division of work by distance gives the force, which after changing units is 5.7×10^{-9} N/bond. The static friction per mer was experimentally determined by chain pull-out experiments, based on the Sambasivam et al. (21,22) dental burr experiments. Both the energy to fracture and the molecular weight reduction were recorded. The fracture toughness in the homopolymer case was determined by the work to form a unit area of polymer.

The fracture toughness, G_c (mixed mode for the dental burr experiment), of the homopolymer, at 300 J/m^2, is about 20 times that of the block copolymer. The reader will note that about 20 times more chains traverse the interface. Dividing G_c by the areal chain density in Table 6.8 yields 1.36×10^{-16} and 1.55×10^{-16} J/chain for the block copolymer and the homopolymer, respectively. This compares with a value of 1.79×10^{-16} J/chain calculated to break the 300 bonds between physical entanglements (see above) at 83 kcal/mol (21). Within reasonable error of experiment or calculation, the three numbers appear to be the same. For comparison, Wool et al. (23) reported a G_{1c} value for molded virgin polystyrene of about 1000 J/m^2. The experimental range reported in the literature goes from about 100 to 1000 J/m^2. For reasons yet to be clarified, molded films apparently have a more random mixing of chains than materials made from latexes, although surface branching or crosslinking has been suspected in some cases.

Since the interfacial tension for the system polystyrene-*blend*-poly(2-vinyl pyridine) has not been reported, the roughly equivalent value for polystyrene-*blend*-poly(methyl methacrylate) was substituted in Table 6.8. Of course, the interfacial tension inside a homopolymer is zero by definition.

Reichert and Brown (24) found that crosslinking between the homopolymers and the block copolymers increased toughness in polystyrene–polyisoprene blends. With a similar system, Creton et al. (25) found that the toughness increased with increasing the degree of polymerization of the block copolymer. Wool (26) emphasizes that the full energy to fracture of amorphous homopolymer plastics is reached at about $8M_c$. In fact, commercial plastics such as polystyrene, poly(vinyl chloride), and polycarbonate are often used at about $7M_c$, or about 90% of possible energy to fracture, see Table 6.9. Apparently, the reason is that the melt viscosity goes as the molecular weight to the 3.4 power. Thus, a slight reduction in molecular weight lowers the melt viscosity, and saves money.

6.4.5. Use of Functionalized Copolymers at Interfaces

Xanthos (12) reviewed the literature regarding the use of functionalized co-polymers, especially block and graft copolymers, to toughen polymer interfaces;

Table 6.9   Selected values of M_c and C_∞

Polymer	M_c (g/mol)	C_∞
Polyethylene	4,000	6.7
Polystyrene	31,200	10.0
Polypropylene	7,000	5.8
Poly(vinyl alcohol)	7,500	8.3
Poly(vinyl acetate)	24,500	9.4
Poly(vinyl chloride)	11,000	7.6
Poly(methyl methacrylate)	18,400	8.2
Poly(ethylene oxide)	4,400	4.2
Poly(propylene oxide)	5,800	—
Polycarbonate	4,800	2.4

Source: Based on R. P. Wool, *Macromolecules*, **26**, 1564 (1993).

see Table 6.10. Compatibilization may involve suitable functionalized polymers capable of enhanced specific interactions and/or chemical reactions such as halogenation, sulfonation, hydroperoxide formation, end capping, and so forth. A popular method of toughening a polyamide in this case involves EPM or EPDM as one of the components, with copolymers containing anhydride or carbonyl functionality serving as the compatibilizers.

In the general case, of course, some block or graft copolymer will form during any blending process involving shear, which tends to degrade all polymers to some extent. Another popular way of bringing about block and graft copolymer formation in situ utilizes the addition of various amounts of monomer to the polymer before it is exposed to the shearing conditions. Then, the free radicals, anions, and cations formed initiate graft and block copolymers (27).

Table 6.10   Compatibility through reactive copolymers

Major Component	Minor Component	Compatibilizer
ABS	PA-6/PA-6,6 copolymer	SAN/MA copolymer
PP or PA-6	PA-6 or PP	EPM/MA copolymer
PE	PA-6 or PA-6,6	Ionomers, carboxyl functional PE's
PP or PE	PET	PP-g-AA, carboxyl functional PE
PA-6,6	Acrylate rubber (hydroxyl functional) or EPM	SMA or EPM-g-MA
PPE/PS	Sulfonated EPDM (zinc salt), or Phosphonate ester of EPM or EPDM	Sulfonated PS (zinc salt) plus zinc stearate

6.4.6. Reactive Processing

Interfacial reactions in polymer blends are most easily carried out industrially utilizing reactive processing in the melt state. Kumph et al. (28) pointed out that many engineering thermoplastics share some common characteristics: They are mostly condensation polymers, they contain functional groups that impart a significant polarity to the molecule, and they often contain aromatic rings in the polymer backbone. Increasingly, chemistry is moved from reacting vessels into processing equipment, referred to as reactive processing. Equipment used includes batch mixers, kneaders, single-screw extruders, and twin-screw extruders. Twin-screw polymer reactors offer the ability to integrate important chemical reaction variables, including improved mixing of high- and low-viscosity raw materials, efficient mass transfer and heat transfer, variable reaction times, devolatilization of those ingredients remaining unreacted, or reaction by-products, and transport of the reaction mass through different process zones.

Figure 6.6 (28) illustrates two types of reactive processing: the production of thermoplastic polyurethanes and compatibilized PA6/PP blends through the

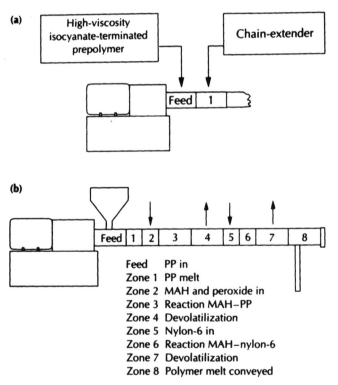

Figure 6.6 Reactive processing in a modern age. (a) Reactive polymerization of a polyurethane, showing the addition of the chain extender to the isocyanate-terminated copolymer; (b) an eight zone extruder for reacting maleic acid anhydride with both polypropylene and nylon 6 to produce a compatibilized blend.

use of maleic anhydride. The use of multiple zones must be noted, which may utilize different screw pitches, temperatures, and the like.

There are two cases to consider in reacting maleic anhydride with polymers. If the polymer contains unsaturation, on heating,

Saturated hydrocarbon polymers, such as polyethylene, polypropylene, or ethylene–propylene–diene monomer (EPDM polymers) react via heat and peroxide,

Block copolymers may be produced via transesterification; see Figure 6.7 (28). Polymers such as poly(*p*-phenylene oxide) may be functionalized with maleic anhydride and may undergo reactions, as also illustrated in Figure 6.7. In each of these two cases, block copolymers are formed.

In recent work, Lambla and co-workers (29,30) studied reactive blends of various polyethylenes and polyamides via melt processing. They used a corotating twin-screw extruder utilizing peroxides and maleic anhydride. They reported that PP/PA6 blends subsequently had good adhesion, while EPR/PA6 blends had a finely dispersed rubber phase and very high impact resistance, see Section 8.6.

6.4.7. Compatibilization with Ionomers

6.4.7.1. Structure of Ionomers   Ionomers are polymers that contain approximately 5–15% ionic groups, depending on the polymer and application (31). These highly useful materials have seen several important applications. These include Surlyn, of du Pont, which is basically a 5% sodium methacrylate

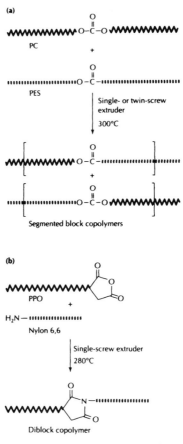

Figure 6.7 Reactive processing of polymer blends. (a) Bond interchange reactions between a polycarbonate and a polyethersulfone, and (b) formation of a diblock copolymer of poly(*p*-phenylene oxide) with a polyamide.

copolymer of polyethylene (32,33). This material is used as a coating for the polyethylene used to package meats in supermarkets, its advantage being sealability in the presence of meat fluids. Another such application is the fluorocarbon based ionomer Nafion, also of du Pont, widely used because of its permeability characteristics as separators in electrochemical applications (31).

As a class, these materials tend to be statistical copolymers of ion-bearing mers in nonpolar polymers. As such, they tend to phase separate into microdomains rich in ionic groups, surrounded by substantially pure ion-free polymer. Thus, they resemble block copolymers with one hard mer; see Chapter 9. The ionic regions tend to increase the melt viscosity in the same way, for example. Although the ionomers, as materials themselves, are beyond the scope of this book, their interactions in the interphases of polymer blends to control compatibility is indeed within the scope of this book.

6.4.7.2. Interphase Control with Ionomers   Compatibility enhancement has been observed in many polymer blends where ionic groups are introduced to one or both components. For example, Smith and Eisenberg (34) incorporated about 5% of styrenesulfonic acid into a polystyrene, and about 5% of 4-vinyl pyridine into a poly(ethyl acrylate) to compatibilize an otherwise incompatible polymer pair. The acid group in the sulfonic acid donates a proton to the amine in the pyridine to produce an exothermic reaction, tending to reduce the interfacial tension and increase stability in the system.

If the level of ionomer interaction is higher, truly miscible materials may be produced. Thus, in stoichiometric blends of poly(styrene-*stat*-styrenesulfonic acid) with poly(styrene-*stat*-4-vinylpyridine), 80–90% of the groups were H-bonded to each other (35). In blending a polyamide-6 with a lithium or sulfonated polystyrene ionomer, the lithium counterions produced miscibility, but the sodium counterions produced only interfacial interactions (36). Similarly, lithium sulfonate groups are more effective than lithium or sodium carboxylate groups in enhancing the miscibility between polyamides and polystyrene (37), thus offering a range of either miscibility or strong interphase interactions, at the will of the investigator.

There are a number of patents that show that blending of polyamides with polyethylene ionomers leads to improved impact resistance of the former (38–40). In this case, the low glass transition of the polyethylene is important. Other patents show that polyethylene ionomers can be used to compatibilize blends of polyamide-6 with polyethylene or polypropylene (41,42). In these latter, the interphase region is probably significantly broadened.

6.5. STRENGTHENING THE INTERFACE THROUGH COPOLYMERS

6.5.1. Wool's Scaling Laws for Polymer Interfaces

The strength of a polymer–polymer interface depends significantly on the extent of interdiffusion of the two polymers. In general, the larger the interphase region, the greater the number of entanglements, and the stronger the interface. Wool and co-workers (26,43–45) developed a theory based on scaling laws that expresses these ideas quantitatively. This theory treats both symmetric interfaces, where the two polymers juxtaposed are identical, and the asymmetric case, where the two polymers are different in some way. While this book is largely concerned with the case where the two polymers are chemically different, and, in general immiscible, a brief treatment of the symmetric case will set the stage for the asymmetric case.

6.5.1.1. Symmetric Case   Scaling laws are primarily concerned with the values of exponents and little concerned with coefficients. They are especially useful for transitions between regimes. In this case, the regimes are inter-diffusion across a symmetric interface and when the system is more or less at equilibrium. As with other branches of mathematics and sciences, scaling laws start with known relationships.

The general scaling law can be written:

$$H(t) = H_\infty (t/T_r)^{r/4} \tag{6.9}$$

$$H_\infty \sim M^{(3r-s)/4} \tag{6.10}$$

where the general dynamic properties, $H(t)$, for the interface are expressed in terms of the static properties, H_∞, and the reduced time, t/T_r, where T_r is the reptation or relaxation time of the system. The quantity T_r depends on the molecular weight to the third power. The quantities r and s are integers with values, 1, 2, 3, ...

Specific solutions of equations (6.9) and (6.10) depend on the quantities of interest. For example, the number of polymer chains intersecting an interface, $n(t)$, can be written:

$$n(t) = n_\infty (t/T_r)^{(1/4)} \tag{6.11}$$

with the equilibrium property given by

$$n_\infty \sim M^{-1/2} \tag{6.12}$$

After substituting equation (6.12) into (6.11), and noting that $T_r \sim M^3$, the following relation emerges:

$$n(t) \sim t^{1/4} M^{-5/4} \tag{6.13}$$

Wool and co-workers (26,43–45) derived similar expressions for the number of bridges crossing the interface, the average mer interpenetration depth, number of mers crossing the interface, and so forth.

The critical fracture energy, G_{1c}, based on reptation of the chains across the weld line of the two parts of polymer is given by

$$G_{1c}(t) \sim G_{1c\infty} (t/T_r)^{1/2} \tag{6.14}$$

In general, Wool (26,43–45) is interested in molecular weights between M_c, the entanglement molecular weight, and $8M_c$. Above that molecular weight, he suggests that mechanical strength should not significantly increase.

6.5.1.2. Asymmetric Case   Helfand and co-workers (46–48) had determined an expression for the equilibrium interpenetration distance, or width of the interphase, as $d_\infty = 2b/(6\chi)^{1/2}$, where b is the statistical segment step length, with a value of approximately 0.65 nm for polystyrene, poly(methyl methacrylate), and similar vinyl polymers. This relation defines the equilibrium strength of the interphase because entanglements are very limited in number for most values of d_∞; see Section 5.4.2. The fracture energy of the interface is given by

$$G_{1c} \approx 2\Gamma (d_\infty/b)^2 \qquad (6.15)$$

where Γ represents the surface tension. Equation (6.15) provides a lower bound relationship. Values of G_{1c} for polymer blend interfaces are frequently in the range of $10–50 \, \text{J/m}^2$, which compares with values of $200–1000 \, \text{J/m}^2$ for virgin homopolymers.

6.5.2. Analysis of the Fractured Interphase

For immiscible (asymmetric) interphases, Foster and Wool (49) and Willett and Wool (50) undertook some interesting experiments to elucidate the path of the crack. X-ray photoelectron spectroscopy (XPS), sometimes called electron spectroscopy for chemical analysis (ESCA; see Section 4.6.1.2) was used. As described before, XPS makes use of the photoelectric effect. Low-energy X-rays bombard the surface of a sample, causing electrons to be ejected. The kinetic energy of the electrons is analyzed, leading to an identification with a particular type of atom, and hence elemental composition of the surface of the sample. One system analyzed was polystyrene interfaced with poly(methyl methacrylate). This pair is incompatible. Some typical atoms and energies analyzed included:

O_{1s} 540 eV
C_{1s} 285 eV
N_{1s} 400 eV

After welding had proceeded for controlled times, temperatures, and pressures, the plates were broken via wedge cleavage fracture mechanics methods. The authors were interested in knowing whether adhesive or cohesive fracture was taking place. The analysis of the results utilized the following reasoning. If, during the fracture process, PMMA were transferred to the PS side, a strong O_{1s} signal should develop on the PS side, whereas if PS were transferred to the PMMA side, a decrease in the O_{1s} signal should occur on the PMMA side. A theoretical ratio of 0.40 (two oxygen atoms and five carbon atoms per mer) would indicate that the PMMA side had no PS transferred. The experimental results are shown in Figure 6.8 (50) for the XPS spectra of the PMMA fracture surface. The O/C ratios for the areas under the peaks were normalized, with values of 0.26 and 0.28 for the 125 and 140°C molding temperatures, respectively, indicating an increase in the carbon content. In contrast, the O/C ratio of as-molded PMMA of 0.38, also shown in Figure 6.8, is in good agreement with the theoretical value of 0.40. The O/C ratios measured on the corresponding PS fracture surfaces were approximately 0.01 or even less. These results indicate

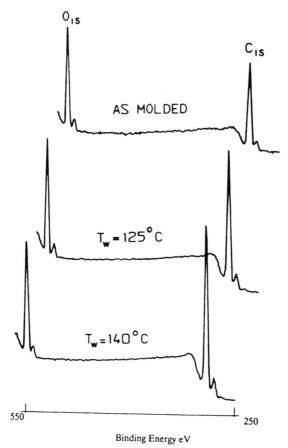

Figure 6.8 XPS spectra of poly(methyl methacrylate) fracture surfaces welded to polystyrene. Range is 250–550 eV. In this and Figure 6.10 the ratio of the oxygen-to-carbon peaks are important in deciding whether adhesive or cohesive failure takes place.

that crack growth at the PS–PMMA interfaces contains both cohesive and adhesive components. Cohesive fracture only occurs on the PS side to the extent that XPS analysis can measure. (Adhesive fracture means that the fracture takes place down the middle of the interphase; cohesive fracture means that fracture took place on the weaker side of the interphase, as indicated.)

Additional analysis by scanning electron microscopy (SEM) on the fractured PS surface revealed a series of ridges of highly deformed materials, with a well-defined spacing of about 50 μm, and hundreds of micrometers in length; see Figure 6.9 (50). These features are indicative of an unstable crack growth pattern commonly called *slip-stick*. The sticking phase probably involves the deformation and rupture of crazelike strands, and the slipping phase leads to the relatively featureless regions between the deformed ridges. The corresponding examination of the PMMA fracture surfaces show both PS residue and rather

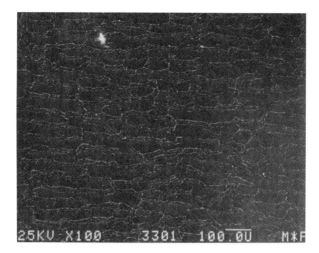

Figure 6.9 SEM micrograph of polystyrene fracture surfaces welded to poly(methyl methacrylate), $T_w = 140°C$.

featureless portions. Interestingly, the resulting fracture energy of the PS–PMMA interface, of about $50 \, J/m^2$, matches 10% ridges of PS having $500 \, J/m^2$, plus 90% interface fracture energy of $0.1 \, J/m^2$ or less to yield the experimental value of about $50 \, J/m^2$. The authors suggest that the cohesive energy contribution associated with the ridges is not the same as that in the virgin components.

In another series of experiments, these same authors substituted styrene–acrylonitrile (SAN) of varying acrylonitrile content for the polystyrene. The acrylonitrile component provides a type of hydrogen bonding between the two polymers; see Table 6.3.

The composition range of greatest compatibility can be estimated through the evaluation of the Flory interaction parameter, χ (51):

$$\chi = \beta\chi_{13} + (1 - \beta)\chi_{23} - \beta(1 - \beta)\chi_{12} \qquad (6.16)$$

where the quantity χ_{ij} represents the interaction parameter between mers i and j, 1 and 2 are the comonomers, 3 represents the homopolymer, and β is the mole fraction of mer 1 in the copolymer, assuming a random mixture of the two polymers.

The minimum in χ can be obtained through differentiation of χ with respect to β in equation (6.16). The result is expressed by β^*:

$$\beta^* = 0.5(\chi_{12} + \chi_{23} - \chi_{13})/\chi_{12} \qquad (6.17)$$

Equation (6.17) implies that the greatest interfacial strength will occur near β^*. From the known values of χ_{ij}, the minimum χ value was -0.005 at $125°C$ for $27 \, wt\%$ acrylonitrile in the SAN. Since χ is negative, actual total miscibility can be predicted for this composition.

Table 6.11 Polymer blend interface strengths, G_{1c} (J/m^2)

Polymer I/ Polymer II	Poly(methyl methacrylate)	Polycarbonate
Polystyrene	50	—
SAN, 5.7% AN	170	28
SAN, 23.4% AN	>240	55
SAN, 37.0% AN	80	42

Source: Based on data from J. L. Willett and R. P. Wool, *Macromolecules*, **26**, 5336 (1993).

Table 6.11 summarizes the maximum energy to fracture, under equilibrium conditions, as determined by Wool and co-workers (50). The value for 23.4% AN SAN shows clear superiority.

The XPS analysis of the PMMA fracture surfaces of the PMMA–SAN interface are shown in Figure 6.10 (50). The O_{1s} peak on the PMMA welded to SAN containing 5.7% AN is significantly reduced, and virtually no O_{1s} signal was measured on the surface welded to SAN containing 23% AN. These results and others show that fracture at these interfaces is almost purely cohesive, although the 37% AN containing material did have an adhesive component.

A third system investigated by Wool and co-workers involved these same SAN samples, but bonded to polycarbonate (50). The value of this last system lies in the practical importance of the rubber toughening of polycarbonate with acrylonitrile–butadiene–styrene (ABS) resins. In this case, the SAN forms a shell material in direct contact with the polycarbonate; see Section 8.5.

The minimum in equation (6.17) occurs at 23% AN for this pair. The energies necessary to fracture these interfaces are also shown in Table 6.10. In this case, only welding below or near the glass transition temperature of the polycarbonate is considered. Only cohesive fracture was indicated by XPS.

Wool (26) found that in most cases interfacial fracture took place in such a manner that a residue of the polymer with the lower entanglement density was found on the fracture surface of the materials with the higher entanglement density. Of the materials considered above, polystyrene has the lowest entanglement density, which increases with increasing AN content. PMMA has a still higher entanglement density, with polycarbonate highest of all. Values of high entanglement density (low values of M_c or M_e) in general correlate with high fracture energies, other factors being equal.

The value of d_∞ for the interphase between polystyrene and poly(methyl methacrylate) has been estimated variously (26,52,53) as being between 2 and 5 nm. Assuming that a value of 3.5 nm is the best overall estimate, the approximate thicknesses for the SAN–PMMA and SAN–PC interfaces can be calculated if the values in equation (6.15) are known. Ideally, the interphase thickness should approach the radius of gyration of the chains making up the interphase, approximately 15–20 nm.

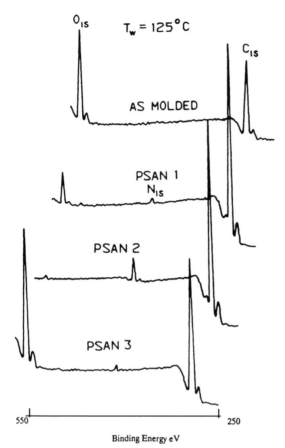

Figure 6.10 XPS spectra of poly(methyl methacrylate) fracture surfaces after molding to styrene–acrylonitrile copolymers with various levels of AN. From top to bottom: as molded, 5.7% AN, 23% AN, 37% AN; $T_w = 125°C$.

While the above represents model experiments on macroscopic surfaces, its value lies in its relation to polymer blend interphase strengths. By synthesizing polymer blend pairs with greater compatibility, the fracture energy of the interfaces can be made to approach that of the virgin materials.

6.6. CRYSTALLINE POLYMER BLENDS

Polymers such as polyethylene, polypropylene, nylon, and the like are all semicrystalline under ordinary circumstances. For most such semicrystalline polymers, the melting temperature replaces the glass transition temperature as the most important landmark in processing and application. In many of these plastics, the amorphous part is above the glass transition temperature

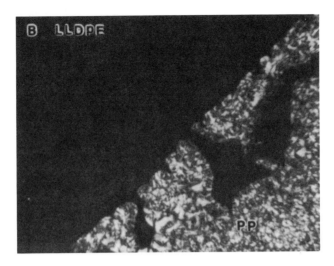

Figure 6.11 Polypropylene/linear low-density polyethylene (LLDPE) interface formed by cooling at 10°C/min from 150°C. Observation at 111°C, while the LLDPE still had not crystallized. Note influxes of LLDPE into the PP side, forming a mechanical interlock.

during engineering use, lending significant flexibility and toughness to the polymer.

When these materials crystallize from the melt, they shrink, sometimes several percent. While interdiffusion at the interphase is still important, the shrinkage at the interphase may dominate because it leads to a pronounced type of mechanical interlocking. Since the first polymer to crystallize shrinks first, the other polymer tends to flow into the spaces provided. Both the kinetics of crystallization as well as the thermodynamic crystallization temperature are important. Frequently, such polymers undercool significantly.

Figure 6.11 (54) illustrates the interface formed when linear low-density polyethylene (LLDPE) (with controlled branching) and polypropylene (PP) are melt blended and cooled. The pear-shaped inclusions of the LLDPE into the PP are caused by shrinkage of the latter polymer during crystallization.

Another effect peculiar to the crystallization of polymer blends at interfaces is the phenomenon of epitaxial crystallization. This may lead to a significant synergism in mechanical behavior. In the system isotactic polypropylene–high-density polyethylene (HDPE), Yan et al. (55) explain that the two types of chains crystallize at a common interface with their chain directions about 50° apart. Their epitaxial relationship has been explained in terms of the alignment of the zig-zag HDPE chains along the methyl group rows of the iPP, with 0.5 nm intermolecular distances. This produces excellent chain-row matching. A combination of bright-field electron microscopy and electron diffraction for slow crystallization led to the model illustrated in Figure 6.12 (55). During an isothermal (or slow) crystallization process, generally three layers of HDPE crystals are formed. The first layer is in direct contact with the iPP surface, and

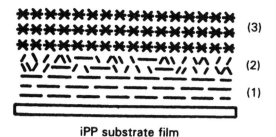

iPP substrate film

Figure 6.12 Cartoon of the crystallization of high density polyethylene onto isotactic polypropylene. The layers indicated are: (1) an epitaxial HDPE layer, (2) oriented crystalline aggregates, and (3) HDPE spherulitic layer.

consists of cross-hatched lamellae grown epitaxially on the oriented iPP substrate. This layer is about 100 nm thick. The second layer consists mainly of oriented crystalline aggregates. The third layer is spherulitic, characteristic of bulk crystallization. For fast crystallization, there are only two layers. The first layer consists of a thicker (250-nm) layer of epitaxial lamellae of HDPE, as above, while the second layer is spherulitic. Under quenching conditions, the epitaxial layer becomes thinner again.

Since polyethylene and polypropylene are the two most important plastics used in the household, they are the ones most often recycled together. The proper understanding of their interface formation will lead to improved properties of the recycled materials, one of the most important problems facing today's polymer scientists and engineers.

6.7. SUMMARY OF INTERFACIAL TOUGHENING METHODS

By way of summary, Koberstein (56) points out that the use of block copolymers or related methods of "modification," that is, improvement, at polymer blend interphases serves two purposes:

1. To decrease the interfacial surface tension
2. To increase chain entanglement within the interphase

Decreasing the interfacial tension makes it easier to decrease the phase domain size. Further, formation of a block copolymer *brush* at the interphase reduces coalescence by steric repulsion effects (57). This latter is of special interest when blending during melt extrusion, for example in an internal batch mixing or a twin-screw extruder.

The use of the term *brush* in the present context is reminiscent of the use of the same term to describe polymer chains that stick up from a colloidal particle into a solution; see Section 5.1. The alternate case was a pancake morphology, where the chains lie down on the colloid particle. In the case of polymer blends containing block copolymers in the interphase, Leibler (58) defined two corresponding brush regimes: wet brush and dry brush. The criteria differentiating the two regions involved the area per block copolymer chain and molecular weights of the individual blocks. For relatively large interfacial areas per block copolymer chain, and/or relatively low homopolymer molecular weights, the wet-brush regime is present, where the homopolymer chains tend to penetrate into the block copolymer brush. At the opposite extreme, when the block copolymer chains are closely packed at the interphase, and/or the molecular weights of the homopolymer are relatively high, a dry-brush regime is present, where there is relatively little interpenetration of the brush by the homopolymer matrix.

Sundararaj and Macosko (57) showed that under melt extrusion conditions, the use of block copolymers is not as efficient in stabilizing morphology as using reactive polymers. (The use of maleic anhydride containing polymers, which react with oxazoline containing polymers to form graft copolymers, was especially interesting.) Working in the dry-brush regime, Sundararaj and Macosko showed that coalescence is suppressed significantly due to the steric effect of block or graft copolymers at the interphase, see Figure 6.13 (57). However, only small amounts of block copolymer, about 1% or even less, can significantly reduce the interfacial tension by 80–90% (59,60).

Koberstein (56) further points out that there are six basic types of polymer interphase, with and without 'modification':

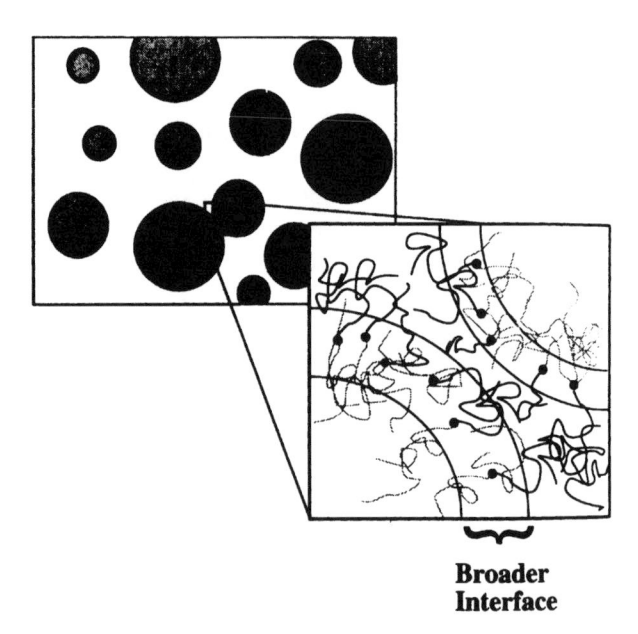

**Broader
Interface**

Figure 6.13 Coalescence is suppressed in polymer blends by the presence of block or graft copolymers at the interfaces.

1. With no block copolymer or reactive polymer present, the equilibrium polymer blend interphase, generally only 2–3 nm in width, with limited entanglement, will have low adhesion strength between the phases.

2. Addition of block copolymer brushes increases interphase entanglements, and interphase width increases to 20–30 nm, with a concomitant rise in interfacial adhesion strength.

3. End-group modifications can work by two different methods: repulsion by its own polymer or favorable interactions with the other polymer (61). Thus, two mers on the same polymer that have strong positive χ interactions may lessen the effect by mixing with another polymer with smaller, although still positive χ interactions. Alternately, there may be direct favorable interactions between an added mer and the other polymer.

4. Formation of interpolymer complexes across the interphase creates a strong physical bond between the two polymers.

5. In the case of composites, polymer brushes may also serve to compatibilize the interfaces through the use of functional group bonding.

6. Through the use of ω-functional block copolymers, a combination of types 2 and 3 above may be accomplished.

Thus, the science of polymer blend interphase modification is seen to be approaching that of a fine art.

Just as this book was being completed, the American Chemical Society symposium, *Interfacial Aspects of Multicomponent Polymer Materials*, was held under the auspices of the Polymer Materials Science and Engineering Division (62). The results of the symposium will appear in the same time frame as this book (63).

REFERENCES

1. E. P. Plueddemann, in *Encyclopedia of Polymer Science and Engineering*, 2nd ed., J. I. Kroschwitz, Ed., Vol. 4, Wiley, New York, 1985.

2. J. L. Koenig and C. Chiang, in *The Role of the Polymeric Matrix in the Processing and Structural Properties of Composite Materials*, J. C. Seferis and L. Nicolais, Eds., Plenum, New York, 1983.

3. P. G. Pape, in *Polymer Modifiers and Additives Div. (SPE) Newsletter*, **X** (3), 1993.

4. M. M. Coleman, J. F. Graf, and P. C. Painter, *Specific Interactions and the Miscibility of Polymer Blends*, Technomic, Lancaster, PA 1991.

5. J. Y. Lee, P. C. Painter, and M. M. Coleman, *Macromolecules*, **21**, 346 (1988).

6. F. M. Fowkes, in *Encyclopedia of Polymer Science and Engineering*, 2nd ed., Supplement, P. 1., J. I. Kroschwitz, Ed., Wiley, New York, 1989.

7. F. M. Fowkes and M. A. Mostafa, *Ind. Eng. Chem. Prod. Res. Dev.*, **17**, 3 (1978).

8. R. S. Drago, G. C. Vogel, and T. E. Needham, *J. Am. Chem. Soc.*, **93**, 6014, (1971).

9. F. M. Fowkes, D. O. Tischler, J. A. Wolfe, L. A. Lannigan, C. M. Ademu-John, and M. J. Halliwell, *J. Polym. Sci., Polym. Chem. Ed.*, **22**, 547 (1984).

10. T. K. Kwei, E. M. Pearce, F. Ren, and J. P. Chen, *J. Polym. Sci., Part B, Polym. Phys.*, **24**, 1597 (1986).

11. M. B. Kaczinski and D. W. Dwight, *J. Adhesion Sci. Technol.*, **7**(3), 165 (1993).

12. M. Xanthos, *Polym. Eng. Sci.*, **28**, 1392 (1988).

13. L. A. Utracki, *Polymer Alloys and Blends*, Hanser, Munich, 1990.

14. Y. Lyatskaya, D. Gersappe, and A. C. Balazs, *Macromolecules*, **28**, 6278 (1995).

15. Y. Lyatskaya, S. H. Jacobson, and A. C. Balasz, *Macromolecules*, **29**, 1059 (1996).

16. R. A. Buckley and R. J. Phillips, *Chem. Eng. Prog.*, **65**, 41 (1969).

17. B. T. Hayes, *Chem. Eng. Prog.*, **65**, 50 (1969).

18. D. B. Xu, C. Y. Hui, E. J. Kramer, and C. Creton, *Mech. Mater.*, **11**, 257 (1991).

19. J. Washiyama, E. J. Kramer, and C. Y. Hui, *Macromolecules*, **26**, 2928 (1993).

20. N. Mohammadi, A. Klein, and L. H. Sperling, *Macromolecules*, **26**, 1019 (1993).

21. M. Sambasivam, A. Klein, and L. H. Sperling, *Macromolecules*, **28**, 152 (1995).

22. M. Sambasivam, A. Klein, and L. H. Sperling, *J. Appl. Polym. Sci.*, **58**, 357 (1995).

23. R. P. Wool, B. L. Yuan, and O. J. McGarel, *Polym. Eng. Sci.*, **29**, 1340 (1989).

24. W. F. Reichert and H. R. Brown, *Polymer*, **34**, 2289 (1993).

25. C. Creton, H. R. Brown, and K. R. Shull, *Macromolecules*, **27**, 3174 (1994).

26. R. P. Wool, *Polymer Interfaces*, Hanser, Munich, 1995.

27. A. Casale and R. S. Porter, *Polymer Stress Reactions*, Academic, New York, Vol. I, 1987, Vol. 2, 1979.

28. R. J. Kumph, J. S. Wiggins, and H. Pielartzik, *TRIP (Trends in Polymer Science)*, **3**, 132 (1995).

29. G. H. Hu, J. J. Flat, and M. Lambla, *Makromol. Chem., Makromol. Symp.*, **75**, 137 (1993).

30. M. Seadan, D. Graebling, and M. Lambla, *Polym. Networks Blends*, **3**, 115 (1993).

31. A. Eisenberg and M. King, *Ion-Containing Polymers: Physical Properties and Structure*, Academic, New York, 1977.

32. J. W. Rees, U.S. Pat. 3,264,272 (1966).

33. J. W. Rees, U.S. Pat. 3,404,134 (1968).

34. P. Smith and A. Eisenberg, *J. Polym. Sci., Polym. Lett.*, **21**, 223 (1983).

35. P. Smith and A. Eisenberg, *Macromolecules*, **27**, 545 (1994).

36. A. Molnar and A. Eisenberg, *Macromolecules*, **25**, 5774 (1992).

37. A. Molnar and A. Eisenberg, *Polymer*, **34**, 1918 (1993).

38. Asahi Chemical Industries Co., Ltd., U.S. Pat. 4,429,076 (1977).

39. Du Pont de Nemours and Co., Eur. Pat. 34,704 (1981).

40. Allied Corp., U.S. Pat. 4,404,325 (1983).

41. R. Armstrong, U.S. Pat. 3,373,222 (1967).

42. Mitsubishi Petrochemical Co. Ltd. Jap. Pat. 57,133,130 (1982).

43. Y. H. Kim and R. P. Wool, *Macromolecules*, **16**, 1115 (1983).

44. R. P. Wool and K. M. O'Connor, *J. Polym. Sci., Polym. Lett. Ed.*, **20**, 7 (1982).

45. H. Zhang and R. P. Wool, *Macromolecules*, **22**, 3018 (1989).

46. E. Helfand and Y. Tagami, *J. Chem. Phys.*, **56**, 3592 (1972).

47. E. Helfand and A. M. Sapse, *J. Chem. Phys.*, **62**, 1327 (1975).

48. E. Helfand, *Macromolecules*, **25**, 1676 (1992).

49. K. L. Foster and R. P. Wool, *Macromolecules*, **24**, 1397 (1991).

50. J. L. Willett and R. P. Wool, *Macromolecules*, **26**, 5336 (1993).

51. J. Kressler, H. W. Kammer, and K. Klosterman, *Polym. Bull.*, **15**, 113 (1986).

52. T. P. Russell, A. Menelle, W. A. Hamilton, G. S. Smith, S. K. Satija, and C. F. Majkrzak, *Macromolecules*, **24**, 5721 (1991).

53. M. L. Fernandez, J. S. Higgins, J. Penfold, R C. Ward, C. Shackleton, and D. Walsh, *Polymer*, **29**, 1923 (1988).

54. B. L. Yuan and R. P. Wool, *Polym. Eng. Sci.*, **30**, 1454 (1990).

55. S. Yan, J. Lin, D. Yang, and J. Petermann, *J. Mat. Sci.*, **29**, 1773 (1994).

56. J. T. Koberstein, *MRS Bull.*, **21**(1), 19 (1996).

57. U. Sundararaj and C. W. Macosko, *Macromolecules*, **28**, 2647 (1995).

58. L. Leibler, *Makromol. Chem, Macromol. Symp.*, **16**, 1 (1988).

59. S. H. Anastasiadis, I. Gancarz, and J. T. Koberstein, *Macromolecules*, **22**, 1449 (1989).

60. W. Hu, J. T. Koberstein, J. P. Lingelser, and Y. Gallot, *Macromolecules*, **28**, 5209 (1995).

61. C. A. Fleischer, J. T. Koberstein, V. Krukonis, and P. A. Wetmore, *Macromolecules*, **26**, 4172 (1993).

62. See *Polym. Mater. Sci. Eng. (Prepr.)*, **75**, 1996.

63. D. J. Lohse, T. P. Russell, and L. H. Sperling, Eds., *Interfacial Aspects of Multi-component Polymer Materials*, Plenum, New York, 1997.

7

SURFACES OF ADVANCED AND COMPOSITE SYSTEMS

There are a wide variety of polymer interface systems not yet discussed. In particular, polymer composite systems involving glass and carbon fibers, carbon-black-reinforced systems, and their respective reinforcement mechanisms will be examined. Langmiur–Blodgett films, self-assembled systems, and organic–inorganic polymer hybrids are playing increasingly important roles in polymer blend and composite materials. Then, there are special application materials, such as surgical implant materials, which by their very nature are composite systems in the body.

7.1. SURFACE SEGREGATION IN MULTICOMPONENT POLYMER SYSTEMS

7.1.1. Systems and Applications

Surface segregation in multicomponent polymer materials is defined as the diffusion to the surface of a thin layer of one component in a polymer blend, block, graft, or interpenetrating polymer networks (IPN). Sometimes the system is composed of a blend of a homopolymer and a block copolymer, one of which is the same as the homopolymer. Surface segregation can be studied most effectively via X-ray photoelectron spectroscopy (XPS), ATR-FTIR, scanning electron microscopy (SEM), FRES, or contact angle measurements; see Section 4.6. A number of systems have been investigated; see Table 7.1. Usually, it is the low surface tension component that rises to the surface. If the polymer blend contains a fluorinated or siloxane polymer, usually it is that polymer that comes to the surface. See Figure 4.4, for example.

Presently, there are several applications, actual or proposed, for having one polymer on the surface, including the medical (1), plastic (2), friction (3), adhesive (4), and soil release (5) fields. For example, Kano et al. (6) found that the tackiness of a poly(2-ethylhexyl acrylate-*stat*-acrylic acid-*stat*-vinyl

Table 7.1   Selected surface segregations in multicomponent polymer materials

System	Surface Polymer	Method	Ref.
Poly(vinylidene fluoride-*stat*-hexafluoro acetone)-*blend*-poly(2-ethylhexyl acrylate-*stat*-acrylic acid-*stat*-vinyl acetate)	Poly(vinylidene fluoride-*stat*-hexafluoro acetone)	XPS FTIR	*a* *b*
[Poly(bisphenol A sulfone) alternating block copolymer with poly(dimethyl siloxane)]-*blend*-polysulfone	Poly(dimethyl siloxane)	XPS	*c*
Polystyrene-*blend-d*-polystyrene	*d*-Polystyrene	Neutron reflection Forward recoil spectometry, FRES	*d* *e*
Polycarbonate-*blend*-[polycarbonate-*block*-poly(dimethyl siloxane)]	Poly(dimethyl siloxane)	ESCA Contact angles	*f* *g*
Poly(ethylene terephthalate)-*block*-perfluoropolyether	Fluoropolymer	ESCA	*h*

[a] Y. Kano, S. Akiyama, and T. Kasemura, *J. Appl. Polym. Sci.*, **50**, 1619 (1993).
[b] Y. Kano, K. Ishikura, S. Kawahara, and S. Akiyama, *Polym. J.* **24**, 135 (1992).
[c] N. M. Patel, D. W. Dwight, J. L. Hedrick, D. C. Webster, and J. E. McGrath, *Macromolecules*, **21**, 2689 (1988).
[d] R. A. L. Jones and E. J. Kramer, *Bull. Am. Phys. Soc.*, **33**, 502 (1988).
[e] R. J. Composto, R. S. Stein, E. J. Kramer, A. L. Jones, A. Mansour, A. Karim, and G. P. Fletcher, *Physica B*, **157**, 434 (1989).
[f] I. Yilgor and J. E. McGrath, *Adv. Polym. Sci.*, **86**, 1 (1988).
[g] D. G. LeGrand and G. L. Gaines, Jr., *Polym. Prepr.*, **11**, 442 (1970).
[h] F. Pilali and M. Toselli, *Macromolecules*, **23**, 348 (1990).

acetate) copolymer was reduced by blending in poly(vinylidene fluoride-*stat*-hexafluoro acetone), the latter component rising to the surface. Sherman et al. (5) coated cotton and polyester fibers with fluorocarbon homopolymers and block copolymers, enhancing their soil release characteristics. In this case, the fluorocarbon copolymer was added after the spinning process as a finish, rather than as a blend in the melt state. The effect may be imagined to be the same, but impossible to do for cottons, and perhaps meeting difficulties in diffusion through crystallizable polymers in general.

7.1.2. Effect of Spinodal Decomposition

Recently, there has been a significant amount of physical polymer science research done on the surface segregation of polymer blends, particularly using blends of polystyrene and deuterated polystyrene (7,8). The two polymers exhibit a positive χ value and phase separate at molecular weights above approximately 1×10^6 g/mol. Studies were carried out using neutron forward recoil spectrometry and neutron reflection. The latter is defined as sending a neutron beam of wavelength λ and grazing incidence θ to the sample's surface and measuring the Fresnel reflectivity **R** as a function of $q = 4\pi \sin \theta/\lambda$. Since the surface tension of d-polystyrene is slightly less than h-polystyrene, the former rises to the surface.

Two cases must be distinguished. If the compositions are outside of the coexistence curve, that is, the mixture is thermodynamically miscible, then three layers must be distinguished in the development of the kinetics: the surface layer, the layer just below the surface, and the bulk of the material. First, the surface-seeking component diffuses from the layer just below the surface up to the surface. The growth of the surface-enriched layer is diffusion limited, with the layer just below the surface trying to stay in local equilibrium with the regions further below the surface, the latter two being depleted of the surface-segregating component (9). The kinetics of the segregation are controlled by diffusion of the surface-seeking polymer down the diffusion gradient from the bulk composition to that in the depleted layer just below the surface. (Meanwhile, the surface-seeking material is increasing in concentration at the surface.) As the enriched surface layer nears equilibrium, the diffusion gradient becomes smaller (the composition gradient becomes larger) and the composition in the depleted layer just below the surface approaches that of the bulk composition, the latter being only slightly depleted. Finally, equilibrium is reached.

The second case involves phase-separated systems, particularly those undergoing spinodal decomposition. Inside of the spinodal region (see Section 2.4, the presence of a surface may alter the course of morphology development because the translational and rotational symmetry of the phase separation is broken. In the pioneering work of Jones et al. (9), the surface of a blend of h-poly(ethylene-*stat*-propylene) with d-poly(ethylene-*stat*-propylene) was investigated by FRES. For a degree of polymerization of both components of approximately 2286, the critical temperature was determined to be about 92.4°C. At 35°C, the system was subjected to spinodal decomposition, noting upper critical solution temperature conditions. When uniform 900-nm films were used, it was found that the spinodal decomposition pattern was directed by the presence of the surface; see Figure 7.1 (9). At short times, the surface enrichment of the d-poly(ethylene-*stat*-propylene) follows a pattern similar to that expected by miscible materials, with a layer of depletion of the d-poly(ethylene-*stat*-propylene) below the surface. However, inside the spinodal the mutual diffusion coefficient is negative, so that rather than gradually being erased as the depth increases, this gradient generates the next layer of enrichment below it. This

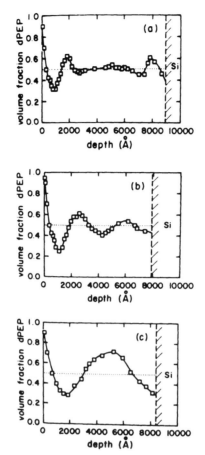

Figure 7.1 Surface enrichment of *d*-poly(ethylene-*stat*-propylene) with time, spinodally decomposing from a 50/50 blend with *h*-poly(ethylene-*stat*-propylene) at 35°C. (a) 19,200 s, (b) 64,440 s, and (c) 172,800 s.

surface-directed spinodal decomposition was the subject of much theoretical interpretation; see for example, the work of Puri and Binder (10).

Of course, the thin film in the preceding discussion has one air interface, and one interface below, where the polymer rests on a support. Since the wavelength of the spinodal decomposition depends on the material parameters, it is possible to generate two distinct spinodal waves originating from the two surfaces, both at early times of decomposition. These two waves interact with each other more and more as the film thickness becomes comparable to the wavelength of the spinodal waves. Constructive and destructive interferences were observed by Krausch et al. (11), which could be modeled by the superposition of two damped oscillatory waves. This picture breaks down as the film thickness becomes smaller than the characteristic wavelength for bulk spinodal decomposition.

Table 7.2 Filler surface chemistry

Filler	Surface Groups	Ref.
Glass fibers	—SiOH	a
Graphite fibers	—OH, —COOH, lactone, N-compounds	b
Carbon blacks	—OH, —COOH, —COOR	c
Aramid fibers	—NH$_2$	d, e

[a] P. K. Mallick, *Fiber-Reinforced Composites*, 2nd ed., Dekker, New York, 1993.
[b] J. P. Wightman, in G. Akovali, ed., *The Interfacial Interactions in Polymeric Composites*, Kluwer, Dordrecht, 1993.
[c] D. Rivin, *Rubber Chem. Technol.*, **44**, 307 (1971).
[d] R. E. Allred, *Polym. Prepr.*, **24**, 223 (1984).
[e] E. M. Liston, in G. Akovali, ed., *The Interfacial Interactions in Polymeric Composites*, Kluwer, Dordrecht, 1993.

7.2. FILLER SURFACE TREATMENTS

7.2.1. Chemistry and Surface Treatments

Very many different fillers are used to make polymeric composites; see Table 7.2. The predominance of oxygen compounds on the surface is obvious. The surface chemistry characteristics fall into three categories during preparation of the material: the natural chemical surface composition, the surface after surface treatments (a wide variety are available), and the surface after finish application (again, a wide variety of finishes are available) (12,13). Each possibility will, of course, have a different composition, each serving different purposes in the final composite.

7.2.2. Glass Fiber Composites

Glass fibers constitute perhaps the most important kind of reinforcing material used to make polymeric composites. Glass fibers are usually 8–16 μm in diameter and exhibit tensile strengths of 2–5 GPa (400,000–700,000 psi). On a strength-to-weight basis, these fibers are among the strongest common structural materials (14). There are several types of glass fibers, usually with letter designations. C-glass, chemical grade, is best for resistance to chemical attack. S-glass, or high silica, is a specialty glass for the highest heat resistance and enhanced structural properties. E-glass, or electrical grade, is used for general structural applications, with good heat resistance and electrical properties. E-glass is the most common grade used.

Untreated glass surfaces have Si–O–Si and Si–OH linkages, among others. Thus, the untreated surface can form esters and ethers with carbonaceous materials such as organic polymers. As a highly polar surface, glass also attracts water and various gases to its surface. Sometimes these chemicals are highly detrimental to the development of high-strength, permanent bonds with polymers.

As the fibers are drawn from the furnace, sizings or surface treatments are applied. These may be either textile or reinforcement sizings. The purpose of these sizings is to protect the fibers from damage during processing. Textile sizings, based on oil–starch emulsions, serve as lubricants that prevent damage to the glass during twisting into yarns and weaving into fabrics. Later, they are removed by heat cleaning or burning before treatment with finishes. Reinforcement sizings, on the other hand, consist of film formers, wetting agents, and surface-active ingredients. These serve to improve the interface or bonding of the fiber with the polymer matrix.

After weaving and heat cleaning, finishes are applied to the fabrics. They promote bonding between the glass and the polymer. Finishes may be based on chrome complexes such as methacrylate chromic chloride. Silanes are an important type of finish; see Section 6.1. Silane coupling agents are used in reinforced plastics, in mineral-filled plastics and elastomers, in coatings, inks, and in adhesives and sealants.

Silanes can also be used as sizings. As such, however, they are rarely used alone. Most commercial silane sizings especially those used in glass/epoxy composites, are proprietary complex compositions. Excluding the solvent or carrier, one epoxy–silane sizing material contained 1–5% of a film-forming resin, 0.1–0.2% antistatic agent, 0.1–0.2% lubricant, and 0.1–0.5% silane coupling agent (15,16). Of course, the constituents of typical sizing systems interact with each other. However, the resulting distribution of these components in the interphase region is not yet well understood, with the formulation of effective sizing systems remaining an empirical art for the most part. Perhaps the most important use of silanes on glass fibers is for unsaturated polyester–styrene layups, where vinyl silane and similar compounds bond the polymerizing styrene to the glass. Of course, it is also crosslinking with the unsaturated polyester, to form an AB-crosslinked copolymer.

7.2.3. Carbon Fibers

Another important fiber is based on carbon, or graphite. One method of preparing carbon fibers is from a polyacrylonitrile (PAN) precursor. The processes include spinning of PAN into fiber form, oxidizing the fiber at 200–300°C, and carbonizing the fiber at 1000–2500°C in an inert atmosphere, surface treating, and eventually sizing (17,18).

According to results obtained via electron spectroscopy for chemical analysis (ESCA) instrumentation, the underivatized top 100 Å of a typical carbon fiber surface contains about 85% carbon, 11% oxygen, 3% nitrogen, 0.2% silicon, and 0.2% sulfur (18). Chemical groups include hydroxy, carbonyl, and carboxylic acid. Thus, its surface composition is quite different than the interior, the latter being nearly 100% carbon.

After the carbon fibers come out of the carbonizing furnace, they are surface treated, which serves several purposes: to remove the outer layer of the carbon fiber surface, believed to be disordered carbon of low shear strength, and to provide a controlled oxidation of the fiber surface, thus supplementing the surface functional groups already on the surface. This latter promotes adhesion of the fiber to the polymer matrix. Possible surface treatment mechanisms include anodization, plasma and flame treatments, solution oxidation, gas phase oxidation, and high-temperature oxidation. Again, the surface finishes act to promote adhesion, involving improvements in fiber wetting, surface chemistry preservation, and surface damage protection.

7.2.4. Use of Plasma and Corona Treatments

Plasma and corona treatments may be given to polymer surfaces, and may also be given to a range of other materials, such as calcium carbonate powder, mica flakes, or other inorganic or organic materials. A plasma may be defined as a partially ionized gas, with equal number densities of electrons and positive ions, in which the charged particles are "free" and possess collective behavior (19). The effect is to provide electrons, ions, and photons with energies sufficient to break any chemical bond. While most commercial plasmas are very complex, a simple plasma treatment can be obtained by slowly passing a Bunsen burner with a Meeker attachment along the surface of the material to be treated. Motion must be fast enough to prevent any charring, of course. The result is an oxidation of most surfaces. This kind of treatment activates both polymeric and nonpolymeric surfaces.

A corona involves a controlled discharge of electrons. Corona discharges may occur on electrodes with sharp edges or small radii of curvature. Flat electrodes may also be used, coated with a dielectric material to prevent arc formation. Under atmospheric conditions, ozone may be generated. The effect of both treatments is to improve bondability of fillers to the polymers.

7.3. FRACTURE OF FIBER COMPOSITES: INTERPHASE EFFECT

7.3.1. Theory of Debonding

Figure 7.2a–c (20) illustrates a crack tip as it intersects a fiber. In Figure 7.2a, the work on a per fiber basis, W_{df}, can be calculated as a function of the interfacial shear stress, sometimes called the interfacial shear strength, τ, the fiber diameter, d, the debonded length, l_d, and the difference in strain between the fiber and the matrix, $\Delta\varepsilon$. The local stresses at the tip can cause fiber–matrix debonding. As the crack tip progresses, the interfacial debonded region becomes extended. Even after debonding, the fiber continues to interact with the matrix through a frictional sliding force.

As shown in Figure 7.2b, the energy stored through deformation up to fracture of the fiber, W_{ff}, can be calculated for each fiber based on the fiber tensile strength, σ_f, and the fiber tensile modulus, E_f. The quantity A represents the fiber cross-sectional area. When W_{ff} equals the fracture energy of the fiber, the energy goes into creating the fracture.

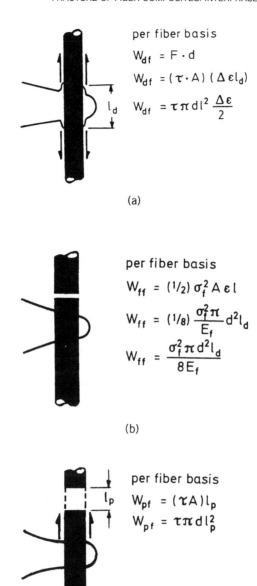

per fiber basis

$$W_{df} = F \cdot d$$

$$W_{df} = (\tau \cdot A)(\Delta \epsilon l_d)$$

$$W_{df} = \tau \pi d l^2 \frac{\Delta \epsilon}{2}$$

(a)

per fiber basis

$$W_{ff} = (1/2)\, \sigma_f^2 A \epsilon l$$

$$W_{ff} = (1/8)\, \frac{\sigma_f^2 \pi}{E_f} d^2 l_d$$

$$W_{ff} = \frac{\sigma_f^2 \pi d^2 l_d}{8 E_f}$$

(b)

per fiber basis

$$W_{pf} = (\tau A)\, l_p$$

$$W_{pf} = \tau \pi d l_p^2$$

(c)

Figure 7.2 Effect of polymer–fiber bonding energies on crack formation and failure. (a) Crack travels through the matrix, reaching a fiber, and initiating debonding. (b) The fiber may deform and break. (c) As the crack opens, the broken fiber may be subject to pull-out.

Table 7.3   Numerical values for fiber–matrix interphase fracture

Symbol	Definition	Value Glass	Value Carbon	Ref.
τ	Interfacial shear stress	55.1 MPa[a]	62 MPa[a]	b
d	Fiber diameter		5–10 μm	c
		10 μm	7 μm	d
l_d	Debonded length	—	—	
$\Delta\varepsilon$	Strain, or difference in strain	0.05	0.01	d
σ_f	Fiber tensile strength	4.30 GPa	4.07 GPa	d
E_f	Fiber tensile modulus	86.9 GPa	248 GPa	d

[a] Fiber–epoxy composite.
[b] P. K. Mallick, *Fiber-Reinforced Composites*, 2nd ed., Dekker, New York, 1993, p. 242.
[c] L. T. Drzal, in *Adv. Polym. Sci., Epoxy Resins and Composites*, K. Dusek, ed., **75**, 3 (1986), Springer-Verlag, New York.
[d] P. K. Mallick, *Fiber-Reinforced Composites*, 2nd ed., Dekker, New York, 1993, p. 18.

A third mechanism operating at the crack tip involves the energy required to pull the broken fiber out of the matrix, W_{pf}; see Figure 7.2c. The work required is the product of the shear strength times the pull-out length, l_p. The total work of fracture is given by the sum of the above contributions:

$$W_t = \tau\pi\, dl_d^2 \Delta\varepsilon/2 + \sigma_f^2 \pi\, d^2 l_d/8E_f + \tau\pi\, dl_p^2 \tag{7.1}$$

7.3.2. Example Calculation: Fiber Pull-Out Length

Table 7.3 shows typical numerical values for equation (7.1). Properties for both glass and carbon-based fibers are shown. Unknown are the debonded length, l_d, and the pull-out length, l_p; see Figure 7.2. It is of some interest to estimate a value of the latter, as a number of SEM photographs are available that show this as an experimental result.

Some assumptions are as follows: The pull-out length probably ranges from $\frac{1}{4}$ to $\frac{1}{2}$ of the debonded length. This assumes that the debonded region goes up and down the fiber for some distance before the fiber actually fractures. A value for l_p of $\frac{1}{2}$ will be assumed here. The work of debonding [first term on the right side of equation (7.1)] cannot be more than that of fiber deformation and fracture (second term on the right), or it is cheaper energetically just to fracture in the plane of the crack. These two terms can be equal as an upper limit, the most desired circumstance, and will be assumed so here. Also, the pull-out energy (third term on the right) must be less than the deformation and fracture (second term), or the fiber would break again rather than pull out. As a desired limit, the two terms should be equal. We are then left with three equal terms in equation (7.1), for the ideal case.

To approximate realistic quantities, the fiber will be assumed to be $10\,\mu$m in diameter and have an interfacial shear stress of 50 MPa. A strain to break of 0.05 (5%) is assumed. A fiber tensile strength of 4 GPa and a fiber tensile modulus of 72 GPa are also assumed.

Equating the first term on the right to the second term yields $222\,\mu$m for l_d, and equating the second term to the third term yields $35\,\mu$m for l_p. While this agreement is only fair, the SEM shown in Figure 1.5 for carbon fibers in poly(aryl ether ether ketone) yields values of $5–40\,\mu$m, spanning much of the distance between the two theoretical values. Similar results are obtained from other literature figures (21,22).

Assuming that the first and third term each equal the second term as the most desired circumstance, and that l_p equals $20\,\mu$m (an intermediate value), W_t has a value of about $2.5 \times 10^2\,\text{J/m}^2$, which compares to values of G_{1c} of $7\,\text{kJ/m}^2$ reported as the sum of the fibers plus the plastic for glass reinforced thermosets (23). In this case, the calculated value, based on the equation given, is too small, providing a kind of lower bound model. Other factors, such as the energy to fracture the matrix (chain pull-out plus scission), and the like come into play, and discussed in Section 6.4.4.

An important concept emerging in polymer interphase research is the controlled development of bonding between two polymers and between polymers and nonpolymers. If the bonding is too weak or nonexistent, then the material will fail prematurely. If, however, the bonding is too tight between the polymer and the filler, failure may take place by inferior energy-absorbing mechanisms. For most systems, there appears to be an optimum interphase bonding level.

7.4. STRESSING AND FRACTURING CARBON-BLACK-REINFORCED ELASTOMER NETWORKS

In Section 5.4.1, the bonding of elastomers (crosslinked polymers above their glass transition temperature) to carbon black was shown to broaden and increase the glass transition temperature of that portion of the elastomer in direct contact with the filler. Now, the importance of the bonding between the carbon black and the polymer in strengthening the elastomer will be explored.

7.4.1. Mullins Effect

The Mullins effect may be defined as the reduction in the modulus of elastomers caused by successive stretching (24). This phenomenon is characteristic of elastomers reinforced with carbon black or silicas but occurs to a lesser extent in gums and nonreinforced systems. In general, the area under the stress–strain curve is reduced, less stress being required for a particular stretch. For samples only modestly strained, the Mullins effect predominates in the strain region already achieved.

Explanations of the Mullins effect include failure of weak linkages, failure of network chains extending between adjacent filler particles, and polymer–filler bond failure, this last a form of dewetting. Considered as a rate process, molecular slippage mechanisms operate at the carbon–rubber interface to redistribute the stress. Thus, the level of stress on the most highly stressed chains is reduced (25). Figure 3.17 summarizes several early models. Note particularly the slippage of the most highly stressed chains to a more relaxed state. With sufficient time, a significant amount of reversibility is noted.

The important point in Figure 3.17 is that weak bonding between the carbon black and the polymer chains, capable of letting go and reforming with chains in more relaxed conformations, is responsible for both the reduction in modulus and toughening the elastomer against fracture.

7.4.2. Particulate Reinforcement Mechanisms

As a matter of well-known fact, carbon-black-reinforced elastomers can be stretched to a much greater extent than the pure vulcanizate. The tensile strength increases from about 3.3 MPa (26) to about 25.5 MPa (27) when styrene–butadiene rubber (SBR) is reinforced with carbon blacks. Hamed (28) recently reviewed the molecular aspects of fatigue and fracture of carbon-black-reinforced rubber. The emphasis was on the fate of chains that scission due to excess stress early in the stress–strain process. He pointed out that in spite of the increased average chain load of reinforced elastomers, noncatastrophic energy-dissipating processes apparently reduce stress magnification at stress-raised points effectively, thus allowing the stress to be somewhat more evenly distributed throughout the network structure. Hamed lists several possible mechanisms:

1. The attachment of a newly broken network chain to the carbon black surface. Thus, the temporarily ruptured chain would regain load-carrying capability rather than irreversibly transmitting its load to neighboring chains. This mechanism is consistent with the need for a high specific surface area in order to have substantial reinforcement with the carbon black or other filler particle. This also emphasizes the importance of filler surface activity.

2. In real networks, there are significant numbers of chains attached to the network at only one end. In the presence of carbon black, the dangling chains may become adsorbed onto the filler surface. Upon application of stress, these chains can now bear load and perhaps the chain may even accommodate the load by slipping reversibly along the filler surface. Such slippage also brings to the fore the possible presence of molecular friction between the two, expending energy.

3. Hamed points out (28) that there may be an optimum in the polymer–filler interactions. A very strong interaction, in which slippage is largely prohibited, may be undesirable. Another idea is that there should be a

threshold for this dissipation of energy, below which dynamic fatigue is minimized. Thus, under ordinary usage, the loss modulus will be small and the system highly elastic. When stressed greatly (in a tire, perhaps hitting a pothole), the energy dissipative mechanisms would kick in, saving the integrity of the object. Carbon black, however, does not particularly reinforce glassy polymers, which lack rapid molecular motion.

Elastomer networks formed through sulfur vulcanization, which then contain the somewhat labile S–S disulfide bonds, are known to be better than peroxide-cured elastomers. This is because the polysulfide linkages can rearrange before the carbon–carbon bonds are broken. Ideas of this nature led to a recent patent by Rauline (29), wherein a highly dispersible silica plus a halogen-containing silane coupler are masticated into the elastomer. The halogen-containing silane then reacts with the filler, the elastomer, and other components such as the sulfur-bearing compounds, providing the labile bond between the silica and the rubber. According to the patent, the rubber composition of the invention confers upon a tread very low rolling resistance, improved adherence on wet ground, improved adherence on snow-covered ground, and very good resistance to wear. The improved adherence aspects, at least, may be due to the more labile bonds formed between the rubber and the filler. Much work remains to be done, however, in optimizing the "kicking-in" of the viscoelastic loss mechanisms at the right stress levels.

7.5. LANGMUIR-BLODGETT FILMS

7.5.1. Introduction

Consider a monomolecular layer of a compound on the surface of a liquid. The monomolecular layer can be soap, ionomer, or other compound where polarity at two points in the molecule differ significantly. The monolayer, or eventually a small number of monolayers of the compound, can be transferred to a solid substrate by dipping the substrate into the liquid. The result is known as a Langmuir–Blodgett film (30–32). A simple Langmiur–Blodgett film can be made by dipping a glass slide into a monolayer of stearic acid on a water surface, which will be transferred in part to the glass slide. A second monolayer film will be formed as the glass slide is withdrawn from the water.

The basic process is illustrated in Figure 7.3 (33). The black dots indicate the polar portions of the molecule, in the case of stearic acid, carboxyl groups. The straight-line portions of the monolayer molecules are the hydrocarbon tails. The carboxyl group is water loving, that is, hydrophilic, and points into the water. The hydrocarbon tail is hydrophobic and sticks out of the water phase. The whole stearic acid molecule is amphophilic (*ampho*- means either). The concept of amphophilic molecules can be immediately generalized, not only to various

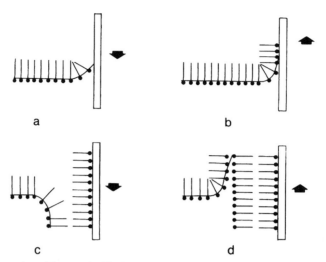

Figure 7.3 Illustration of Langmuir–Blodgett film formation and deposition of multilayers. (a) First immersion; (b) first withdrawal; (c) second immersion, and (d) second withdrawal of a substrate entering and leaving a liquid with a surface active layer.

kinds of molecules in water but also to molecules exposed to various kinds of solvents.

When the surface concentration of the amphophilic molecules is low, and the interactions between the molecules small, the surface can be regarded as a two-dimensional gas. Under these conditions, the surface monolayer has relatively little effect on the water's surface tension, $73\,\mathrm{mNm}^{-1}$, or $73\,\mathrm{ergs/cm}^{-2}$. If the amphophilic molecules are in a Langmuir trough, they can be compressed by movement of a barrier system, which reduces the surface area available to the amphophilic molecules. The pressure rises to an equilibrium spreading pressure, which for stearic acid on an acidified water subphase is about $58\,\mathrm{mNm}^{-1}$ $(5.8\,\mathrm{J/m^2})$ at $0.22\,\mathrm{nm^2/molecule}$. At higher pressures, the monolayer collapses, and forms a bilayer. Most of the early work has involved small molecules, such as soaps and detergents.

The field of Langmuir–Blodgett film formation and application today involves both organic molecules and polymer molecules. Using polymers usually results in a stronger film, less easily removed from the substrate. However, because of the high viscosity of the film, transfer is necessarily slower. The polymer-containing Langmuir–Blodgett films are all a type of composite, bearing polymer on a (usually) nonpolymer substrate.

A multilayer film can be formed by three possible deposition modes, called X, Y, and Z (33,34). These three depositions modes can be realized by immersing (or emersing) a substrate either perpendicular or parallel to the liquid surface. *Immersing* means going into the system, *emersing* means going out. The more common method involves vertical deposition. In the X type, the film is deposited only when the substrate enters the subphase. Correspondingly, the Z type film forms only when the substrate leaves the subphase. Then, in the X type, the polar groups are facing outward only, away from the substrate, and in the Z type they are always facing inward, toward the substrate. In the Y type, layers are deposited each time the substrate moves across the phase boundary, up and down. These layers then have alternating polarity direction. It will be shown below that the properties of these three deposition types are significantly different.

It is also possible to transfer one monolayer, structure A, by immersion, and another, structure B, by emersion. This produces alternating layers of A and B, usually of the Y type.

7.5.2. Methods of Deposition

Monomers for Langmuir–Blodgett films must be amphophilic (35). Early choices were monomers related to stearic acid but containing double bonds for vinyl polymerization. Table 7.4 summarizes some important choices.

Alternately, polymers can be added directly to a substrate via Langmuir–Blodgett technology. The main structural requirement of polymeric Langmuir–Blodgett materials relates to having hydrophilic groups regularly distributed at short intervals along the main chain. If long sections of the chain exist without such groups, then these will tend to form ill-defined loops, and the like, clear of the water surface. Subsequent films on the substrate will suffer concomitant structural defects. Table 7.4 also describes typical polymer choices.

Table 7.4   Selected polymers for Langmuir–Blodgett films

Polymer	Comments	Ref.
A. Monomer to Polymer		
Poly(vinyl stearate)	^{60}CO γ-irradiation	a
Poly(α-octadecylacrylic acid)	Electron beam irradiation	b
Poly amphilic dienes	Photopolymerization	c
Poly diacetylenic amphiphiles	Alternating double and triple bonds	d
B. Preformed Polymers		
Poly(octadecyl methacrylate)	$M_n = 92,200\,\text{g/mol}$	e
Poly(vinyl alkylals)	100 layer films	f
Polyimide precursers	Thermal stability	g

Source: Based on A. Ulman, *An Introduction to Ultrathin Films*, Academic, Boston, 1991.
[a] A. Cemal, T. Fort, Jr., and J. B. Lando, *J. Polym. Sci., A-1*, **10**, 2061 (1972).
[b] G. Fariss, J. B. Lando, and S. Rickert, *Thin Solid Films*, **99**, 305 (1983).
[c] A. Laschewsky and H. Ringsdorf, *Macromolecules*, **21**, 1936 (1988).
[d] D. Day and H. Ringsdorf, *Makromol. Chem.*, **180**, 1059 (1979).
[e] S. J. Mumbly, J. F. Rabolt, and J. D. Swalen, *Macromolecules*, **19**, 1054 (1986).
[f] M. Watanabe, Y. Kosaka, K. Oguchi, K. Sanui, and N. Ogata, *Macromolecules*, **21**, 2997 (1988).
[g] M. Uekita, H. Awaji, and M. Murata, *Thin Solid Films*, **160**, 21 (1988).

Table 7.5 Actual and proposed applications of Langmuir–Blodgett films

Application	Polymer	Remarks	Ref.
Second-harmonic generation (nonlinear optics)	CH_2Cl — $(CH-CH_2-O)_x$ $(CH-CH_2-O)_{1-x}$ — CH_2-O — (phenylene–CH=CH–pyridinium) $N^+-C_{12}H_{25}$, Br^-	True noncentrosymmetric Y-type	a
Third-harmonic generation (nonlinear optics)	[thiophene polymer] $CH_2(CH_2)_{10}CH_3$	$\chi^{(3)} = 1 \times 10^{-9}$ esu	b
Third-harmonic generation (nonlinear optics)	Polydiacetylenes	$\chi^{(3)} = 2 \times 10^{-10}$ esu	c
Pyroelectric device	Side-chain liquid crystalline copolymers containing nitrobiphenyl	Type X horizontal deposition	d

Application	Structure	Property	Ref.
Photoinduced electron	$(CH-CH)_n$ (acenaphthylene)	Easily prepared	e
Photochromic effect	(azobenzene with piperazine-2,5-dimethyl amide)	cis–trans photoisomerism	f
Resists	Poly[styrene-alt-(maleic anhydride)]	250 Å resolution	g
Lubricants	Poly(perfluoropropylene oxide)	First monolayer chemisorbed, >10,000 sliding cycles	h

[a] B. L. Anderson, R. C. Hall, B. G. Higgins, G. Lindsay, P. Stroeve, and S. T. Kowel, *Synth. Metals*, **28**, D683 (1989).

[b] P. B. Logsdon, J. Peleger, P. N. Prasad, *Synth. Metals*, **26**, 369 (1988).

[c] P. N. Prasad, *Thin Solid Films*, **152**, 275 (1987).

[d] S. H. Ou, V. Percec, J. A. Mann, J. B. Lando, L. Zhou, and K. D. Singer, *Macromolecules*, **26**, 7263 (1993).

[e] T. Murakata, T. Miashita, and M. Matsuda, *Macromolecules*, **21**, 2738 (1988).

[f] H. S. Blair and C. B. McArdle, *Polymer*, **25**, 1347 (1984).

[g] R. Jones, C. S. Winter, R. H. Tredgold, P. Hodge, and A. Hoorfar, *Polymer*, **28**, 1619 (1987).

[h] V. Novotny, J. D. Swalen, and J. D. Rabe, *Langmuir*, **5**, 485 (1989).

Table 7.6   Definition of terms

Second-harmonic generation is a combination of two photons of frequency ω to give one photon at frequency 2ω.

Third-harmonic generation: Three photons of frequency ω combine to give one photon of frequency 3ω.

Piezoelectric device measures small changes in (applied) mass.

Pyroelectric device detects small temperature changes.

Photo-induced electron transfer involves excitation of an electron donor from its π to π^* orbital.

Photochromic effect involves the reversible change of a molecule between two states having different absorption spectra.

7.5.3. Actual and Proposed Applications

The Langmuir–Blodgett films are currently being examined for a host of possible applications, principally in the optical and electrical areas. These include nonlinear optics, pyroelectric devices, and photochromic effects, among others; see Table 7.5. Some of the terms are defined in Table 7.6.

A noncentrosymmetric structure, such as deposited in the X or Z mode, is essential for a material to be able to possess nonlinear optical effects, pyroelectricity, and piezoelectricity. Unfortunately, only a few materials have been reported to produce genuine X- or Z-type multilayers. At the time of writing this is a very active research area.

7.6. SELF-ASSEMBLY OF BLOCK COPOLYMERS

The concept of self-assembly in polymers is closely related to the concept of Langmuir–Blodgett films; see Section 7.5. However, the routes for their preparation are much broader in scope. Indeed, the search for polymers that will self-assemble is just beginning.

Russell and co-workers (36–39) showed that films of symmetric diblock copolymers will, in general, self-assemble into a multilayered morphology with alternating layers of the two components oriented parallel to the surface of the substrate. On proper annealing, a near perfect orientation of microphase-separated layers will result from two forces: One of the segments will preferentially segregate to the substrate, and the other will preferentially segregate to the air surface.

Symmetrical diblock copolymers of polystyrene-*block*-poly(methyl methacrylate) were spin coated from toluene solutions onto a silicon substrate. The films were then dried and annealed at 170°C for various periods of time. Instrumentation included neutron reflectivity (36) and secondary ion mass spectroscopy (SIMS) (37). For these experiments, one block or the other was deuterated.

Neutron reflectivity involves measuring the reflection properties of these multilayered films at very low (grazing) incidence angles. The as-cast films were

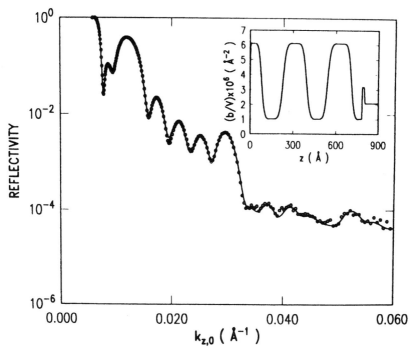

Figure 7.4 Organization of a *d*-polystyrene-*block*-poly(methyl methacrylate) thin layer on a surface as indicated by a neutron reflectivity profile. Film annealed at 170°C for 24 h under vacuum. While the circles represent the experimental data, the solid line was calculated from the scattering length density profile; see inset. The theory matches the experiment excellently well in this case.

found to exhibit no preferential orientation of the microdomain morphology with respect to the surface. Annealing of the block copolymers for various lengths of time produced dramatic orientation of the microdomains parallel to the surface of the film; see Figure 7.4 (36). The poly(methyl methacrylate) (PMMA) block was preferentially located at the copolymer–substrate interface, while the polystyrene (PS) was preferentially located at the polymer–air interface. The arrangement is such that the first layer from the substrate is one block of PMMA thick, followed by a two-block thickness of polystyrene. In this case, the two-block thickness arises from two different block copolymers, one on the bottom and one on the top. This double thickness profile lasts until the block at the surface is reached, which is again a single layer. However, if the silicon wafer substrate is coated with a chemically attached layer of PS, then the PS of the block copolymer prefers the substrate (36). In either case, the film thickness is forced to be quantized according to the repeat period dictated by the molecular weights of the two blocks. If the total amount of polymer deposited per unit area does not meet these requirements, then either islands or holes in the surface will be formed, each with a step height again dictated by the molecular weights.

7.7. INTERFACIAL ASPECTS OF SURGICAL IMPLANTS

Today, there are very many surgical implants in use, ranging from heart valves to hip replacements to breast protheses. Some implants are polymeric in nature, some metallic, some ceramic, and some a combination of the above. In all cases, the surfaces of implants interface with biological fluids and living tissue (40,41).

Cells usually do not recognize synthetic polymers and do not generally interact with the surface of an implanted material. Rather, within seconds of being implanted, a layer of proteins from the surrounding fluids adsorbs on the implant surface. Of course, the proteins that adsorb are themselves polymers. The cells recognize the proteins, inducing a series of hierarchial reactions that may lead to growth of cellular structures that migrate to the surface by diffusive and active processes.

The factors important in protein adsorption are given in Table 7.7 (40). This list is reminiscent of the factors important in polymer adhering to colloids from dilute solution; see Section 5.3. The exact conformation on the implant surface surely influences subsequent cell growth.

7.8. ORGANIC–INORGANIC POLYMER HYBRIDS

The organic–inorganic polymer hybrids are a new type of composite, where the inorganic portion is so finely dispersed as to form transparent materials. These microheterogeneous polymer composites utilize combinations of metallic oxides or carbonates with basic organic groups such as polyamides (42).

Table 7.7 Adsorbed protein film influence on implant performance

Effect	Comments
Total amount of protein	Depends on surface activity of the polymer
Which proteins are present	Fractionation of the some 200 blood and plasma proteins
Accessibility of the adsorbed protein molecules to the surface	In multilayer films, the proteins beneath the surface layer may exert less influence
Orientation of the protein molecules at the surface	Do the chains assume brush, pancake, or random conformations, etc.?
Activation of the proteins by the surface	Does the plastic surface induce reactivity?
Patchiness or "island" formation of the adsorbed protein	What is the effect of bare spots?
Exchange of adsorbed proteins with surrounding media	Kinetics of weakly vs. strongly bonded proteins
Surface-induced denaturation	The activity of many proteins depends on their conformation; this may be altered by the surface

Three basic types of organic–inorganic polymer hybrids exist:

1. Those based on hydrogen bonding between the inorganic and organic phases
2. Those based on covalent bonding, such as through silane reactions
3. Combinations of (1) and (2)

The formation reactions frequently utilize the sol–gel process, starting with a metal alkoxide in the presence of an organic polymer in a homogeneous reaction mixture. Useful metal alkoxides include $Si(OR)_4$, $Al(OR)_3$, or $Ti(OR)_4$. On reaction, secondary bonds form with functional groups on the polymer.

Because these hybrid materials have great clarity and high modulus, they are useful as scratch-resistant coatings.

7.9. TYPES OF SURFACES AND INTERFACES

There are several kinds of surfaces and interfaces that are important in polymer science broadly, and in multicomponent polymer materials in particular. Surprisingly, all of them have several features in common.

First, there is the polymer A–polymer A interface. If the polymers are linear, amorphous, and above their glass transition temperature, the polymers will diffuse across the interface, eventually obliterating it.

In the polymer A–air interface, the polymer A–polymer B interface, and in the polymer A–nonpolymer interface, assuming only physical interactions, the ends of the chains are vertical to the plane of the interface, while the center portions of the chains lie parallel to the interface. Because the interface tends to be rich in chain ends, the next layer of polymer molecules down may be partly denuded of chain ends.

Several types of chemical interactions have been identified as being important. These include hydrogen bonding, acid–base interactions, and direct covalent bonding between the two kinds of molecules making up the interface. These can include the addition of special molecules that either develop physical bonds, as in block copolymers, or chemical bonds, as in silanes. Of course, both types of interaction may be present.

Since the region between the two phases has a finite thickness, it has become popular to talk about the interphase. The interphase region may be between 1 and 100 nm thick. If there is little bonding between the two phases making up the interphase, the material may be abnormally low in density and mechanically weak. In many cases, there may be an optimum area average of bonding between the two phases that allows a maximum amount of energy to be spent in preventing fracture of the material as a whole. Many new methods of orienting polymers at surfaces are being developed, including Langmiur–Blodgett films and self-assembled systems.

REFERENCES

1. A. Takahara, J. Tashita, T. Kajiyama, M. Takayanagi, and W. J. MacKnight, *Polymer*, **26**, 978 (1985).
2. J. M. Xie, M. Matsuoka, and K. Takemura, *Polym. Prepr. Jpn.*, **40**, 2800 (1991).
3. M. Owen and J. Thompson, *Br. Polym. J.*, **4**, 297 (1972).
4. M. Tachikake, *Koubunshi*, **41**, 348 (1992).
5. P. O. Sherman, S. Smith, and B. Jahannsessen, *Textile Res. J.*, **39**, 449 (1969).
6. Y. Kano, S. Akiyama, and T. Kasemura, *J. Appl. Polym. Sci.*, **50**, 1619 (1993).
7. R. A. L. Jones and E. J. Kramer, *Bull. Am. Phys. Soc.*, **33**, 502 (1988).
8. R. J. Composto, R. S. Stein, E. J. Kramer, A. L. Jones, A. Mansour, A. Karim, and G. P. Fletcher, *Physica B*, **157**, 434 (1989).
9. R. A. L. Jones, L. J. Norton, E. J. Kramer, F. S. Bates, and P. Wilzius, *Phys. Rev. Lett.*, **66**, 1326 (1991).
10. S. Puri and K. Binder, *Phys. Rev.*, **46**, R4487 (1992).
11. G. Krausch, C. A. Dai, E. J. Kramer, J. F. Marko, and F. S. Bates, *Macromolecules*, **26**, 5566 (1993).
12. L. T. Drzal, *SAMPE*, Sept./Oct., 7 (1983).
13. L. T. Drzal, in *Treatise of Adhesion and Adhesives*, R. L. Patrick, Ed., Marcel Dekker, New York, 1989.
14. D. Rosato, in *Encyclopedia of Polymer Science and Engineering*, 2nd ed., J. I. Kroschwitz, Ed., Wiley, New York, 1988.
15. E. K. Drown, H. A. Moussawi, and L. T Drzal, *J. Adhesion Sci. Technol.*, **5**, 865 (1991).
16. Dow Corning Corporation, *A Guide to Dow Corning Silane Coupling Agents*, 1985, p. 15.
17. J. P. Wightman, in *The Interfacial Interactions in Polymeric Composites*, G. Akovali, Ed., Kluwer, Dordrecht, The Netherlands, 1993.
18. E. Bayramli, in *The Interfacial Interactions in Polymeric Composites*, G. Akovali, Ed., Kluwer, Dordrecht, the Netherlands, 1993.
19. J. E. Klemberg-Sapieha, L. Martinu, S. Sapieha, and M. R. Wertheimer, in *The Interfacial Interactions in Polymeric Composites*, G. Akovali, Eds., Kluwer, Dordrecht, the Netherlands, 1993.
20. L. T. Drzal, in *Adv. Polym. Sci.*, Vol. 75, *Epoxy Resins and Composites II*, K. Dusek, Ed., Springer, Berlin, 1986.
21. E. Amendola, L. Nicolais, and C. Carfagna, in *The Interfacial Interactions in Polymeric Composites*, G. Akovali, Ed., Kluwer, Dordrecht, The Netherlands, 1993, pp. 400 and 404.
22. M. J. Folkes, in *Multicomponent Polymer Materials*, I. S. Miles and S. Rostami, Eds., Longman, Essex, England, 1992.
23. A. J. Kinloch and R. J. Young, *Fracture Behavior of Polymers*, Applied Science, Elsevier, Essex, England, 1983.
24. L. Mullins, *Rubber Chem. Technol.*, **42**, 339 (1969).
25. J. A. C. Harwood, L. Mullins, and A. R. Payne, *J. IRI*, **1**, (Jan/Feb) 17 (1967).

26. G. S. Whitby, Ed., *Synthetic Rubber*, Wiley, New York, 1954.

27. D. P. Tate and T. W. Bethea, in *Encyclopedia of Polymer Science and Engineering*, 2nd ed., J. I. Kroschwitz, Ed., Wiley, New York, Vol. 2, 1985, p. 576.

28. G. R. Hamed, *Rubber Chem. Technol.*, **67**, 529 (1994).

29. R. Rauline, U.S. Pat. 5,227,425 (1993).

30. I. Langmuir, *J. Am. Chem. Soc*, **39**, 1848 (1917).

31. K. A. Blodgett, *J. Am. Chem. Soc*, **57**, 1007 (1935).

32. K. A. Blodgett, *Phys. Rev.*, **51**, 964 (1937).

33. R. A. Hann, in G. Roberts, Ed., *Langmiur–Blodgett Films*, Plenum, New York, 1990.

34. S. H. Ou, V. Percec, J. A. Mann, J. B. Lando, L. Zhou, and K. D. Singer, *Macromolecules*, **26**, 7263 (1993).

35. A. Ulman, *An Introduction to Ultrathin Organic Films*, Academic, San Diego, 1991.

36. T. P. Russell, A. M. Mayes, and P. Bassereau, *Physica A*, **200**, 713 (1993).

37. G. Coulon, T. P. Russell, V. R. Decline, and P. F. Green, *Macromolecules*, **22**, 2580 (1989).

38. S. H. Anastasiadis, T. P. Russell, S. K. Satija, and C. F. Majkrzak, *Phys. Rev. Lett.*, **62**, 1852 (1989).

39. S. H. Anastasiadis, T. P. Russell, S. K. Satija, and C. F. Majrzak, *J. Chem. Phys.*, **92**, 5677 (1989).

40. B. D. Ratner, in *Comprehensive Polymer Science*, Vol. 7, *Specialty Polymers and Polymer Processing*, S. L. Aggarwal, Ed., Pergamon, Oxford, England, 1989.

41. T. A. Horbett, in *Proteins at Interfaces: Physicochemical and Biochemical Studies*, T. A. Horbett and J. L. Brash, Eds., American Chemical Society, Washington, DC, 1987.

42. T. Saegusa, presented at the IUPAC Macro Seoul '96, August 4–9, 1996, Seoul, Korea, Abstracts, p. 329.

GENERAL READING

Adamson, A. W., *Physical Chemistry of Surfaces*, 5th ed., Wiley-Interscience, New York, 1990.

Akovali, G., Ed., *The Interfacial Interactions in Polymeric Composites*, Kluwer, Dordrecht, the Netherlands, 1993.

Chaudhury, M. J., *Mat. Sci. Eng.* **R16(19)**, No. 3, 97 (1996).

Feast, W. J. and H. S. Munro, Eds., *Polymer Surfaces and Interfaces*, Wiley, New York, 1987.

Feast, W. J., H. S. Munro, and R. W. Richards, Eds., *Polymer Surfaces and Interfaces II*, Wiley, New York, 1993.

Fleer, G. J., M. A. Cohen Stuart, J. M. H. M. Scheutjens, T. Cosgrove, and B. Vincent, *Polymers at Interfaces*, Chapman & Hall, London, 1993.

Garbasi, F., M. Morra, and E. Occhiello, *Polymer Surfaces: From Physics to Technology*, Wiley, Chichester, 1994.

Garton, A., *Infrared Spectroscopy of Polymer Blends, Composites, and Surfaces*, Hanser, Munich, 1992.

Israelachvili, J., *Intermolecular and Surface Forces*, 2nd ed., Academic, New York, 1992.

Krim, J., *Sci. Am.*, **275**(4), 74 (1996).

Miles, I. S. and S. Rostami, Eds., *Multicomponent Polymer Systems*, Longman Scientific & Technical, Avon, Great Britain, 1992.

MRS Bulletin, **21**(1), (1996), special issue, *Polymer Surfaces and Interfaces.*

Napper, D. H., *Polymeric Stabilization of Colloidal Dispersions*, Academic, New York, 1983.

Noda, I. and D. N. Rubingh, *Polymer Solutions, Blends, and Interfaces*, Elsevier, Amsterdam, 1992.

Sanchez, I. C., Ed., *Physics of Polymer Surfaces and Interfaces*, Butterworth-Heinemann, Boston, 1992.

Urban, M. W., *Vibrational Spectroscopy of Molecules and Macromolecules on Surfaces*, Wiley-Interscience, New York, 1993.

Vigo, T. L. and B. J. Kinzig, Eds., *Composite Applications: The Role of Matrix, Fiber, and Interface*, VCH, New York, 1992.

Wool, R., *Polymer Interfaces: Structure and Strength*, Hanser, Munich, 1995.

Wu, S., *Polymer Interface and Adhesion*, Marcel Dekker, New York, 1982.

STUDY PROBLEMS

1. What silane chemical structure would you choose to bond glass fibers to the following?
   a. An epoxy resin
   b. An unsaturated polyester–styrene fiber layup
   c. Nylon-66

2. How would you determine the specific surface area of a polymer composite using light scattering? What is the theoretical and experimental interpretation of the data?

3. Common polymers such as polyethylene and polystyrene make poor blends because they are *incompatible*. What method(s) would you select to remedy the problem? Present details as to molecular weights, concentrations, processing conditions, and the like as needed.

4. A continuous, oriented glass fiber system is embedded in a crosslinked epoxy matrix. Because the modulus of the epoxy is significantly lower than the glass fiber, a very significant strain is imposed on the fiber–matrix interface, even though the total stress on the composite is less than the fracture stress.
   a. If it is determined that the radius of gyration of the epoxy chain is 10 nm, and that the shear strain of the matrix relative to the fiber is 20 nm near

the surface, what bonding energy at 25°C is required to hold the chains onto the fiber?

b. Assume that the fibers (S-Glass) are discontinuous, short fibers of 1000 μm length and of 0.05% w/w concentration, but still highly oriented. Noting Tables 3.1 and 7.3, what total strain on the system is required to achieve a shear strain of the matrix relative to the fiber of 20 nm? Please make any assumptions required.

5. In a line or two of words, and/or preferably with a drawing, how do the following methods work, and what do they measure?
   a. Dynamic light scattering
   b. ESCA
   c. Contact angle determination
   d. SANS

6. Polymers at interfaces sometimes have lower than average densities, and sometimes higher than average densities.
   a. What are the nature of the forces that control the interfacial density?
   b. Which type of interaction is preferred for improved mechanical strength? Why?
   c. What is the conformation of the polymer chains at polymer blend interfaces?

7. Fiber glass is to be embedded in polymer matrices of polystyrene, epoxy, or poly(ethylene terephthalate). The interface is supposed to be toughened with silanes. Finding that the exact silane structures of Table 6.1 are "patented" and unavailable to your company, what new ones can you invent, and how will they work?

8. Silanes have become highly important as surface treatments, especially for silica or glass-based composite systems.
   a. How do silanes bond to glass and silica?
   b. How do they work to strengthen polymer composites containing glass or silica? Give two examples, one for glass fibers in an unsaturated polyester–styrene mix undergoing polymerization and for glass fibers embedded in a nylon matrix.

9. How would you design a can of nondrip latex paint by adding polymers? On a molecular level, how does the nondrip feature work?

10. The dilute solution–colloid people tend to talk about adsorption of polymer chains on fillers, the bulk polymer people tend to talk about adhesion of the polymer to the filler.
    a. What similarities and differences exist between the two systems, besides vocabulary?

b. What is the approximate concentration profile of the polymer chains as a function of distance from the surface of a colloid particle and a filler? Separate out the concentrations of bound polymer and unbound polymer. Include the concept of entropic repulsion from the wall, if applicable.

11. At a polymer blend interphase, the behavior of polymer chains is quite different from that of the bulk material of either phase.
    a. What happens to the density of each polymer and that of the mix as the interphase is crossed? (A drawing would help.)
    b. The quantity $-d\gamma/dT$ is positive for most polymer blend interfaces. What does this mean in terms of molecular structure or orientation?
    c. If there are small molecules present in the blend, where would you expect to find them?

12. A latex is forming a film under conditions that the relaxation time, $T_r \approx 1\,\text{min}$.
    a. What new relaxation time would be obtained if the molecular weight was doubled?
    b. What new relaxation time would be obtained if the temperature of interdiffusion was raised from 70 to 100°C?

13. According to the Lake and Thomas theory, what work is required to fracture an infinite molecular weight linear polystyrene? Please report the answer in terms of energy per unit of surface area created.

14. What fractions of chain scission and chain pull-out should be expected theoretically for the fracture of 100,000 g/mol poly(methyl methacrylate)?

15. What is known about the shape of a polymer chain near
    a. a polymer blend interphase, or
    b. a nonpolymer solid surface, or
    c. at an air surface?

16. Will poly(dimethyl siloxane) spread over polychloroprene at 140°C?

17. A 50/50 blend of two polymers has a Debye correlation distance, a, of 10.5 nm. What is the internal specific surface area?

18. Be inventive! What monomer or polymer not in Table 7.4 of your own choice/discovery/invention/(literature?: give reference!) would be useful for Langmuir–Blodgett film formation?

19. In 18 above, how will you apply the material to a substrate (X, Y, or Z methods) to get an asymmetric film? What possible uses would such a material have?

PART III

SELECTED ENGINEERING
POLYMER MATERIAL

8

RUBBER TOUGHENING OF ENGINEERING PLASTICS

All modern plastics are light compared to metals and ceramics; many of them are relatively inexpensive. While some of them are intrinsically "strong" as homopolymers, many are brittle, or fracture relatively easily under rapid loading conditions. There are two leading solutions to this problem: fillers, particularly fibrous fillers, may be added. The second solution involves the toughening of engineering plastics with elastomers, the subject of the present chapter.

8.1. HISTORICAL ASPECTS

The history of rubber-toughened plastics stretches back at least to the beginning of the twentieth century. In 1914, Thomas Edison's chief chemist, Jonas Aylsworth, was working on the new Bakelite materials for phonograph records. Of course, Edison had invented the phonograph himself. Bakelite, basically a thermoset plastic based on phenol and formaldehyde, had just been invented by Leo Baekeland in Yonkers, New York, in 1906. The Edison research and manufacturing complex was across the Hudson River in West Orange, New Jersey. While the new material was a definite improvement for the recording industry over the waxes that preceded it, simple phenol–formaldehyde plastics were extremely brittle, and the phonograph records of that day needed to be physically thick and heavy to prevent fracture on being handled or dropped.

At that time, Aylsworth added many items to Bakelite, forming a long series of blends and composites detailed in the patent literature. The most successful was the addition of natural rubber and sulfur to the phenol and formaldehyde mix, and simultaneously polymerizing and crosslinking, thus making a simultaneous interpenetrating network (IPN) (1,2). According to the patent dates listed on the phonograph record sleeves, natural-rubber-toughened phenol–formaldehyde materials were in regular production until the end of the 1920s, when the West Orange operation was shut down; see Section 10.12.

The next invention was by Ostromislensky in 1927 (3). He added natural rubber to styrene monomer, which was then polymerized quiescently. The end result was technically an adhesive, since the rubber remained the continuous phase. Due to grafting reactions between the rubber and the polystyrene, the product was more or less insoluble and difficult to process. As a rubber-toughened plastic, apparently even the best compositions were poor.

A very simple method of blending rubber with plastics such as polystyrene involves equipment such as rollers or extruders, thus effecting the mechanical blending of the two polymers in the molten state (4). While this method still finds practice today, the increase in impact resistance of the final plastic is minimal. The main problem stems from the rather coarse dispersion of the rubber in the plastic, that is, the domains are far larger than optimum. Clearly, the method also does not take into account the interfacial issues discussed in Chapters 4–7.

8.2. HIGH-IMPACT POLYSTYRENE (HIPS)

8.2.1. Synthesis

Homopolymer polystyrene is a very brittle plastic with a glass transition temperature of about 100°C. The first modern high-impact polystyrene was developed by the Dow Chemical Co. in the late 1940s by Amos et al. (5). Basically, the elastomer, usually polybutadiene, is mixed with styrene monomer, and the mass polymerized *with stirring*. Subjecting the reacting mass to a shearing agitation both breaks up crosslinking portions and causes a phase inversion [equation (2.15)]; see Section 2.3. The basic process has been the subject of many improvements (6–8) and has been reviewed many times (9,10). The general process of manufacturing HIPS is as follows:

1. 5–10% linear polybutadiene is dissolved in styrene monomer, along with peroxide, and a bulk thermal polymerization is initiated, with shearing agitation, and cooling.
2. At first, the growing amount of polystyrene (PS) and the polybutadiene (PB) are mutually soluble in the styrene monomer.
3. When phase separation occurs, first the polybutadiene-rich phase is the continuous phase. Both the polybutadiene-rich phase and the polystyrene-rich phase are highly swollen with styrene monomer.
4. With continued shearing, a phase inversion occurs, and the polystyrene-rich phase becomes continuous. For many commercial preparations, phase inversion takes place after about 5–15 percent conversion of polystyrene in the system. As stated above, equation (2.15) governs the composition at which phase inversion occurs. Of course, it is the viscosity and volume fraction of the phase as a whole, rather than the quantity and viscosity of the polymer that must be measured. The particle size

and particle size distribution are largely controlled by the applied shear rate.

5. At this point, with both phases still highly swollen with styrene monomer, there is a fork in the path. The bulk polymerization may be stopped, and the whole mass dispersed in water as a suspension. The polymerization is continued and completed as a suspension polymerization. Important reasons include the increasing viscosity of the polymer system and the need for heat removal. Alternately, the material may be continued as a bulk polymerization. Amos (9) describes a process using two or three continuous polymerizers in series, with 85–90% polymerization, followed by devolatilization. The remaining monomer is recovered and recycled, and the molten polymer converted into pellets.

6. The styrene in the polystyrene-rich phase continues to make more polystyrene. The styrene in the polybutadiene-rich phase, on continued polymerization, phase separates again. By this time, the polybutadiene and the polystyrene are grafted together significantly, and the polybutadiene is actually crosslinked, forming polybutadiene-*cross*-polystyrene.

7. The kinetics of phase separation of the polystyrene from the polybutadiene in this latter stage is thought to follow a modified spinodal decomposition mechanism.

8. To a significant extent, the toughness of the high-impact plastic that results increases with the volume fraction of polystyrene trapped in the polybutadiene droplets. Another feature that is important includes the distance between the domains. The glass transition temperature of the polybutadiene as well as that of the polystyrene are substantially the same as in their respective homopolymers, so very little mixing of the components takes place.

9. The total amount of grafting, in the chemical sense, is small, of the order of a few percent, but highly important in increasing interfacial bonding.

The importance of the Amos et al. patent (5) must be emphasized. More than a full column of U.S. Patent No. 2,694,692 is devoted to the necessity of agitation with shearing. In part:

It is important that the solution of the rubber and the polymerizable monovinyl aromatic compound, e.g., styrene, be agitated, preferably with a shearing action throughout its mass, during the early stages, or first part, of the polymerization reaction in order to obtain homogeneous linear interpolymerization products which are free, or substantially free, of cross-linked or highly branched-chain interpolymer molecules....

The tendency toward the formation of cross-linked interpolymers appears to be greatest during the early part of the polymerization, e.g., when 10 per cent by weight or less of the starting materials have been polymerized. During this stage of the polymerization, the interpolymer molecules appear to be attached to each other, or held together by relatively weak forces, i.e., the cross-linked interpolymers or highly branched-chain polymers are apparently agglomerates of polymer molecules which are held together by only a few or by relatively weak bonds. . . . The tendency toward the formation of such cross-linked interpolymer molecules, which cause inhomogeneities in the polymeric product, can be prevented or substantially reduced by application of a shearing action to the polymerizing mass, i.e., by agitating the polymerizing mass, particularly during the early or first stages of the polymerization.

In those days, the term *interpolymer* was used to mean a chemically reacted mixture of two or more polymers, such as block and graft copolymers. The important point is that Amos et al. (5) recognized that without shearing, the product formed an intractable and insoluble gel. This latter had already been invented by Ostromislensky (3). While the Amos et al. (5) patent does not mention phase inversion, agitation with shearing during the first 10% of conversion is critical to obtaining the correct morphology. Thus, a material with a continuous, crosslinked rubber phase is avoided, and a continuous polystyrene phase containing small rubber gel particles appears instead; see Section 8.3.2.

The synthesis and the processes for making HIPS are substantially the same as they were in the early 1970s. However, HIPS is widely manufactured today on a global scale because of its relatively low price and good impact properties.

8.2.2. Example Calculation

A general problem in the type of solution polymerization described above relates to the partition of the last 20% or so of the remaining monomer (11). In this case, it is the partition of styrene between the polystyrene-rich phase and the polybutadiene-rich phase. Of course, the partition will be governed by the minimum in free energy of the system. The partition coefficient is given by

$$K_{A/B} = A/B \tag{8.1}$$

where A and B represent the concentration of monomer in phases A and B. There are two parts to the calculation: the heat of mixing and the entropy of mixing.

The free energy of mixing, ΔG_M, can be expressed as the difference between the enthalpy, ΔH_M, and the entropy of mixing, ΔS_M, times the absolute temperature, T:

$$\Delta G_M = \Delta H_M - T\Delta S_M \tag{8.2}$$

The enthalpy of mixing is given by

$$\Delta H_M = V_M(\delta_1 - \delta_2)^2 \nu_1 \nu_2 \tag{8.3}$$

where V_M represents the total volume of the mixture, and δ and ν represent the solubility parameters and the volume fractions, respectively, of components 1 and 2, where by tradition, 1 represents the solvent, and 2 the polymer. In this case, there are two polymers involved, A (PS) and B (PB).

The entropy of mixing is given by

$$\Delta S_M = -k(N_1 \ln \nu_1 + N_2 \ln \nu_2) \tag{8.4}$$

where k is the Boltzmann constant, and N is the number of molecules of species 1 and 2, respectively. Of course, the volume selected for equations (8.3) and (8.4) must be identical.

Assume 20% of remaining styrene(s) in the system, and polymerization being slow relative to diffusion, so that the system can be considered in quasi equilibrium, and 1 cm^3 of material selected. The solubility parameters required are given in Tables 2.4 and 2.6, that is, $\delta(\text{PB}) = 8.4$, $\delta(\text{PS}) = 9.1$, and $\delta(\text{S}) = 9.3 \, (\text{cal}/\text{cm}^3)^{1/2}$. A 50/50 mix of polystyrene and polybutadiene is assumed. The value of ΔH for mixing 20% of styrene with 80% of polybutadiene is given from equation (8.3):

$$\Delta H_M = 0.5 \, \text{cm}^3 (9.3 - 8.4)^2 \times 0.2 \times 0.8 \, \text{cal}/\text{cm}^3 = 0.064 \, \text{cal}$$

A similar calculation involving mixing styrene with polystyrene yields $0.0032 \, \text{cal}$, a much smaller number. Those stopping there may conclude that the styrene is much more soluble in the polystyrene.

The calculation for the entropy of mixing for the styrene with both phases is identical. The value of N, for the styrene, is 1.05×10^{21} molecules per cm^3, assuming 20% concentration. If the degree of polymerization (DP) of the chains is 1000, the second term in equation (8.4) is small, and

$$\Delta S_M = -3.27 \times 10^{-24} \, \text{cal/K} \times 1.05 \times 10^{21} \times \ln 2 = 5.5 \times 10^{-3} \, \text{cal/K}$$

Assuming room temperature, 298 K, the total entropic contribution is 1.64 cal, much larger than the enthalpic contribution. Calculations assuming all the styrene is in either phase yields a total entropic term calculation of 0.937 cal, much smaller. It is easy to show that the monomer should be nearly evenly distributed, as outlined by the calculation given below:

The calculation scheme in general involves several steps:

1. The proportion of A/B must be selected.
2. Define the volume fraction of monomer to be added, for example, 20%.
3. Assume a range for possible $K_{A/B}$ values, ranging from, say, 0.1 to 10.
4. For each $K_{A/B}$, the amount of monomer dissolving in each phase must be calculated.
5. For each $K_{A/B}$, ΔG_M for each phase must be calculated.

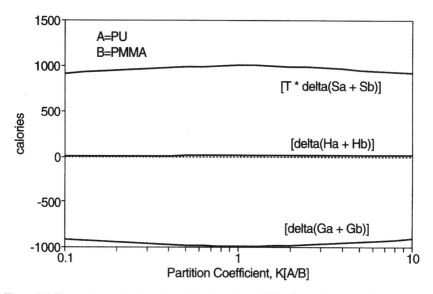

Figure 8.1 Thermodynamic calculation of the location of 20% of methyl methacrylate monomer in an SIN (or related material) of 50/50 polyurethane and poly(methyl methacrylate). The lowest free energy is when the partition coefficient is close to unity.

6. Finally, a plot of $\Delta G_{\text{total}} = \Delta G_{MA} + \Delta G_{MB}$ vs. $K_{A/B}$ is determined. The minimum in ΔG_{total} predicts the value of $K_{A/B}$.

The results are shown in Figure 8.1 (11), worked out for the polymerization of methyl methacrylate (MMA) in the presence of polyurethane (PU), in the form of an SIN (12,13). Clearly, the entropy of mixing dominates, suggesting that $K_{A/B}$ must be close to unity, over a wide range of ΔH_M values. Thus, Figure 8.1 predicts that the last remaining monomer in this polymerization will be nearly evenly partitioned between the two phases, in nearly equal concentrations.

The conclusion is that for most bulk solution polymerizations where monomer II is polymerized in the presence of polymer I, as well as SIN or IPN polymerizations, the monomer will be distributed substantially evenly between the two phases. A few possible exceptions include crystalline polymers, very densely crosslinked polymers, and the few cases where monomer II is almost totally insoluble in one polymer or the other.

This also bears on the results of Jin et al. (14), and many others. In the Jin et al. study, an almost identical PU/PMMA SIN system as shown in Figure 8.1 was prepared. It was proposed that in the latter stages of polymerization, the remaining MMA reaction takes place primarily at the interphase, where MMA arrives by diffusion. The locus of polymerization of the last traces of MMA cannot be in the PMMA-rich phase because it is vitrified. Analysis such as performed above for both the remaining monomer and the remaining initiator suggests that both would be expected to be nearly evenly partitioned between both phases.

This has strong implications for the locus of the polymerization of the last portions of the styrene monomer. It must be noted that the polystyrene becomes glassy in the last stages of the polymerization, when the glass transition temperature of the plasticized polymer rises above the temperature of the polymerization, usually about 70°C. Thus, the polymerization in the polystyrene phase is slowed very significantly. It may be that the last traces of styrene are polymerized within the polybutadiene-rich phase, and in the interphase between the two polymers. The latter is suggested by the continued random diffusion of the monomer out of the polystyrene-rich phase to the interphase, where it is polymerized.

It must be noted that a similar thermodynamic analysis of the remaining peroxide also suggests that it must be nearly evenly distributed between the two phases.

This may be responsible for the occasional finding of a "fine structure" secondary phase separation in such polymerizations, where the secondary domain sizes may be only 10 nm in diameter (15).

8.2.3. Morphology

An excellent way to study the morphology of HIPS involves ultrathin sectioning, staining with osmium tetroxide, and examining the features with transmission electron microscopy. See Section 1.4 for a review of these methods. Typical results are shown in Figure 8.2 (10), which shows typical morphologies for solution-polymerized (bulk process) HIPS, mechanically blended with a butadiene–styrene block copolymer in polystyrene. The solution process produces the spherical cellular particles with relatively large, uniform domains of polystyrene occluded inside. These rubber domains are of the order of $2\,\mu$m in diameter. The block copolymer rubber domains are of the order of $5\,\mu$m in size, with the polystyrene occlusions being much smaller. The dimensions of the block copolymer polystyrene domains are controlled in significant measure by the molecular weight of the blocks; see Section 9.3. Note the similarity between the morphologies in Figure 8.2 and in Figure 2.5, the latter having castor oil rubber instead of polybutadiene rubber.

Grafting between the rubber and the polystyrene clearly plays an important role in terms of interphase strength. However, originally it was thought that the actual grafting level was much higher, about 50% or more (16,17). It is now believed that the products of such graft copolymerizations contain many fewer graft sites than originally believed to be the case. The older conclusion originated with two observations: (1) Standard separation techniques based on extraction and precipitation showed it was not possible to separate much of the polystyrene from the rubber. (2) The relatively high chain transfer constants (from polymerization kinetics) known for this system supported the possibility of such high grafting.

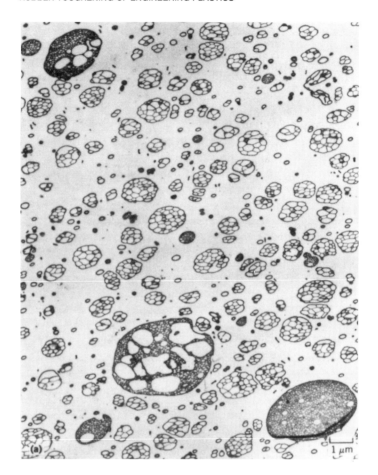

Figure 8.2 Morphology of an osmium tetroxide stained high-impact polystyrene. The finely divided morphology is a polystyrene–polybutadiene block copolymer mechanically blended in for extra toughness.

The high grafting level conclusion was reached before transmission electron microscopy. The polymer blend shown in Figure 8.2 (for the pure HIPS component) may be considered to be comprised of three components: a plastic phase, much of which is far removed from any elastomer, but partly contained within the rubber phase itself, and the graft copolymer. Note that in all cases, the domain size is large compared with the size of individual polymer chains. In a now classical experiment, Wagner and Cotter (18) showed in 1971 that ultrasonic vibration could be employed to disrupt the cellular structure. Then the occluded polystyrene could be extracted. Thus, the actual level of grafting is now thought to be only a few percent (19), albeit a very important few percent.

Table 8.1   Izod impact strength of polymeric materials

Polymer	Impact strength, J/m^a
Polystyrene	16–24
HIPS	48–100
ABS	100–450
Poly(vinyl chloride)	50
Rubber-toughened poly(vinyl chloride)	800
Nylon 6,6	240
Super Tough Nylon	1100–1200
Epoxy	25–50
CTBN toughened epoxy	125–200
Polycarbonate	800
Polycarbonate/ABS	400–500
Polysulfone	60
Polysulfone/polyacrylate	600
Poly(methyl methacrylate)	16
Poly(methyl methacrylate)/rubber	80

^{a} 1 ft-lb/in notch $= 2.5$ ft-lb/in$^2 = 5.25$ kJ/m$^2 = 52.5$ J/m notch.

8.2.4. Mechanical Behavior

8.2.4.1. Measures of Toughness   The most important reason for using HIPS [or acrylonitrile–butadiene–styrene (ABS) as described below], rather than polystyrene homopolymer, is the much greater toughness of the material. It absorbs more energy on loading, delaying, or preventing fracture. This is true for a variety of test methods, including stress–strain, impact loading, and fatigue.

Although the stress to break decreases, the energy to fracture in stress–strain increases with these materials. Usually, the stress–strain curves show well-defined yield points. Of course, the area under the curves is a measure of the energy required to break the sample.

Izod impact strengths are shown in Table 8.1 for HIPS and other polymers. The rubber in HIPS was originally thought to act primarily by increasing the crazing level, each craze absorbing energy. However, cavitation is now known to play a role, as will be described below.

In Section 3.1, the fracture energy, G, was introduced. The critical energy to fracture, G_c, can be written as

$$G_c \preceq -(1/b)(\partial U/\partial a)_1 \qquad (8.5)$$

The quantity b is the sample thickness, and the quantity a represents the crack length. While the fracture energy is a general term, it most easily characterizes materials that obey linear elastic fracture mechanics (LEFM). An improved method of characterizing the fracture toughness of ductile and toughened

Table 8.2. J-integral toughness of polymers

Part A. HIPS J Data via Three Methods

	10	12.5	15
Specimen thickness, mm			
ASTM E813-81 method			
J_c (kJ/m^2)	3.8	3.7	3.6
$dJ/d\Delta a$ (MPa)	5.0	3.7	3.5
J at $\Delta a = 1$ mm, kJ/m^2	8.0	6.7	6.2
ASTM E813-81 by neglecting the blunting line J_c (kJ/m^2)	3.21	3.24	3.24
ASTM E813-87 method J_c (kJ/m^2)	4.9	4.3	4.3
ASTM E813-87 with 1 mm offset line J_c (kJ/m^2)	3.8	3.3	3.5
Hysteresis energy method J_c (kJ/m^2)	3.3	3.2	3.0

B. J-Integral Toughness of Other Polymers, J_c (kJ/m^2)

Nylon 66	18.5	a
Super-tough nylon	27–32	a
Polycarbonate	4–5	b
Rubber-toughened polycarbonate	8–10	b

[a] D. D. Huang and J. G. Williams, *J. Mater. Sci.*, **22**, 2503 (1987).
[b] C. B. Lee and F. H. Chang, *Polym. Eng. Sci.*, **32**, 792 (1992).

polymers and blends involves the *J*-integral approach (20). The quantity J_c, (or in mode I, J_{Ic}), represents the energy required to initiate crack growth. In general, it can be written (21)

$$J_c = -(1/b)(\partial U/\partial a)_1 \qquad (8.6)$$

so that in general, J_c may be expected to be slightly larger than G_c.

The quantity J_c is a material property. However, it tends to vary somewhat with the method of measurement. Table 8.2 (22) shows values in the range of 3000–4000 J/m^2, based on the ASTM and hysteresis methods described. An earlier impact loading test by Plati and Williams (23) gave approximately 15,000 J/m^2 for another HIPS material. These values compare to values of G_c for homopolymer polystyrene in the range of 100–1000 J/m^2, typical values being around 250 J/m^2. Thus, by this measure, rubber-toughened polystyrene in the form of HIPS is several times tougher than the corresponding homopolymer.

8.2.4.2. Morphology Changes In general, the following physical observations must be explained to understand the reinforcing role of rubber in HIPS (24):

1. A several-fold increase in impact strength over unmodified polystyrene
2. A 10-fold increase in elongation

3. The whitening that appears on application of a stress
4. The improvements that arise through the use of block copolymers and/or grafting
5. The advantage of having the reinforcing rubber crosslinked
6. The optimum mechanical behavior observed with the rubber particles being about 2–5 μm in size
7. The desirability of having occluded polystyrene so as to extend the effective volume of the rubber phase

Using osmium tetroxide staining, Keskkula et al. (25) examined the failure processes in rubber-toughened polystyrenes. They found that multiple crazes initiated from some of the rubber particles; see Figure 8.3a. Several reasons for the existence of these multiple craze initiations include multiple stress concentration regions on particles with a "raspberry surface," stress field interactions in polymers with a high volume fraction of particles, and particle separation from the matrix polymer.

Another significant feature is failure within the rubber particles. Holes, or cavities, as they are now called, are formed, which seem more or less independent of the crazes; see Figure 8.3b. This last figure was taken with scanning electron microscopy of a highly damaged HIPS, after bending it in a U shape. Cavitation is also visible in Figure 8.3a; see arrows.

Using real-time small-angle X-ray scattering techniques, Bubeck et al. (26) demonstrated that noncrazing mechanisms, predominantly due to rubber particle cavitation and associated ligament bending of the surrounding glassy matrix, occur before crazing in HIPS and ABS. They found that crazing accounts for only about half of the total plastic strain in HIPS and ABS.

Cieslinski (27) studied real-time deformation of HIPS and ABS in the transmission electron microscope (TEM). For ABS materials (see below), with a bimodal particle distribution, 20% 0.1 μm and 80% 0.2–1.0 μm particles, he found that under tensile loading both rubber-induced shear planes and particle cavitation occurred. The shear planes initiated at the larger rubber particles and propagated through the material along aligned bands of the smaller particles. He thought that the cavitation relieves the initial thermal stress around the particles and the hydrostatic pressure generated within the particles during shearing. This leads to a larger volume of plastic deformation and greater toughness.

According to Henton (28), a popular current view is that in HIPS fracture, the rubber particle cavitates first. Then a craze is initiated at the edge of the rubber particle. The craze is then propagated to other rubber particles, where it may be further divided. Keskkula (29) summarized the current state of knowledge as follows:

> It is well agreed that in brittle polymers, as well as in the ductile ones most of the impact energy is absorbed by the matrix polymer. In polystyrene it is due to crazing. In ductile polymers, however, the impact energy is principally absorbed

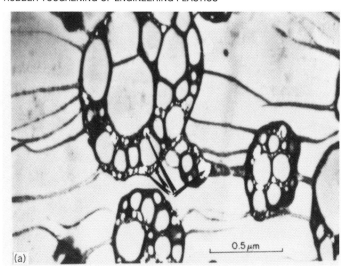

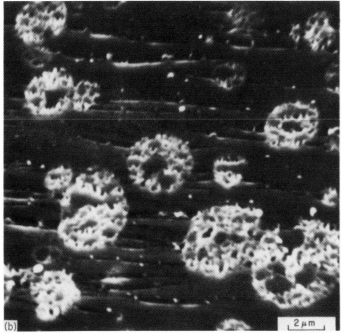

Figure 8.3 Electron microscopy of crazing and fracture in HIPS. Upper figure: Transmission electron microscopy. Note multiple crazes initiating from the rubber particles. Lower figure: Scanning electron microscopy. Note the presence of cavities in the rubber domains.

through shear yielding of the matrix polymer. In this case, the rubber particles relieve the triaxial stress by cavitation that will contribute to the initiation of shear yielding of the matrix.

It appears that there are thus two important mechanisms of rubber toughening: cavitation and crazing. In HIPS, and also in ABS (Section 8.3), crazing seems to be playing the more important role. However, in rubber-toughened polycarbonates (Section 8.5) and similar modern engineering plastics, cavitation followed by shear yielding is the most important toughening mechanism. These materials may not craze at all but rather undergo a shear deformation.

8.3. ABS PLASTICS

8.3.1. Synthesis

ABS plastics constitute the next step up on the toughness ladder. Similar to HIPS, these materials were fully developed and commercialized in the 1940s and 1950s and have undergone only minor improvements since. However, the ABS plastics remain a major global product.

ABS plastics are basically styrenic in nature. However, instead of the polystyrene homopolymer in HIPS, a copolymer containing 5–66% acrylonitrile is used. Many ABS materials have between 20 and 30% acrylonitrile in the styrene–acrylonitrile (SAN) copolymer (30). The acrylonitrile contributes both oil resistance and a higher glass transition temperature to the plastic, in addition helping to increase its toughness.

There are three commercial polymerization processes for the manufacture of ABS. These are the emulsion, suspension, and bulk processes. Each of the three manufacturing methods result in the formation of a material composed of discrete rubber particles dispersed in a rigid matrix. Basically, the suspension and bulk processes are similar in nature to that discussed in Section 8.2.1 for polystyrene and will not be further discussed here.

The emulsion polymerization, however, is quite different. A typical ABS emulsion process is described in Figure 8.4 (31). First, a seed latex of polybutadiene or styrene–butadiene rubber (SBR) is synthesized (32–35). Sometimes, statistical copolymers of butadiene and acrylonitrile are used (36). As synthesized, the polybutadiene is crosslinked, important for future integrity during processing. Thus, the ABS polymers are technically semi-1 IPNs, with the first synthesized polymer crosslinked. The size of the seed latex particles is also important. In a second step, styrene and acrylonitrile are added, and polymerized with significant grafting. A SAN latex can also be polymerized separately and blended with the high rubber ABS, thus providing another degree of control over the product composition and structure (16).

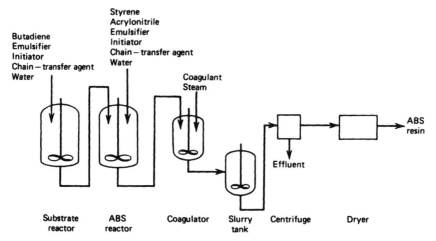

Figure 8.4 Illustration of the emulsion ABS process. The final ABS resin or plastic is a multiphase material.

8.3.2. Morphology

The morphology of suspension and bulk polymerized ABS is also similar to that of HIPS; see Figure 8.5 (30). Here, a TEM of a thin section, with the rubber domains stained by osmium tetroxide is illustrated. Again, there is a phase

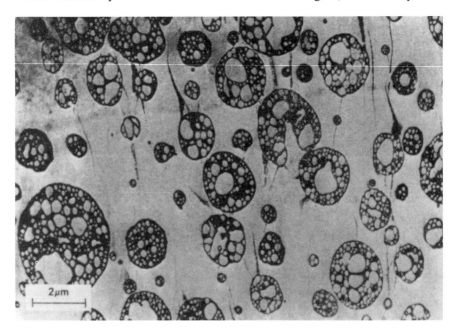

Figure 8.5 Osmium tetroxide stained suspension prepared ABS, as seen by transmission electron microscopy. Note occluded SAN copolymer.

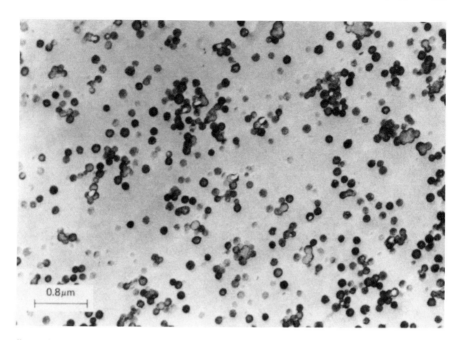

Figure 8.6 Emulsion prepared ABS, via transmission electron microscopy. This material has little occluded SAN copolymer.

within a phase within a phase, with the SAN polymer continuous and the rubber domains containing occluded SAN.

The emulsion-synthesized ABS, however, is quite different; see Figure 8.6 (30). Figure 8.6 shows little occluded SAN inside the seed latex portion, the more usual case. However, other preparations contain a quite distinctive morphology of occluded SAN; see Figure 8.7 (37). The particle sizes are

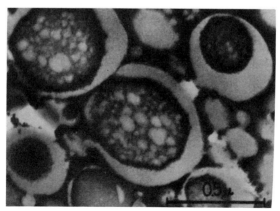

Figure 8.7 Transmission electron micrograph of epoxy-embedded type G ABS latex particles. In contrast with those shown in Figure 8.6, these particles contain a considerable amount of occluded SAN.

significantly larger than those in Figure 8.6. In Figures 8.5, 8.6, and 8.7, the styrene and acrylonitrile monomers were originally dissolved in the rubber-rich phase. On polymerization, however, phase separation occurs. Since the rubber is crosslinked, the occluded polymer cannot escape, and forms a cellular domain structure. Analysis of the morphology strongly suggests a spinodal decomposition type of phase separation. These developments are also true for the HIPS polymers.

At least some occluded rubber seems important for the development of good mechanical behavior, especially impact resistance. By increasing the volume fraction of space occupied by the rubber-continuous phase, the effectiveness of the domains in multiplying the appearance of crazes, and hence increasing the toughness of the product, increases. These will be the subjects of the following sections.

8.3.3. Mechanical Behavior

While the general pattern of stress–strain and impact behavior of HIPS and ABS is quite similar, the ABS plastics are much tougher; see Table 3.1 and Table 8.1.

More recently, Aoki and co-workers (38–44) investigated the dynamic viscoelastic behavior of model ABS polymers made via emulsion polymerization. In studies on these polymers in the molten state, they found that there are two kinds of ABS polymers: one kind exhibits a pseudoequilibrium shear modulus that does not depend strongly on frequency at low frequencies, and the other kind does not have such a plateau, the modulus continuing to decrease with decreasing frequency. The former was found for systems in which rubber particles agglomerate in the matrix phase, forming three-dimensional network structures, while the latter was found in finely dispersed systems without such an agglomerated structure. Opposing agglomeration is an entropic repulsion force between neighboring rubber particles. This entropic repulsion leads to a more random distribution of the particles (but not a regular array of the particles).

For example, a series of transmission electron microscope experiments were performed on ABS materials with 10% rubber content but of variable surface grafting level before and after shearing. The samples were stained with osmium tetroxide, which stains preferentially the double bonds of the rubber. In Figure 8.8 (44), samples at 41.3% grafting remain dispersed, while higher levels agglomerate. The degree of agglomeration also increases if the samples are heated for extended periods of time, that is, for 170°C or higher for 30 min. In this case, the shearing energy needed for agglomeration is replaced by thermal energy.

It was also found that the appearance of agglomeration depended upon the composition of the poly(styrene-*stat*-acrylonitrile) in the latexes. When the difference in the percent of acrylonitrile between the grafted (i.e., cellular domain structure) and the matrix (i.e., shell) poly(styrene-*stat*-acrylonitrile) was small, the rubber particles were finely dispersed without agglomeration, and

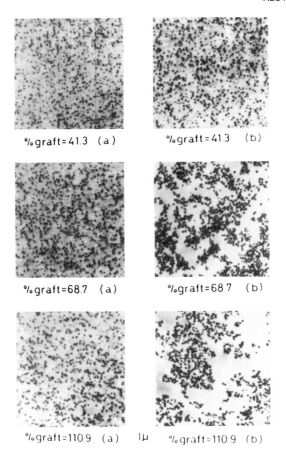

%graft=41.3 (a)   %graft=41.3   (b)

%graft=68.7 (a)   %graft=68.7   (b)

%graft=110.9 (a)   1µ   %graft=110.9 (b)

Figure 8.8 Electron micrograph of ABS materials containing variable amounts of grafting and subsequent agglomeration.

no low-frequency modulus plateau was found. However, the mismatching of the acrylonitrile composition caused the agglomeration of the particles and led to the observation of the low-frequency plateau modulus in these materials.

The low-frequency plateau in those materials exhibiting this phenomenon is caused by the appearance of a three-dimensional network (29). The nature of the network apparently depends on the structure of the ABS particles. For low degrees of grafting, the ABS cannot form stable colloid particles. When the temperature is raised, the viscosity of the matrix poly(styrene-*stat*-acrylonitrile) decreases. The unstable rubber particles are then relatively free to agglomerate. For high degrees of grafting, the grafted chains are stretched inside the latex particles. The matrix poly(styrene-*stat*-acrylonitrile) is expelled. If the composition between the two kinds of poly(styrene-*stat*-acrylonitrile) differs, gross phase separation between the two types of plastic ensues, and the rubber particles agglomerate because, again, they form an unstable dispersion. The

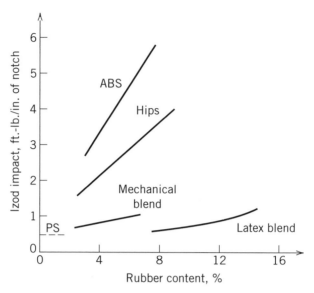

Figure 8.9 Impact resistance of several elastomer-containing styrenics. Commonly, polystyrene homopolymer has an impact resistance of around 0.3–0.4 ft-lb/in. of notch, lower left.

rubber particles of ABS, then, form stable dispersions only at intermediate grafting states.

If aspects such as grafting and particle sizes are roughly equal, the impact strength of the ABS depends significantly on the extent of agglomeration of the rubber particles. For small particles, for example, 0.1 μm in diameter, an increase in grafting will first reduce the cluster dimensions, resulting in an increase in impact strength. If grafting is increased further, the particles will be prevented from clustering and impact strength will drop (45). The toughness of the product also increases if a bimodal rubber particle size distribution is used (46).

Of course, toughness and impact behavior depend on the quantity of rubber incorporated (47,48); see Figure 8.9. Of the styrenics, ABS is seen to be the best. Mechanical blends and latex blends (simple blends of polybutadiene latex and polystyrene latex, precipitated, dried, and molded) exhibit poor impact resistance because the interfacial bonding between the rubber and the plastic is poor.

Impact resistance is also a function of temperature. The now classical finding of Bucknall and Street (49) (Figure 8.10) that ABS (and HIPS also) exhibits three distinct regions of increasing toughness as the temperature is raised has led to important new advances in the understanding of the rubber toughening mechanism (50). In region I, the low-temperature region, the rubber particles are glassy at times up to 10^{-4} s, leading to brittle crack initiation and propagation. The impact strength is low. In region II, the moderate temperature region, the rubber particles exhibit their glass transition temperatures near 10^{-4} s. The material is tough in the face of slow-moving cracks, but as cracks accelerate,

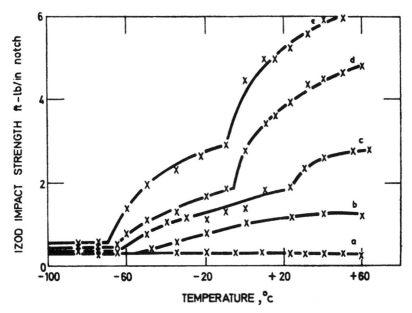

Figure 8.10 Izod impact resistance as a function of temperature for ABSs of different rubber contents: (a) 0%, (b) 6%, (c) 10%, (d) 14%, and (e) 20%. Note three temperature zones: I, brittle; II, tough to slow deformation but brittle to fast deformation; III, tough even at impact loading rates.

impact resistance is lost. This is the region of notch sensitivity, where the shape of the notch or crack, and hence its geometry, is important. In region III, the high-temperature region, the rubbery particles remain rubbery under conditions of fast crack propagation. Great amounts of stress whitening will be observed for styrenic polymers.

The results of Figure 8.10 can be modeled, by considering the motion of a crack or craze as it approaches a rubber domain (14). As illustrated in Figure 8.11, a crack (or craze) of 100 nm radius (estimated PS crack radius) moves with a velocity of 620 m/s through the plastic under impact loading conditions. This is governed by the modulus of the plastic, usually in the range of 3×10^9 Pa. During the time that the crack progresses from point A to point C, a bit of matter, slightly off center of the crack tip, moves from near point A to point B. The velocity of the crack in the elastomer is much slower, about 25 m/s. Now, instead of considering the crack as always moving forward (reality), consider the imaginary experiment where the crack tip moves back and forth from A to C in a sinusoidal motion of average velocity 620 m/s. This generates a frequency of approximately 10^9 Hz. Noting that the glass transition temperature is usually classically recorded at 10 s measurement, or approximately 10^{-1} Hz, the total increase in frequency is 10 decades. The Williams, Landel, Ferry (WLF) equation states that the T_g of a polymer increases about 6–7°C per decade of frequency. Thus, the T_g of the elastomer domain in Figure 8.11 is increased

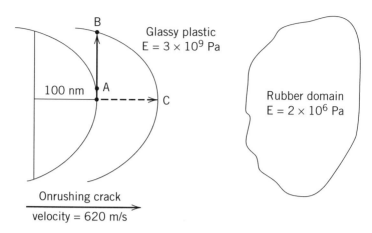

Figure 8.11 Motion of a crack (or craze) through a plastic. As the crack moves from point *A* to point *C*, a bit of matter slightly displaced from point *A* moves to point *B*. This crack or craze is moving toward a much softer rubber domain.

about 60°C. Examination of Figure 8.10 shows that the temperature change between the region I–region II intersection and the region II–region III intersection is approximately 60°C, the width of region II.

When the crack (or craze) strikes the rubber domain (see Figure 8.11), the velocity of the crack is suddenly reduced from about 620 m/s to approximately 25 m/s. The effect is to divide the crack (or craze) into several smaller cracks, with each propagating further into the sample. The crack takes about 5 μm to accelerate back to 620 m/s, thus providing the optimum distance between rubber domains, and hence a measure of the optimum rubber concentration. Each of these cracks will then intersect new rubber domains and redivide. The energy is absorbed through the creation of large amounts of new surface area, with concomitant large numbers of chain scissions and chain pull-outs, and the creation of regions of highly oriented polymer, especially where crazes predominate.

For rubber-toughened plastics intended for use at room temperature (ca. 20°C), the T_g of the rubber phase must be −40°C or lower. Thus, elastomers such as polybutadiene and polyisoprene work excellently well for rubber toughening. Elastomers such as poly(*n*-butyl acrylate), $T_g = -54$°C, work fairly well, while poly(ethyl acrylate), $T_g = -24$°C, works relatively poorly. For lower temperature applications, such as ice cube trays, ABS plastics still work excellently well because the T_g of the polybutadiene is around −90°C.

The HIPS materials are frequently blended with poly(2,6-dimethyl-1,4-phenylene oxide) (PPO) to produce still tougher materials. As described in Section 2.6, PS and HIPS are miscible. When PPO is added to HIPS, a transition from brittle to ductile behavior occurs between 20 and 50% PPO, indicating that the deformation mechanism in fracture changes from crazing to shear yielding. The impact resistance is raised accordingly, making these materials true engineering plastics. The HIPS/PPO blends are manufactured under the trade name Noryl^R; see Section 8.9.1.

8.4. TOUGHENING OF POLY(VINYL CHLORIDE)

As a homopolymer, poly(vinyl chloride) (PVC) is a glassy, slightly crystalline polymer, with an impact resistance in the range of 1 ft-lb/in. of notch (51). Applications of PVC include piping and other items of construction. However, it is better known to the public in two modified forms: plasticized, used as flexible seat covers and the like, and rubber toughened, serving a host of applications such as bottles.

In the field of rubber-toughening PVC, as well as with some other polymers, it is common to talk about *impact modifiers*. This term may be misleading to the uninformed because what is always meant is *impact improvement*. Therefore, the term impact modifiers will be used only sparingly.

PVC is particularly sensitive to thermal degradation and other problems. Therefore, a range of additives such as stabilizers, lubricants, processing aids, flame retardants, fillers, heat distortion improvers, and pigments are often added. Many of these additives, if used in their most effective concentration range, cause significant reductions in the impact resistance of the materials. A useful solution in many of these cases involves addition of special materials to boost the impact resistance of the formulation to the desired level.

Some of the newer impact improvers include the core–shell type of ABS and MBS (methacrylate–butadiene–styrene) (52). Such additives can boost the impact resistance up to the range of 16 ft-lb/in. of notch or higher (32,33,53). One of the major functions of core–shell impact improvers of this type is to act as stress concentrators located thoughout the polymer matrix. These stress concentrations are produced by the large difference that exists between the modulus of the rubbery portion of the particles and the polymer matrix. Such sites provide multiple sites at which shear yielding and/or crazing of the PVC can be initiated simultaneously on impact. The result is a structure with a large number of small crazes and/or shear bands that absorb energy. These impact improvers work by elongation and/or by cavitation. It has long been suggested that such impact improvers work by bridging the crazes; the modfier is able to elongate significantly without completely debonding from the matrix. Thus, it helps keep crazes from becoming cracks. A more modern theory proposes that the ability of the rubber to cavitate plays an important role in initiating shear yielding. The formation of a free surface within the rubber particles relieves the triaxial stress state of the surrounding polymer matrix significantly (33).

Stevenson (52) points out that these core–shell ABS and MBS particles work especially well in the presence of those additives that ordinarily reduce impact resistance. One reason why many additives reduce impact resistance relates to the fact that polymer is replaced by the additive, reducing the total ability of the material to dissipate the stress. A second reason is that certain additives,

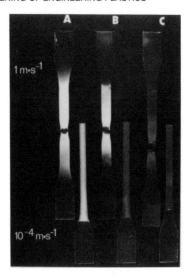

Figure 8.12 Dumbbell specimens of poly(vinyl chloride) blends with commercial toughening agents, "modifiers" A, B, and C, drawn at strain rates of 1 m/s and 1×10^{-4} m/s.

especially those that are miscible with the PVC, may hinder the local chain motions of the polymer chains, sharply decreasing the ability of the polymer to absorb energy. An example of that is the use of poly(methyl methacrylate) to increase the T_g of the system, hence raising the heat distortion temperature (32).

PVC toughened with ABS and similar materials stress whitens significantly; see Figure 8.12 (54). The ABS rubber particles were noted to exhibit significant cavitation, and the cavitated rubber particles were accumulated in bands that corresponded to shear bands in the PVC matrix. This was the principal cause of the extensive stress whitening. It was concluded that the rubber particles improved the impact strength of the PVC by initiating the shear bands, and not by generating crazes.

The materials in Figure 8.12 were prepared by blending ABS with suspension-polymerized PVC. The morphology of the materials, then, depends on the temperature of the blending process, as illustrated in Figure 8.13 (54), higher temperatures yielding more uniform blends. However, the notched Charpy impact strength for Figure 8.13a, b, and c, reads 42, 26, and 8 kJ/m^2, respectively, showing that some nonuniformity in the dispersion of the rubber particles yields tougher materials.

More recently, Tse et al. (55) blended chlorinated PVC with PVC. The profuse microshear banding visible in the damage zone was thought to nucleate subsequently from the voids formed in the chlorinated PVC portion.

The most important conclusion of these works is that the rubber particles cavitate under stress. The cavitation leads to a conversion of triaxial stresses in the matrix into a plane shear state, favorable for shear band initiation. The cavitation theory will be explored in the following sections.

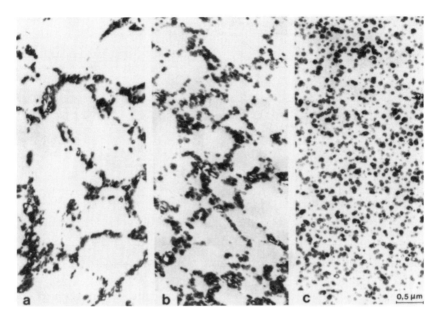

Figure 8.13 Transmission electron micrographs of PVC blends with ABS impact improver, mixed at various temperatures: (a) 140°C, (b) 160°C, and (c) 185°C. Outlines of the suspension-sized PVC particles can be seen in (a), much less so in (b), and not at all in (c), indicating the greater melting and mixing of the PVC at higher temperatures.

8.5. TOUGHENING OF POLYCARBONATES

Modern studies recognize three major mechanisms of toughening of plastics by rubber particles: by inducing crazing, shear yielding, and cavitation. While the older methods (see previous sections on HIPS or ABS, e.g.) concentrated on polybutadiene toughening of styrenics, today's research centers on the toughening of many other plastics, epoxies, nylons, and polycarbonates, to name a few. Frequently, the "rubber" particles blended into the plastic are themselves complex. The use of ABS particles in toughening polycarbonates has already been commercial for some years (56,57). This section will explore recent results and theories of toughening engineering plastics.

8.5.1. Properties of Polycarbonate Plastics

While the term polycarbonate covers a range of different materials (58), the principal material is that based on bisphenol A. The chemical structure of this polycarbonate is:

$$\left(O-\overset{\displaystyle O}{\underset{\displaystyle \|}{C}} - \left\langle \bigcirc \right\rangle - \overset{\displaystyle CH_3}{\underset{\displaystyle CH_3}{C}} - \left\langle \bigcirc \right\rangle - \right)_n \qquad (8.7)$$

Table 8.3 Basic properties of bisphenol A polycarbonate

Property	Value
Density, g/cm^3	1.20
Linear thermal expansion, per °C	7×10^{-5}
Refractive index, n_D^{20}	1.586
T_g, °C	150
T_m, °C	260^a
Brittle temperature, °C	-10
Tensile strength, MPa	70
Elongation to break, %	120
Notched izod impact strength, J/m	—
3.2 mm thickness	850
6.4 mm thickness	120

^{a} Bisphenol A polycarbonate, difficult to crystallize, is usually amorphous.

In this text, unless noted differently, "polycarbonate" will mean bisphenol A polycarbonate. Desirable properties include transparency, heat distortion resistance, toughness, and good electrical resistance. A few basic properties are summarized in Table 8.3 (58). The low brittle temperature for slow deformation is caused by a broad G'' maximum at -100°C. This is due to a γ relaxation based on the onset of rotation about the bond to the carbonate group. With a heat deflection temperature of 131°C, polycarbonates enjoy a broad range of useful temperatures.

Sample thickness affects the impact resistance significantly in polycarbonate. For polycarbonates in thin cross-section, and many other engineering plastics, the ductile state is characterized by stable crack propagation, with the formation and propagation of a macroscopic plastic zone and accompanying shear deformation. Brittleness occurs in thicker samples when a plastic zone with shear deformation no longer can form because the relaxation process is no longer able to eliminate the three-dimensional stress state that develops behind the notch. In this case, microcavities (crazing) occurs behind the notch, resulting in unstable macroscopic crack propagation.

In thin samples, the polymer is subjected to plane stress and thus can participate freely in the shearing mechanism, resulting in great toughness. In thick samples, the material exhibits plane strain, inhibiting the onset of shear deformation. This, however, facilitates the onset of crazing, resulting in brittle behavior. The brittle–ductile transition is shifted toward higher temperatures as the sample thickness increases. Under impact loading conditions, the brittle–ductile transition temperature is about 10°C for materials thin enough to exhibit plane stress. As will be shown, the critical thickness of polycarbonates can be increased by the addition of various kinds of impact modifiers, particularly complex rubber particles in latex form. Here, the term impact modifier, of course, means an agent that increases the impact resistance of the plastic.

8.5.2. ABS in Polycarbonate

ABS is particularly useful for toughening polycarbonates for two reasons: first, the glass transition temperature of the polybutadiene portion, usually in the range of -90 to $-70°C$, is excellent for crazing and shear yielding initiation mechanisms, and second, the SAN portion, usually in the shell of the poly-butadiene/poly(styrene-*stat*-acrylonitrile) particles, is relatively compatible with polycarbonate. In fact, the interfacial adhesion between polycarbonate and SAN has been found to be near its maximum at 25% acrylonitrile (59–61), close to that widely used in ABS materials. Thus, the level of interfacial adhesion between polycarbonate and SAN is sufficient such that useful products can be formed without the aid of any additional compatibilizer, contrary to experience with blends of ABS with other polymers such as the nylons or polyesters.

Transmission electron microscopy studies of emulsion-polymerized ABS materials dispersed in polycarbonate are shown in Figure 8.14 (56). Samples were first stained with OsO_4, which makes the rubber appear black, and then with RuO_4, making the SAN appear gray in the TEM photomicrographs. The particular ABS, ABS GRC, is manufactured by Dow Chemical. The polycarbonate, Caliber 200-3, is also a Dow product. ABS GRC contains 50% rubber in the latex, with 25% of acrylonitrile in the SAN portion. The polycarbonate forms the continuous phase for the 90/10 and 70/30 compositions, but appears

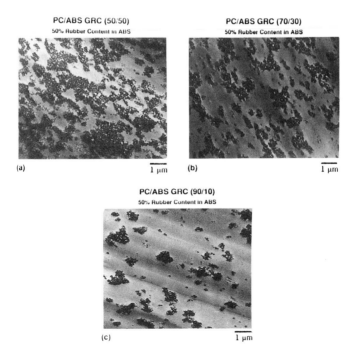

Figure 8.14 Transmission electron micrographs of polycarbonate–ABS blends at three compositions as indicated. Note the agglomeration of the ABS.

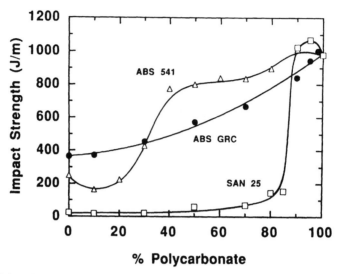

Figure 8.15 Polycarbonate blends with ABS and SAN may show decreases in standard notch Izod impact strength as the ABS or SAN is added.

to form a co-continuous morphology in the 50/50 mixture. The rubber particles appear agglomerated in all three of these blends, with the ABS domains having irregular shapes and jagged boundaries. Again, agglomeration is thought to be caused both by a low degree of grafting between the rubber and the SAN, and SAN–SAN interparticle interactions. In practice, a controlled degree of agglomeration provides technological advantages, such as greater toughness.

Figure 8.15 (56) illustrates the impact resistance of three ABS-type materials in polycarbonate. The SAN 25 contains no rubber and is a copolymer of styrene (75%) and acrylonitrile (25%). The ABS 541 is a mass-produced ABS, made by a bulk process similar to that of HIPS. The bars were 3.18 mm thick, conforming to ASTM D 256. In this thickness range, rubber toughening is used for lowering the ductile–brittle transition temperature range rather than raising the impact resistance. While the SAN is not very effective by itself, the two ABS materials produced high impact strengths at a level of 30% ABS or higher.

While the value of ABS additions to polycarbonate is not obvious from a study of Figure 8.15, a standard notch was used. When a new razor blade was pressed into the center of the machined standard notch (producing a sharp notch), the resistance to impact loading becomes more obvious (Figure 8.16) (56). Thus, the effect of notch radius, termed notch sensitivity, is seen to be very important in determining polycarbonate impact resistance (62).

The temperature also plays an important role. Under impact loading conditions (56), the ductile-to-brittle transition temperature for polycarbonate containing ABS is lowered from $-10°C$ to approximately $-25°C$. Additions of ABS at levels of 10% or higher lowers the ductile–brittle transition temperature to between -50 and $-60°C$, near the T_g of the rubber component.

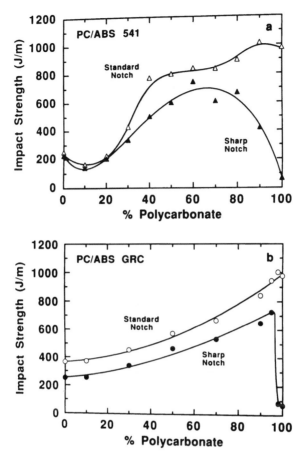

Figure 8.16 Differences between a standard notch and a sharp notch Izod impact strength of poly-carbonate blends with various ABS materials. Polycarbonate is especially sharp notch sensitive.

Importantly, ABS plastics are added to polycarbonates primarily because they allow lower melt viscosities and easier processing. (Thus the molecular weight of the polycarbonate employed can be lower, and still maintain reasonable mechanical properties.) For many commercial compositions, the actual toughening of the already tough polycarbonate homopolymer is minimal. As noted above, polycarbonate is notch sensitive, especially to very sharp notches. Polycarbonate (PC) also loses toughness in thicker sections. ABS and similar toughening elastomer compositions help in both of these situations.

8.5.3. Role of Cavitation in Toughening Polycarbonates

Many polymers suffer from notch sensitivity, that is, they fracture much more easily if a sharp notch is imposed on the sample. Among such polymers are the ductile thermoplastics poly(vinyl chloride), nylon, and polycarbonate.

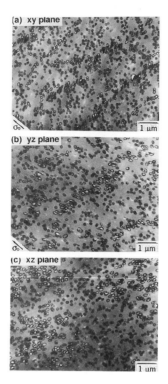

Figure 8.17 Osmium-tetroxide-stained MBS particles in polycarbonate. Note rows of cavitated particles.

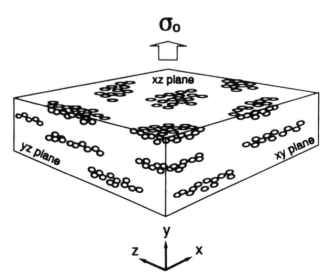

Figure 8.18 Model of the cavitated particle patterns, showing their organization in space.

Under stress, an internal craze initiates ahead of the notch, where the triaxial stresses are highest. The presence of dispersed rubber particles in the matrix provides a cavitation mechanism to relieve such triaxial stresses before an internal craze can initiate or propagate. Thus, an important new toughening mechanism for elastomers in such plastics involves cavitation of the rubber domains. In particular, cooperative cavitation among the rubber particles is very important (63,64). Once the rubber particles cavitate, the ductile matrix deforms in a shear mode. The primary energy absorption mechanism in toughened PC involves shear deformation of the matrix, rather than cavitation of the rubber particles per se.

In the experiments described here, Cheng and co-workers (63,64) used MBS rather than ABS because of its greater adhesion to the polycarbonate. The MBS used was Paraloid EXL-3607® (Rohm and Haas), which is composed of substantially monodisperse particles of 0.2 μm in diameter. The matrix poly-carbonate was Sinvet 251 (EniChem SpA, with a molecular weight of about 32,000 g/mol). A 1-mm radius semicircular notch was used. For electron microscopy, samples were stained with osmium tetroxide, and thin sections were examined in a JEOL 100 SX transmission electron microscope.

At low magnifications, sections at the far notch region of the stress-whitened zone show numerous parallel dark, wavy lines that are oriented approximately perpendicular to the loading direction. At higher magnification, the wavy dark lines were shown to consist of rows of cavitated particles; see Figure 8.17 (63). Cuts in all three planes are shown. The cavitated arrays were thin with a broad range in length. Typically, they incorporated 1–4 particles in the thickness direction and from 8–35 cavitated particles in the length direction. The shapes of these domains are modeled in Figure 8.18 (63). Sometimes the cavitated particles in such arrays exhibited distortion, suggesting that locally, the matrix was highly deformed. Interestingly, most of the particles outside of the arrays were not cavitated, nor were they distorted. For 5 and 10% MBS particles in the polycarbonate, the fraction of particles that cavitated was about 0.22 in the near notch region and about 0.50 in the cavitated arrays.

The size and shape of the cavitated regions depend on the position of the particles in the stress-whitened zone (SWZ), as modeled in Figure 8.19 (64). These regions go from single clumps to long arrays, then back to single clumps at the end of the SWZ.

The mechanism of cavitated array formation was explored via a microme-chanical approach. Cavitated arrays form from strings of particles. The particle separation distance in these strings is of the order of 0.05 μm. The initial event involves cavitation of a single particle in response to the local mean stress field; see Figure 8.20 (64). Cavitation alters the local stress state, with the increased size of the particle being accommodated by formation of a small plastic zone in the matrix. The process is repeated, with extension of the PC ligaments between cavitated particles leading to large local strains.

The finding that strings of particles form cavitation arrays has important implications for the blending process. It suggests that an overmixing of the

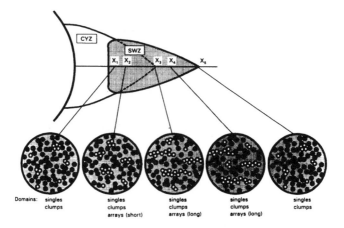

Figure 8.19 Development of the cavitated regions in the stress-whitened zone. Long arrays appear at X_3 and X_4.

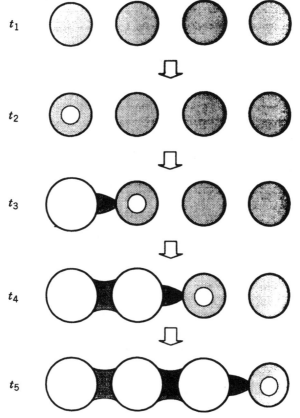

Figure 8.20 Development of cavitation arrays with time.

particles to achieve total randomness is not desired. However, a distribution of rubber particle domain regions, of controlled size and extent of mixing, are important for greatest toughening.

8.5.4. Model Microlayer Composites

The phenomena associated with crazing and shear yielding in rubber-toughened polycarbonates can be modeled with microlayer composites of polycarbonate and SAN (64–66). This micromechanics approach has proved valuable for the understanding of certain aspects of particle cavitation in polymer blends. In particular, the critical cavitation condition and the subsequent effect of particle cavitation on matrix shear yielding (67) are amenable to such analysis.

The experiment involves microlayer composites in the form of coextruded sheets of polymer. The microlayer composites are composed of alternating layers of two immiscible polymers, in this case polycarbonate and SAN. The total number of layers may be in the hundreds, and the thickness of each layer of the order of a few micrometers, with both the total number of layers and the thickness of each layer being variables. The samples were clamped in a micro-tensile tester, and mounted on the stage of an optical microscope, so that the deformation processes could be photographed as the specimen was deformed.

The results are shown in Figure 8.21 (65) for a PC/SAN microlayer composite with individual layers of 5.9/7.9 μm thickness, respectively. The crazes are forming in the SAN layer, for identification of the layers. The corresponding stress–strain curve is shown in Figure 8.22 (65), where the strain is seen to increase up to about 5% for optical micrographs a–d in Figure 8.21.

Initial crazing took place at about 2% strain, position a. The initial deformation event was random crazing throughout the SAN layers. As the strain increased, occasional craze doublets were formed. The formation of such doublets and growth of the doublets to triplets was halted when micro-shearbands appeared at the craze tips (Figure 8.21c). The micro-shearbands coalesced just before yielding and extended across the PC layers while some shearbands propagated through both the PC and SAN layers.

From such experiments, it was concluded that SAN layer thickness was not a factor in formation of arrays and doublets; formation of craze doublets and craze arrays depended only upon PC layer thickness. During deformation, crazing advances primarily from single crazes to multiple crazes to craze arrays as the PC layer thickness decreases to less than about 1.3 μm. The zone that forms in the PC layer consists of a colinear plastic zone, together with a pair of micro-shearbands that grow at an angle of about 45°. When the PC layer thickness is less than 1.3 μm, the length of the colinear plastic zone is comparable to the PC layer thickness; formation of a craze at the point of impingement on the neighboring SAN layer leads to craze arrays with many aligned crazes in neighboring layers. If the layers are thin enough, these micro-shearbands grow through the PC layers and extend into adjacent SAN and PC layers, producing a change in deformation mechanism from craze opening to shear yielding. This

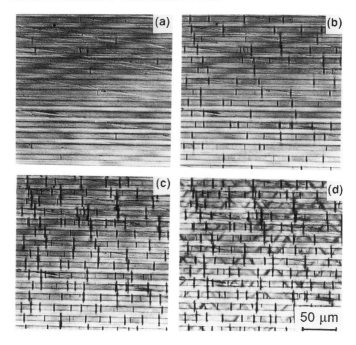

Figure 8.21 Development of crazes, craze doublets, and micro-shearbands with increasing strain (a)–(d) for a microlayer composite of PC/SAN (5.9/7.9 μm).

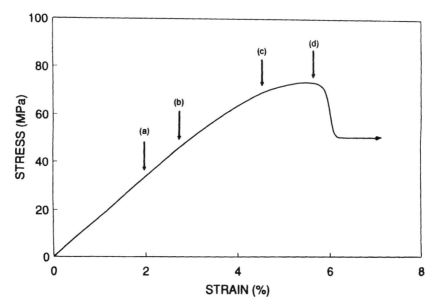

Figure 8.22 Stress–strain curve for a 194-layer PC/SAN (5.9/7.9 μm) composite. Points (a)–(d) refer to strains in Figure 8.21.

change in deformation mechanism of the SAN layers is responsible for the increased ductility and toughness in composites with thinner layers. While these microlayer composites are important in their own right, the length of the colinear plastic zone in the PC gives valuable insight as to the maximum distance between rubber or other particles in PC, to develop improved toughness and impact resistance.

PMMA-based plastics can be toughened by the addition of core-shell or multilayer acrylic latex particles of about 0.3 μm in diameter. See P. A. Lovell and D. Pierre, in *Emulsion Polymerization and Emulsion Polymers*, P. A. Lovell and M. E. El-Aasser, Eds., Wiley, Chichester, 1997.

8.6. COMPATIBILIZED POLYAMIDE BLENDS

The use of maleic anhydride and similar materials for reacting with hydrocarbon polymers for subsequent grafting to polyamides for purposes of compatibilization was already discussed in Section 6.4. Such materials have been extensively used to toughen polyamide 6 and polyamide 66, for example. Common toughening polymers include polypropylene ethylene–propylene–diene monomer (EPDM) and styrene–ethylene–butylene–styrene (SEBS) triblock copolymers.

Commonly, the maleic anhydride is reacted with the hydrocarbon polymer in the presence of a decomposing organic peroxide. This is subsequently reacted with the amine on the polyamide, thus effecting graft copolymer molecules that serve as compatibilizers (68–73). Compatibilization, having the meaning of providing useful interfacial bonding, is considered to occur between the chemical linkage of the anhydride and the polyamide end groups. Both amide and imide graft linkages develop, as shown (72):

$$(8.8)$$

Anhydride   Polyamide amine
end group

Amide linkage

$$\left.\begin{array}{c} H \quad\quad O \\ | \quad\quad \nearrow \\ C{-}C \\ | \quad\quad \searrow \\ | \quad\quad\quad N{+}CH_2{+}_n C{+}N{+}CH_2{+}_x C{+}_{n-1} OH \\ C{-}C \\ | \quad\quad \searrow \\ H_2 \quad\quad O \end{array}\right.$$

Imide linkage

where X runs from one to 11 (usually 5), and n represents the degree of polymerization. Other polyamides react similarly.

Ide and Hasegawa (68), for example, studied maleic anhydride-compatibilized polypropylene as toughening agents for polyamides, finding that the impact strength was increased significantly depending on the quantity of maleic anhydride and the fraction of polypropylene added.

Commercial materials by du Pont are known as Super Tough Nylon, under the Zytel trade mark (71). These materials are among the most impact resistant plastics made so far, see Table 8.1, and energy absorbing materials, see Table 8.2.

Of course, both polypropylene and the polyamides are crystallizable polymers (72). Duvall et al. (73) modeled the morphology of a low-anhydride compatibilizer (LAC) and a high-anhydride compatibilizer (HAC) containing PP; see Figure 8.23 (73). For the LAC (Figure 8.23a), the reacted PP and the PP homopolymer are able to cocrystallize, while for the HAC, cocrystallization is thermodynamically unfavorable, and the two polypropylenes are separated (Figure 8.23b). In both cases, points of chemical linkage to the polyamide are indicated. Of course, in the cases of SEBS and EPDM or EPM, crystallization of the hydrocarbon component does not occur. In each of these cases, toughening depends on the decreased rubber domain size and increased adhesion between the rubber and the polyamide, both augmented by the presence of the maleic anhydride (or similar) grafting reaction.

Oshinski et al. (74) studied SEBS triblock copolymers, partly grafted with maleic anhydride. As the fraction of the maleated material increased from zero to 100%, the rubber domain sizes decreased from about 4 μm to about 0.04 μm, a factor of 100! The optimum Izod impact resistance was with a rubber domain size of about 0.4 μm, yielding up to 1000 J/m values.

Bucknall et al. (75) investigated the mechanical behavior of ethylene–propylene (EP) rubbers in nylon 66. In stress–strain experiments, they showed that there was up to a 6% increase in the sample volume with extension. This they ascribed to the formation of voids (cavitation) in the rubber particles, leading to fibrillation of the nylon matrix at high strains.

In almost all of the above experiments, part of the elastomer was treated with maleic acid anhydride, providing graft sites for reaction with the amine group of the nylon. This reacted portion of the rubber was then blended with a larger portion of unreacted rubber of the same type. This blend, mostly forming a

COCRYSTALLIZATION MODEL

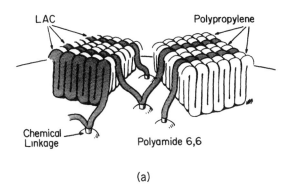

(a)

SEPARATE CRYSTALLIZATION MODEL

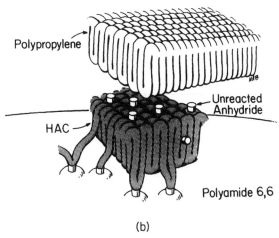

(b)

Figure 8.23 Schematic representation of co-crystallization of polyamide 6,6 with polypropylene. (a) Conditions favor co-crystallization, but (b) conditions favor separate phase crystallization.

singe phase, is then further blended with the nylon under reactive processing conditions. By this route, optimum interfacial bonding, domain size, and toughening can be achieved.

8.7. RUBBER-TOUGHENED EPOXIES

8.7.1. General Characteristics of Epoxy Resins

Epoxy resins are crosslinked, glassy polymers, most often prepared from the diglycidyl ether of bisphenol A (DGEBPA). The structure of the monomer

DGEBPA is

$$CH_2-CH-CH_2-O \underset{\diagdown O \diagup}{} \left\langle\!\!\bigcirc\!\!\right\rangle - \overset{\overset{\textstyle CH_3}{|}}{\underset{\underset{\textstyle CH_3}{|}}{C}} - \left\langle\!\!\bigcirc\!\!\right\rangle - O-CH_2CH-CH_2 \quad (8.9)$$

showing two epoxy groups, one at each end. The DGEBPA monomer is often cured with agents such as triethylenetetramine or N-aminoethylpiperazine, or anhydrides such as phthalic anhydride (76). The choice of resin and curing agent depends on the application and on required handling characteristics, such as viscosity, pot life, and gel time, the curing temperature and time, and the use properties such as mechanical, chemical, thermal, or electrical. The glass transition temperature of epoxies depends on the choice of resin and curing agent also, but commonly ranges from 40 to 120°C, most often near the upper end of this range for tough and impact resistant plastics, and near the lower end for adhesives.

(The reader will note the common use of archaic nomenclature common in the field of epoxies. Resin translates as polymer, epoxy translates as oxirane, and curing agent as crosslinker.)

Epoxies are generally brittle plastics with K_{IC} in the range of 0.4–1.2 $MNm^{-3/2}$ (77). Izod impact values are shown in Table 8.1. The work of fracture, G_{IC}, varies from 50 to 200 J/m^2, for many brittle plastics (78,79). Values of G_{IC} provide important engineering design information.

8.7.2. Types of Rubber-Toughened Epoxies

There are two general types of elastomers that are used to toughen epoxies. The older and more important commercially is based on carboxy-terminated butadiene nitrile (CTBN) and similar compositions,

$$HO-\overset{\overset{\textstyle O}{\|}}{C}(CH_2-CH=CH-CH_2)_n(CH_2-CH)_m\overset{\overset{\textstyle O}{\|}}{\underset{\underset{\textstyle CN}{|}}{C}}-OH$$

These polymers are synthesized by anionic polymerization with a dianion living polymer initiator. The polymerization is terminated with carbon dioxide, to yield carboxyl groups at each end of the chain. The CTBN rubber is in a prepolymer form, with molecular weights of the order of a few to several thousand. Usually, the CTBN is mixed with the DGEBA under controlled conditions, often preheated to form graft copolymers, and then the total mass is cured. The use of CTBN elastomers to toughen epoxies can sometimes lead to a 10-fold or more increase in G_{IC}. McGarry and co-workers (80,81) were the

Table 8.4 Fracture toughness values of some rubber-modified epoxies[a]

Formulation	Particle Size (μm)	K_{IC} (MPa m$^{1/2}$)	G_{IC} (Jm^{-2})
DGEBA/PIP	0 (neat resin)	0.80	180
DGEBA/ PIP/CTB-162	$\leq$200	1.10	410
DGEBA/PIP/CTB/CTBN	1–2, 100–200	1.95	1275
DGEBA/PIP/CTBN-8	1–2	2.10	1440
DGEBA/PIP/CTBN-31	1–2, 10–20	2.00	1300
DGEBA/PIP/MBS	~0.2	2.90	2725
DGEBA/PIP/MBS/CTBN	0.2, 1–2	2.75	2465

[a] All rubber-modified epoxies contain 10 phr rubber.

first to explore this field in the late 1960s, incorporating rubber particles of approximately 1 μm.

The newer types are based on latexes, often of the core–shell type. Here, the elastomer forms the core, while the shell provides protection for the core and sites for bonding to the matrix. The latex is then blended into the unreacted epoxy system, forming a dispersion. Of course, proper interfacial bonding between the shell polymer and the epoxy monomer mix is essential. Both of these types will be explored below. Of course, epoxies are often used in composites as well, in particular with continuous glass fibers.

8.7.3. Effect of Particle Size and Size Distribution

Pearson and Yee (82) explored the use of both CTBN elastomers and MBS latex particles in toughening epoxies; see Table 8.4. The epoxy was based on DER 331 resin, basically a DGEBA sold by Dow Chemical Co. The "liquid rubbers" were based on Hycar CTBN 1300x8 and others, sold by B. F. Goodrich Co. The MBS latex particles were Acryloid® KM 653, sold by Rohm and Haas. The curing agent was piperidine. Note that the last composition in Table 8.4 contains both kinds of particles. Pearson and Yee (82) remark that the CTBN by itself often forms a bimodal distribution of particle sizes. This may be due to phase separation characteristics: the large particles were phase separated from the start, and the smaller particles separated during the polymerization due to the increased immiscibility of the two components.

The fracture toughness appears to increase with decreasing particle sizes. The MBS particles, at 0.2 μm, second from the bottom in Table 8.4, provide a 15-fold increase in G_{IC}. One explanation of the results is that large particles tend to span the two crack surfaces, whereas small particles cavitate in a process zone in the vicinity of the crack tip. Merely spanning the two crack surfaces only provides a small energy absorption mechanism, while cavitation, as will be described below, relieves stresses triaxially.

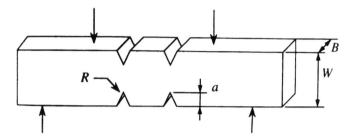

Figure 8.24 Schematic of a symmetrical double-edge double-notched special, where $W = 19.1\,\text{mm}$, $B = 6.3\,\text{mm}$, $a/W = 0.25$, and $R = 0.5$ or $1.5\,\text{mm}$.

8.7.4. Cavitation in Elastomer-Toughened Epoxies

The most important mechanism in rubber toughening of epoxies is now thought to be cavitation, which attributes the fracture toughness increase to cavitation of the CTBN or latex rubber particles and the induced dilational deformation and shear yielding in the surrounding epoxy matrix. In support of this mechanism, Li and co-workers (83) showed that cavitation in CTBN-rubber-modified epoxies could be suppressed by the application of hydrostatic pressure. Upon the application of 30–38 MPa, relatively low pressures, the rubber particles were unable to induce massive shear yielding in the epoxy matrix, and the fracture toughness was no higher than that of the unmodified epoxy.

Yee et al. (84) used a four-point bending test to examine their epoxy materials; see Figure 8.24. The advantage of this sample geometry is that it can provide both fracture and subfracture information. It also allows the comparison of the behavior of the rubber particles in tensile and compressive regions.

The cavitation of the rubber particles was studied via scanning electron microscopy; see Figure 8.25 (84). These workers noted that rubber cavitation

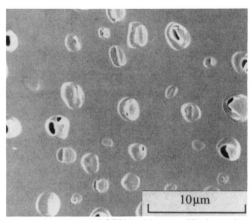

Figure 8.25 Scanning electron microscopy of CTBN rubber-modified epoxies, showing cavitation of a microtomed suface.

precedes plastic yielding. At room temperature and at moderate strain rates, the CTBN appears to cavitate well before shear yielding in the matrix. However, in MBS-type core–shell latex particles, cavitation occurs just before matrix shear yielding. It was also observed that the cavitation zone size decreased as the rubber content increased. Also, the elastic-plastic boundary, where the mean stress is a maximum, is closer to the notch tip as the rubber concentration increases. An alternative explanation suggests that the effect could be due to a change of the fracture mode from plane strain to plane stress.

A model of the deformation mechanisms is illustrated in Figure 8.26 (84). The following sequence of microscopic events seems to be taking place: (a) The rubber particles cavitate after a critical mean stress value is reached; (b) these cavitated rubber particles can no longer support load, hence constraint relief is

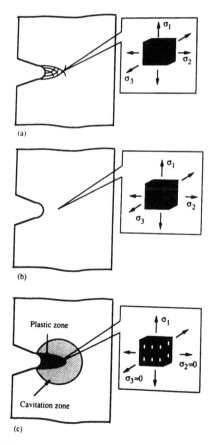

Figure 8.26 Modeling of the deformation mechanisms in blunt-notched epoxy specimens, pure and containing CTBN. (a) Pure epoxy. Triaxial tensile stresses build up in front of the notch. (b) CTBN-containing epoxy. Before rubber particle cavitation, behavior is similar to pure epoxy. (c) CTBN-containing epoxy after rubber particle cavitation. The constraints in the transverse directions are relieved, and the matrix epoxy deforms in the plane stress state.

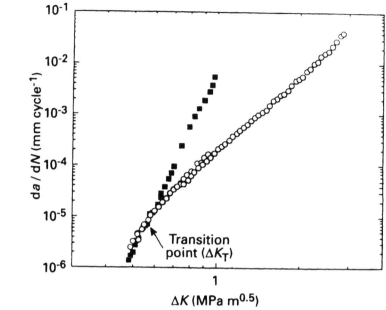

Figure 8.27 Fatigue propagation behavior of epoxy polymers. Solid squares: unmodified. Open circles: 10% CTBN. Note that at low crack propagation rates, the rubber does not increase crack resistance.

effected within the material; and (c) with a sufficiently large number of cavities formed, constraint relief causes the stress state to change toward a plane stress state.

Azimi et al. (85) studied the fatigue crack propagation of similar materials. They found that at relatively low crack driving forces, ΔK, the crack growth rate, da/dN, was nearly the same for CTBN-toughened materials as for unmodified epoxy; see Figure 8.27 (85). Below a transition value, ΔK_T, these particles are not effectively interacting with the crack tip. When the size of the plastic zone at the crack tip is large enough for the rubber particles to be engulfed in the plastic zone, they begin to interact with the crack tip stress field, and the toughening mechanisms become active. According to this hypothesis, smaller particles should result in higher fatigue crack resistance. Preliminary results on epoxies modified with MBS particles, at $0.2\,\mu m$, show that the fatigue crack growth resistance of these materials in the threshold regime is improved by one order of magnitude when compared to that of epoxies toughened by $3\,\mu m$ CTBN rubber particles (86).

Several mechanisms responsible for toughening are illustrated in Figure 8.28 (85). The most commonly observed mechanisms include localized shear yielding, which refers to shear banding in the matrix occurring between the rubber particles, cavitation in the rubber matrix, and rubber particle bridging behind the crack tip. Other mechanisms, thought not so important here, include rubber particle stretching, microcracking, and crack deflection.

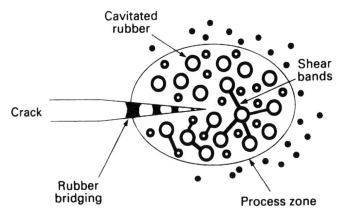

Figure 8.28 Illustration of several toughening mechanisms known to take part in rubber-toughened epoxy polymers.

8.8. MODELS FOR PARTICLE CAVITATION

The preceding sections illustrate that cavitation plays a key role in the rubber toughening of many plastics. This includes both thermoplastics and thermosets, commodity polymers and engineering plastics. Lazzeri and Bucknall (87) recently developed a new treatment of rubber particle cavitation based on an energy balance approach. The underlying principle states that the particle will cavitate when the energy released during cavitation is greater than the energy needed to form the void. The energy stored within the particle itself was emphasized.

Bucknall et al. (88) considered the effect of introducing additional terms for the energy input from the matrix and surroundings, and examined the relationship between cavitation resistance and the morphological and mechanical behavior of the rubber, as predicted by the model. Basic assumptions emphasized that the largest defects within a typical rubber particle are microvoids with dimensions of the order of a nanometer, and that these microvoids expand only if the resulting release of stored strain energy is sufficient to increase the surface area of the void, and to stretch the surrounding layers of rubber biaxially. It is important that the rubbers not only have very low shear moduli but also high bulk moduli relative to the plastic they are toughening.

While cavitation in itself cannot be regarded as an important energy-absorbing process, it reduces the resistance of the polymer to volumetric expansion in response to the dilational applied stress fields that occur, especially at crack tips. Depending on the loading conditions, particular particle sizes provide optimum void formation. The reduced void size, r/R, ratio of void to rubber domain size, varies from about 0.1 to about 0.4, depending on the amount of energy supplied to the system.

Thus, if the plastic contains 20% of elastomer, each particle of which increases in volume by 30%, the volume of the sample (or that portion of the sample under stress) will increase in volume by 6%. This allows relief of the triaxial stresses developed in the material before catastrophic failure.

8.9. BLENDS OF DIFFERENT POLYETHYLENES

Today, polyolefins need not be the simple homopolymers or copolymers as was known in the past. While straight linear polyethyene is known as high-density polyethylene (HDPE), a wide variety of copolymers are prepared today (89). These are made via catalytic copolymerization of ethylene with a comonomer such as butene-1 or hexene-1, commonly called linear low-density polyethylenes (LLDPE). These materials must be contrasted with free-radical-initiated homopolymerization of ethylene, called low-density polyethylene (LDPE), which contains short alkyl branches as well as long branches. In general, the more branches the polymer has, long or short, the lower the density is. This is a result of the lower degree of crystallinity of the final product. Some of the nomenclature and blending characteristics of these materials were treated in Section 2.3.

Polyethylene blends, such as higher and lower density materials, are not impact resistant as such. However, they are widely used as films for packaging, especially for plastic shopping bags, because of their high tear strength. They are also quite puncture resistant. The modulus and mechanical properties can be engineered to meet a wide variety of needs via such blending.

Important questions relating to these materials involves their phase relations. There are two aspects: mixing in the melt state and whether given compositions will cocrystallize. Some of the methods used to examine the melt state include optical and electron microscopy, differential scanning calorimetry (DSC), small-angle neutron scattering (SANS), and small-angle X-ray scattering (SAXS). These experiments indicate that HDPE and LDPE mixtures form a single phase for all concentrations (90). By contrast, highly branched LLDPE, with more than eight branches per 100 backbone carbons, can phase separate in the melt (91).

In the crystalline state, the fraction of cocrystallization that takes place depends on both the cooling rate and the composition of the components. Rapid cocrystallization encourages cocrystallization. Also, high concentrations of the linear component in HDPE and LLDPE allow for more complete cocrystallization, as indicated by single melting behavior and the inability to extract the pure copolymer at its dissolution temperature (92). The extent of segregation of the crystals is driven by the crystallization kinetics of each of the components, the linear materials crystallizing more rapidly, and finally melting at a higher temperature. The amorphous portions of the material remain homogeneous. Again, technologically important issues are the exact extent of crystallization, as well as the size and perfection of the crystals, all taking part in determining the modulus and mechanical behavior of the final film. It may be that two kinds of crystals, differing in size or perfection, and hence melting temperature, contribute to such properties as high tear strength.

As a side issue, those materials that crystallize in separate crystal states, but mix in the amorphous portions, constitute a special type of thermoplastic IPN; see Chapter 10.

8.10. APPLICATIONS OF RUBBER-TOUGHENED PLASTICS

8.10.1. Selected Trade Names

Rubber-toughened plastics come in many kinds and grades. At the bottom of the line are materials such as polystyrene with several percent of rubber mechanically blended in. HIPS and ABS form a kind of staple material, relatively inexpensive, yet having reasonable properties. Higher up the line are such materials as ABS-toughened polycarbonate and HIPS-toughened poly(2,6-dimethyl-1,4-phenylene oxide), the latter sometimes known as poly (phenylene ether). Obviously, properties other than impact resistance need to be factored into any application, as well as cost. Table 8.5 summarizes some of the more important trade names and applications (89–91). Many of the products have applications in the transportation industry, ranging from automotive instrument panels to aircraft seating. Another class of applications includes business equipment housings, such as computers, other office appliances, and telephones. Sporting goods and bicycles are also important. Interestingly, HIPS is now being used for food packaging, including plastic lids for food containers.

8.10.2. Selected Patents

Rubber-toughened plastics constitute a major fraction of polymer patents (see Table 1.3, Chapter 1). Some patent literature in the field of rubber-toughened plastics includes the use of ABS in polycarbonates. The first U.S. patent was issued in 1964, to Grabowski (92). This was followed by a number of patents covering novel compositions, processes, and applictions (93–98). For example, Henton and O'Brian (99) examined the role of acrylic agglomerating agents on the behavior of the final ABS. The particle size distribution in elastomeric latex preparations was advantageously controlled and improved by treatment with such agglomerating agents. Blends of polycarbonate with poly(ethylene terephthalate) (PET) or poly(butylene terephthalate) (PBT) and an elastomer have also been described (100).

The field of rubber-toughened epoxies is also replete with patents. Of course, epoxies are usually sold in the monomer or prepolymer stage. Table 8.5 delineates some of the commerical products. While large amounts of epoxy are used without rubber toughening, the rubber, where used, is usually added just before the curing reactions. While CTBN elastomers such as Hycar® of B. F. Goodrich are often used, other elastomers are frequently employed.

Table 8.5   Rubber-toughened plastics: selected trade names and compositions

Trade Name	Company	Composition	Applications
HIPS			
Limera®	Dainippon Ink and Chemicals	PS/Elastomer	Automative, electrical equipment, applicances, food packaging, business machines, lids
Styron®	Dow Chemical	PS/PB	
Polystyrol®	BASF AG		
Styroplus®	BASF AG		
ABS			
Lacqram®	ATO	PB/SAN	Applicances, automotive, pipes, telephones, business machine housings
Teluran®	BASF		
Novodur®	Bayer		
Magnum®	Dow Chemical		
Cycolac®	G.E.		
Lustran ABS®	Monsanto		
Urtal®	Montedison		
Toughened Polycarbonates			
Arloy®	Arco	PC/P(S-co-MA)	Automotive instrument panels
Bayblend®	Miles, Inc.	PC/ABS	Automotive, vacuum cleaners, computer and business equipment,
Cycloy®	G.E.		
Lupoy®	LG Chemical, Ltd.		
Pulse®	Dow Chemical		
Xenoy®	G.E.	PC/PBT/EPDM	automotive and transportation industries
Toughened Nylons			
Akuloy®	Du Pont	PA-6 or PA-66/PP or EPDM	Automotive underhood, pump housings, and gears
Zytel®	Du Pont	PA-6/ionomer or PA-6 or PA-66/elastomer	Outdoor applications
Celanese Nylon®	Hoechst-Celanese Engineering Resins	PA-66/TPU	Automotive applications
Poly(vinyl chloride)			
Hostalite Z®	Hoechst	PVC/CPE (30%)	Roofing, wire and cable, hoses
Daisolac®	Osaka Soda		
Elasten®	Showa Denko		
Levepren®	Bayer	PVC/EVA	Window frames
Elvaloy®	Bayer		
Vyna®	USI		

Table 8.5   *(continued)*

Trade Name	Company	Composition	Applications
Hycar	B. F. Goodrich	PVC/NBR	Wire and cable,
Krynac	Polysar		hose and belting
Rubber-Toughened Epoxies			
DER	Dow Chemical	Epoxy/CTBN	Coatings, printed
Epoxin	BASF	or	circuit boards,
Levepox	Bayer	Epoxy/Elastomer	encapsulation, or
Araldit	CibaGeigy		structural plastics
Hostapox	Hoechst		
Epotuf	Reichhold		
Epon	Shell		
Epikote	Shell		
Polyolefins			
Ferro Flex	Ferro Corp.	PP/rubbery olefinic polymers	Automotive and appliance
Hostalen GC	Hoechst	HDPE/LDPE	Photographic paper, transportation crates
Kelburon	DSM	PP/EPDM	Bumpers, suitcases
Paxon Pax Plus	Allied	PE/Elast.	Film applications
Unsaturated Polyester–Styrene Resins			
MR 13006	Aristech Corp.	Typical: polypropylene	Automotive, boat hulls
Aropol	Ashland Chemicals	glycol, phthalic anhydride, maleic	
Freefix	DSM Resins BV	anhydride, 35–45% styrene[a], often	
NPG	Eastman Chemicals	reinforced with glass fibers	
Other			
Lumax	LG Chemical, Ltd.	PBT/ABS	Electronic components, automotive, office, appliances
Noryl	GE Plastics	PPO/HIPS	Electronic components
Rynite	Du Pont	PET/elastomer	Furniture, motor housings, bicycles
SBRE	Dow	PC/Polyester	Lawn mower covers
Prevail	Dow Chemical	ABS/PU	Fascia

[a] The styrene polymerizes across the double bonds of the unsaturated polyester, creating a tightly crosslinked network of the AB-crosslinked polymer type.

For example, core–shell polybutadiene-based elastomers with styrenic or methacrylic shells have been employed (101). Dispersed acrylic elastomers can also be used (102,103). Since epoxies are crosslinked, IPNs or semi-IPNs will also be formed on the addition of a second polymer. For example, systems have been prepared by dissolving thermoplastics such as polysulfones, polyethersulfones, or polyetherimides into an epoxy resin and subsequently polymerizing the solution under conditions such that the epoxy resin phase separates from the thermoplastic during polymerization and crosslinking (104–106).

It must be pointed out that except for a very few commodity polymer applications, nearly all commercial polymer materials do not use chemically "pure" polymers. Additives range from simple antioxidants to plasticizers, these usually forming one phase with the matrix material. This chapter has emphasized the use of judicial amounts of rubber to toughen the material.

8.11. LATEX BLENDS

Latex blends are defined as combinations of two or more chemically distinct types of latex particles. Since latexes are usually composed of particles only a few thousands of Ångstroms in size, latex blends will abound in interior surface areas.

Frequently, latex blends are prepared by selecting polymers of different modulus at use temperature, resulting in bumpy compositions. Recent research has emphasized atomic force microscopy to study the morphology of films made from these materials. For example, Patel et al. (107) studied blends of harder and softer latex particles, finding that in blends where the soft component is present in amounts larger than 40%, smoothed bumps were observed which appear larger than either the hard or the soft particles. The smoothness of each bump, supported by other evidence, suggests that the soft particles coalesced into a continuum at the surface. The underlying hard particles cause the overall surface unevenness. These hard particles were submersed because of the lower surface energy of the soft polymer.

Gilicinski and Hegedus (108) identified hard and soft domains by supplementing atomic force microscopy topographic data with mechanical property maps. These reflected phase identity in the blend coatings. Geurts et al. (109) used minimum film formation temperatures to study blends of two monodisperse sizes via atomic force microscopy. Usually, the addition of increasing amounts of small particles into larger particles leads to a disruption of the particle matrix until the point at which the matrix changes from a continuous matrix of large particles into a continuous matrix of small particles, a phenomenon easily studied via atomic force microscopy.

A major application of these latex blends is in the coatings industry. A mixture of hard and soft latex particles provides a film with a higher modulus than otherwise easily achievable. However, a wide range of patents have been issued for latex blends. Poly(vinyl chloride)/poly(vinyl acetate) latex blends

(110) containing 0.3–0.5% of (meth)acrylic acid and a fraction of hydroxy-methyl diacetone acrylamide were found to confer improved wet primed adhesion, for example. More recently, styrene–acrylic and SBR latex blends were suggested for toner aggregation processes (111). Combinations cf SBR and NBR latexes have been patented for use in foam rubber backings (112). In each case, the two latexes complement each other in some way, such as hardness–adhesiveness and so on.

8.12. MODERN VIEW OF TOUGHENING PLASTICS

Michler (113) summarized the micromechanical mechanisms of toughening polymers as observed in the electron microscope. He divided these into three groups. In group I, basically homopolymers and one-phased copolymers, there were the stretching of molecular chains during early stages of crack growth, microyielding, craze formation, shear bands, and plastic zone development.

In group II, various materials are added to the plastic to increase toughness. This involves the initiation of a large number of microcracks by inorganic particles, short fibers, or similar materials. If the particles are ductile, they may be plastically stretched across the crack tip, the so-called bridging mechanism. If the particles are rubbery, the main focus of the present chapter, there is the initiation of a large number of small plastic zones (crazes or shear bands), induced by stress concentration at soft structural heterogeneities (the rubber particles). Cavitation may occur either inside or at the surface of the particles, with subsequent stretching of the polymeric ligaments between the microvoids.

Group III mechanisms typically involve the propagation of the crack into a local volume of higher strength, for example, oriented chains, crystalline regions, particles or fibers with a higher strength, and the like. The crack tip may be blunted by rapid relaxation processes or by propagating the crack into a weak second phase, another microcrack, a craze, or a shear band. Alternately, the stress at the crack tip may be reduced by propagation of the crack into a locally unloaded region between other cracks or near strong fibers. Other mechanisms that prevent rapid crack propagation utilize changing the crack propagation direction at phase boundaries or other structural details, or by thermal effects at the crack tip.

Some of the major properties of rubber-toughened plastics may be summarized:

1. The toughening elastomer should have a glass transition temperature at least 60°C below ambient. For room temperature, this means below about $-40°C$.

2. The domain size must be of the order of a fraction of a micrometer. For many materials, the optimum diameter seems to be about 0.3 μm.

3. The spacing between the rubber particles should be of the order of 1–5 μm. This allows for both reacceleration of the craze or shearing

mechanism before the next particle is encountered, and maximizing the number of particles. Also, the microfailure mechanism may change if the distance between the particles is too large. Proper particle spacing encourages cooperative cavitation.

4. The modulus of the elastomer must be low enough that it can cavitate easily. Since most types of such elastomer particles are crosslinked, care must be taken to achieve high enough crosslinking so that the particles do not deform unduly during processing, but are able to cavitate under externally imposed stresses later.

5. There must be good adhesion between the elastomer particle and the plastic. Since the mechanism of cavitation around the particle (as well as within the particle) is often important, there may be an optimum to the interfacial bonding level. Adhesion may be brought about through chemical bonding, such as grafting, or through the use of complex particles, such as core–shell latexes.

6. Large rubber volumes produce greater toughness. Clearly, the modulus of the material decreases with large amounts of rubber. However, incorporating plastic domains within the elastomer particle apparently counts toward the rubber volume, while not decreasing the modulus.

7. Multiple sizes of rubber particles, or mixtures of different kinds of particles (glass beads or fibers, etc.), allow for alleviating different kinds of stresses, such as impact stresses, stress–strain deformations, or fatigue.

8. Clearly, industrial needs also involve financial criteria. The choice of elastomer type and concentration may be governed by the cost as well as the intended use.

REFERENCES

1. J. W. Aylsworth, U.S. Pat. 1,111,284 (1914).
2. L. H. Sperling, *Polym. News*, **12**(12), 332 (1987).
3. I. Ostromislensky, U.S. Pat. 1,613,673 (1927).
4. M. Matsuo, *Jpn. Plastics*, **2**(July), 6 (1968).
5. J. L. Amos, J. L. McCurdy, and O. R. McIntire, U.S. Pat. 2,694,692 (1954).
6. A. B. Finestone and R. C. Westphal, U.S. Pat. 3,781,283 (1973).
7. D. E. Carter and R. H. M. Simon, U.S. Pat. 3,903,202 (1975).
8. A. Finberg, U.S. Pat. 3,644,587 (1972).
9. J. L. Amos, *Polym. Eng. Sci.*, **14**, 1 (1974).
10. M. E. Soderquist and R. P. Dion, *Encyclopedia of Polymer Science and Engineering*, Vol. 16, 2nd ed., J. I. Kroschwitz, Ed., Wiley, New York, 1989, p. 88.
11. V. Mishra, D. A. Thomas, and L. H. Sperling, *J. Polym. Sci., Part B: Polym. Phys. Ed.*, **34**, 2105 (1996).
12. V. Mishra, F. E. Du Prez, E. J. Goethals, and L. H. Sperling, *J. Appl. Polym. Sci.*, **58**, 347 (1995).

13. L. H. Sperling and V. Mishra, *Polym. Adv. Technol.*, **7**, 197 (1996).

14. S. R. Jin, J. M. Widmaier, and G. C. Meyer, *Polymer*, **29**, 346 (1988).

15. A. A. Donatelli, D. A. Thomas, and L. H. Sperling, in *Recent Advances in Polymer Blends, Grafts, and Blocks*, L. H. Sperling, Ed., Plenum, New York, 1974.

16. M. Fels and R. Huang, *J. Appl. Polym. Sci.*, **14**, 537 (1970).

17. H. A. J. Battaerd and G. W. Tregear, *Graft Copolymers*, Interscience, New York, 1967.

18. E. R. Wagner and R. J. Cotter, *J. Appl. Polym. Sci.*, **15**, 3043 (1971).

19. J. A. Manson and L. H. Sperling, *Polymer Blends and Composites*, Plenum, New York, 1976.

20. M. L. Lu and F. C. Chang, *J. Appl. Polym. Sci.*, **56**, 1065 (1995).

21. A. J. Kinloch and R. J. Young, *Fracture Behavior of Polymers*, Applied Science, Essex, England, 1983.

22. C. B. Lee, M. L. Lu, and F. C. Chang, *J. Appl. Polym. Sci.*, **47**, 1867 (1993).

23. E. Plati and J. G. Williams, *Polym. Eng. Sci.*, **15**, 470 (1975).

24. R. F. Boyer and H. Keskkula, in *Encyclopedia of Polymer Science and Technology*, 1st ed., Vol. 13, N. Bikales, Ed., Wiley, New York, 1970, p. 128.

25. H. Keskkula, M. Schwarz, and D. R. Paul, *Polymer*, **27**, 211 (1986).

26. R. A. Bubeck, D. J. Buckley, Jr., E. J. Kramer, and H. R. Brown, *J. Mat. Sci.*, **26**, 6249 (1991).

27. R. C. Cieslinski, *J. Mat. Sci. Lett.*, **11**, 813 (1992).

28. D. Henton, seminar at Lehigh University, April, 1995.

29. H. Keskkula, private communication, April 28, 1995.

30. G. H. Fremon and W. N. Stoops, U.S. Pat. 3,168,593 (1965).

31. D. M. Kulich, P. D. Kelley, and J. E. Pace, in *Encyclopedia of Polymer Science and Engineering*, 2nd ed., J. I. Kroschwitz, Ed., Vol. 1, Wiley, New York, 1985, p. 388.

32. C. W. Childers and C. F. Fisk, U.S. Pat. 2,820,773 (1958).

33. W. C. Calvert, U.S. Pat. 3,238,275 (1966).

34. A. A. Collyer, Ed., *Rubber Toughened Engineering Plastics*, Chapman & Hall, London, 1994.

35. T. A. Clayton, Br. Pat. 649,166 (1951).

36. S. A. Harrison and W. E. Brown, U.S. Pat. 2,614,089 (1952).

37. K. Kato, *Jpn. Plastics*, **2**(April), 6 (1968).

38. Y. Aoki, *Soc. Rheol. Jpn.*, **7**, 20 (1979).

39. Y. Aoki, *J. Rheol.*, **25**, 351 (1981).

40. Y. Aoki and K. J. Nakayama, *J. Soc. Rheol. Jpn*, **9**, 39 (1981).

41. Y. Aoki and K. Nakayama, *Polym. J.*, **14**, 951 (1982).

42. T. Masuda, A. Nakajima, M. Kitamure, Y. Aoki, N. Yamauchi, and A. Yoshioka, *Pure App. Chem.*, **56**, 1457 (1984).

43. Y. Aoki, *J. Non-Newt. Fluid Mech.*, **2**, 91 (1986).

44. Y. Aoki, *Macromolecules*, **20**, 2208 (1987).

45. A. Echte, in *Rubber-Toughened Plastics*, C. K. Riew, Ed., Adv. in Chem. Ser. No. 222, ACS Books, Washington, DC, 1989, p. 48.

46. L. Morbitzer, D. Kranz, G. Humme, and K. H. Ott, *J. App. Polym. Sci.*, **20**, 2691 (1976).

47. R. N. Haward and J. Mann, *Proc. R. Soc.*, **282A**, 120 (1964).

48. D. A. Lannon and E. J. Hoskins, in *Physics of Plastics*, P. D. Richie, Ed., Plastics Institute, London, 1965.

49. C. B. Bucknall and D. G. Street, *Advances in Polymer Science and Technology*, Soc. Chem. Ind. Monograph No. 26, London, 1967, p. 272.

50. C. B. Bucknall, *Toughened Plastics*, Applied Science, London, 1977.

51. G. H. Hoffman, in *Polymer Blends and Mixtures*, D. J. Walsh, J. S. Higgins, and A. Maconnachie, Eds., Martinus Nijhoff, Dordrecht, The Netherlands, 1985.

52. J. C. Stevenson, *PMAD* (Polymer Modifiers and Additives Division SPE Newsletter), **12**(2), 11 (1994).

53. C. M. Gruber and J. R. Patterson, *J. Vinyl Technol.*, **11**(1), 23 (1989).

54. H. Breuer, F. Haff, and J. Stabenow, *J. Macromol. Sci., Phys.*, **B14**, 387 (1977).

55. A. Tse, E. Shin, A. Hiltner, E. Baer, and R. Laakso, *J. Mater. Sci.*, **26**, 2823 (1991).

56. B. S. Lumbardo, H. Keskkula, and D. R. Paul, *J. Appl. Polym. Sci.*, **54**, 1697 (1994).

57. L. A. Utracki, *Polymer Alloys and Blends*, Hanser, Munich, 1990, pp. 7–8, 275–281.

58. D. Freitag, U. Grigo, P. R. Muller, and W. Nouvertne, *Encyclopedia of Polymer Science and Engineering*, 2nd ed., Vol. 11, J. I. Kroschwitz, Ed., Wiley, New York, 1988, p. 648.

59. T. A. Callaghan, K. Takakuwa, D. R. Paul, and A. R. Padwa, *Polymer*, **34**, 3796 (1993).

60. V. Janarthanan, R. S. Stein, and P. D. Garrett, *J. Polym. Sci., Polym. Phys. Ed.*, **31**, 1995 (1993).

61. J. D. Keitz, J. W. Barlow, and D. R. Paul, *J. Appl. Polym. Sci.*, **29**, 3131 (1984).

62. T. W. Cheng, H. Keskkula, and D. R. Paul, *J. Appl. Polym. Sci.*, **45**, 531 (1992).

63. C. Cheng, A. Hiltner, E. Baer, P. R. Soskey, and S. G. Mylonakis, *J. Appl. Polym. Sci.*, **55**, 1691 (1995).

64. C. Cheng, A. Hiltner, E. Baer, P. R. Soskey, and S. G. Mylonakis, *J. Mater. Sci.*, **30**, 587 (1995).

65. D. Haderski, K. Sung, J. Im, A. Hiltner, and E. Baer, *J. Appl. Polym. Sci.*, **52**, 121 (1994).

66. K. Sung, D. Haderski, A. Hiltner, and E. Baer, *J. Appl. Polym. Sci.*, **52**, 147 (1994).

67. A. Lazzeri and C. B. Bucknall, *J. Mater. Sci.*, **28**, 6799 (1993).

68. F. Ide and A. Hasegawa, *J. Appl. Polym. Sci.*, **18**, 963 (1974).

69. S. Hosoda, K. Kojima, Y. Kanda, and M. Aoyaki, *Polym. Networks Blends*, **1**, 51 (1991).

70. S. J. Park, B. K. Kim, and H. M. Jeong, *Eur. Polym. J.*, **26**, 131 (1990).

71. F. M. Sweeney, Ed., *Polymer Blends and Alloys: Guidebook to Commercial Products*, Technomic, Lancaster, PA, 1988, pp. 206–220.

72. J. Duvall, C. Sellitti, C. Myers, A. Hiltner, and E. Baer, *J. Appl. Polym. Sci.*, **52**, 195 (1994).

73. J. Duvall, C. Sellitti, C. Myers, A. Hiltner, and E. Baer, *J. Appl. Polym. Sci.*, **52**, 207 (1994).

74. A. J. Oshinski, H. Keskkula, and D. R. Paul, *Polymer*, **33**, 268 (1992).

75. C. B. Bucknall, P. Heather, and A. Lazzeri, *J. Mater. Sci.*, **24**, 2255 (1989).

76. L. V. McAdams and J. A. Gannon, in *Encyclopedia of Polymer Science and Engineering*, Vol. 6, 2nd ed., J. I. Kroschwitz, Wiley, New York, 1986.

77. A. J. Kinloch and R. J. Young, *Fracture Behavior of Polymers*, Applied Science, London, 1983.

78. R. A. Gledhill and A. J. Kinloch, *Polym. Eng. Sci.*, **19**, 82 (1979).

79. R. A. Gledhill, A. J. Kinloch, and S. J. Shaw, *J. Mater. Sci.*, **14**, 1769 (1979).

80. F. J. McGarry and A. M. Willner, *Org. Coat. Plast. Chem.*, **28**, 512 (1968).

81. N. J. Sultan, R. C. Laible, and F. J. McGarry, *J. Appl. Polym. Sci.*, **6**, 127 (1971).

82. R. A. Pearson and A. F. Yee, *J. Mater. Sci.*, **26**, 3828 (1991).

83. D. Li, A. F. Yee, I. W. Chen, S. C. Chang, and K. Takahasi, *J. Mater. Sci.*, **29**, 2205 (1994).

84. A. F. Yee, D. Li, and X. Li, *J. Mater. Sci.*, **28**, 6392 (1993).

85. H. R. Azimi, R. A. Pearson, and R. W. Hertzberg, *J. Mater. Sci. Lett.*, **13**, 1460 (1994).

86. H. R. Azimi, Ph.D. Dissertation, Lehigh University, 1994.

87. A. Lazzeri and C. B. Bucknall, *J. Mater. Sci.*, **28**, 6799 (1993).

88. C. B. Bucknall, A. Karpodinis, and X. C. Zhang, *J. Mater. Sci.*, **29**, 3377 (1994).

89. L. Mandelkern, R. G. Alamo, G. D. Wignall, and F. C. Stehling, *TRIP*, **4**(11), 377 (1996).

90. R. G. Alamo, J. D. Londono, L. Mandelkern, F. C. Stehling, and G. D. Wignall, *Macromolecules*, **27**, 411 (1994).

91. G. D. Wignall, J. D. Londono, R. G. Alamo, L. Mandelkern, and F. C. Stehling, *Macromolecules*, **29**, 5332 (1996).

92. R. G. Alamo, R. H. Glaser, and L. Mandelkern, *J. Polym. Sci., Polym. Phys. Ed.*, **26**, 2169 (1988).

89. H. Ulrich, *Introduction to Industrial Polymers*, 2nd ed., Hanser, Munich, 1993.

90. L. A. Utracki, Ed., *Encyclopedic Dictionary of Commercial Polymer Blends*, ChemTec, Toronto, 1994.

91. L. A. Utracki, *Polym. Eng. Sci.*, **35**, 2 (1995).

92. T. S. Grabowski, U.S. Pat. 3,130, 177 (1964).

93. H. S. Chao, U.S. Pat. 5,229,169 (1993).

94. D. L. Dufour and J. P. St. Denis, U.S. Pat. 5,061,754 (1991).

95. R. Dujardin, J. Schoeps, and M. Wandel, U.S. Pat. 5,177,145 (1993).

96. T. Eckel, D. Wittmann, D. Freitag, U. Westeppe, and K. H. Ott, U. S. Pat. 5,137,970 (1992).

97. J. D. Fischer, R. B. Darmstadt, U. Numrich, and W. Siol, U. S. 5,232,986 (1993).

98. S. K. Gaggar, U. S. Pat. 5,128,409 (1992).

99. D. E. Henton and T. M. O'Brian, U.S. 4,419,490 (1983).

100. J. Y. J. Chung and M. W. Witman, Eur. Pat. Appl. 135,904 (1985).

101. D. E. Henton, D. M. Pickelman, C. B. Arends, and V. E. Meyer, U.S. Pat. 4,778,851 (1988).

102. K. D. Hoffman and C. B. Arends, U.S. Pat. 4,708,996 (1987).

103. D. K. Hoffman and C. B. Arends, U.S. Pat. 4,789,712 (1988).

104. H. Jabloner, B. J. Swetlin, and S. G. Chu, U.S. Pat. 4,656,207 (1987).

105. S. G. Chu, H. Jabloner, and B. J. Swetlin, U.S. Pat. 4,656,208 (1987).

106. J. K. Bard, U.S. Pat. 4,680,076 (1987).

107. A. A. Patel, J. Feng, M. A. Winnik, and G. Vancso, *Polymer*, **37**, 5577 (1996).

108. A. G. Gilicinski and C. R. Hegedus, in *Film Formation in Waterborne Coatings*, T. Provder, M. A. Winnik, and M. W. Urban, Eds., ACS Books, Washington, DC, 1996.

109. J. M. Geurts, M. Lammers, and A. L. German, *Colloids Surf., A*, **108**, 295 (1996).

110. R. G. Nickerson, R. T. Bouchard, P. J. C. Hurtubise, and E. A. Duchesneau, Jr., U.S. Pat. 3,935,151 (1976).

111. M. D. Croucher, F. D. Patel, G. E. Kmiecik-Lawrynowicz, M. A. Hopper, and B. Grushkin, U.S. Pat. 5,496,676 (1996).

112. G. Peltier, M. Longuet, and C. A. Midgley, U.S. Pat. 4,500,591 (1985).

113. G. C. Michler, *TRIP* (Trends in Polymer Science), **3**, 124 (1995).

GENERAL READING

Arends, C. B., Ed., *Polymer Toughening*, Marcel Dekker, New York, 1996.

Collyer, A. A., Ed., *Rubber Toughened Engineering Plastics*, Chapman & Hall, London, 1994.

J. Karger-Kocsis, Ed., *Polypropylene: Structure, Blends, and Composites, 2. Copolymers and Blends*, Chapman & Hall, London, 1995.

Kinloch, A. J. and R. J. Young, *Fracture Behavior of Polymers*, Applied Science, Essex, England, 1983.

Lipatov, Y. S., *Polymer Reinforcement*, ChemTec, Toronto, 1994.

Michler, G. H., *Kunstoff-Mikromechanik: Morphologie, Deformations-und Bruchmechanismen*, Carl Hanser, Munich, 1992.

Riew, C. K., Ed., *Rubber-Toughened Plastics*, ACS Adv. Chem. Ser. No. 222, American Chemical Society, Washington, DC, 1989.

Ulrich, H., *Introduction to Industrial Polymers*, 2nd ed., Hanser, Munich, 1993.

Utracki, L. A., Ed., *Encyclopaedic Dictionary of Commerical Polymer Blends*, ChemTec, Toronto-Scarborough, 1994.

Williams, J. G., *Fracture Mechanics of Polymers*, Ellis Horwood, Chichester, England, 1984.

9

BLOCK COPOLYMERS AND THERMOPLASTIC ELASTOMERS

9.1. INTRODUCTION TO BLOCK AND GRAFT COPOLYMERS

9.1.1. Concept of Thermoplastic Elastomers

Thermoplastic elastomers (TPEs) and their polyurethane subset, thermoplastic urethanes (TPUs), share an important concept. All of these materials are composed of "hard" and "soft" blocks, phase separated. The soft blocks are usually the larger volume fraction and are elastomeric. The hard blocks are either glassy or crystalline. At use temperature, these hard blocks constitute physical crosslinks, holding the elastic material together. At some higher temperature, these hard blocks go through their glass transition temperature, or melt. Then, the physical crosslink holding power of the once hard blocks is much diminished, and the material can flow. Thus, on heating, the material is reversibly transformed from an elastomer to a thermoplastic melt, hence the term *thermoplastic elastomers*. It must be pointed out that thermoplastic elastomers are a somewhat broader field than that concerned directly with block and graft copolymers. The thermoplastic interpenetrating polymer networks (IPNs), Section 10.7.2, and even some polymer blends, with dual-phase continuity and containing a glassy or crystalline polymer, exhibit this behavior.

In the same sense that a chemically crosslinked elastomer must have two or more crosslinks per chain to exhibit elastomeric properties, so a thermoplastic elastomer must contain at least two hard blocks. These may be arranged in space in several different ways.

9.1.2. Topology of Block and Graft Copolymers

Block and graft copolymers have the general architectures shown in Figure 9.1. Block copolymers are distinguished by having chain portions, or *blocks*, linked end on end. Graft copolymers have a backbone, from which one or more side chains emanate. Thus, Figure 9.1a illustrates a simple AB diblock copolymer,

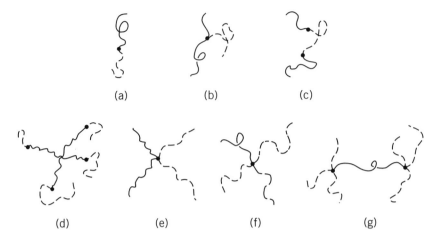

(a)   (b)   (c)

(d)   (e)   (f)   (g)

Figure 9.1 Some block and graft copolymer structures. See text for definitions.

while Figure 9.1b shows a simple AB graft copolymer, with A as the backbone chain, and B as the side chain. Figure 9.1c shows an ABA triblock copolymer, composed of identical end blocks and a central block. A variation on this structure is the ABC triblock copolymers, where there are three nonidentical blocks. Figure 9.1d shows an $(AB)_4$ star block copolymer, illustrated with four arms. One may make many other variations, however. Figures 9.1e, f, and g are the creations of Hadjichristidis and co-workers (1–3). Figure 9.1e is called the 4-miktoarm star copolymer of the A_2B_2 type, Figure 9.1f is the 4-miktoarm star quaterpolymer of the ABCD type, and Figure 9.1g is called the super-H block copolymer of the B_3AB_3 type. Not shown is the $(AB)_2(BA)_2$ inverse 4-miktoarm block copolymer, in which two of the arms are as in Figure 9.1d, and two of the arms are reversed in synthetic order.

Another class of block copolymer structures arises through a continuation of the Figure 9.1 a and c topologies to make a linear multiblock copolymer of the type $(AB)_n$. In addition, a number of tapered structures have been made, where the A mers in polyA are gradually replaced by B mers down the chain, eventually becoming all polyB. Such a tapered structure tends to broaden the interphase, thus giving another degree of control over properties. Many other possibilities exist. As Figure 9.1 illustrates, further, there are some structures that can be interpreted multiple ways. For example, Figure 9.1e might be a graft copolymer joining two previously synthesized chains, as well as considered a relative of the star block copolymers.

Block and graft copolymers have similar sets of properties. Conceptually, all are thermoplastic in behavior, in that they can flow on heating, and almost all exhibit phase separation if the molecular weights are high enough, that is one block or graft is immiscible in the other. This latter feature, of course, is critical to the proper development of most polymer blends, as illustrated in Chapter 8, for example. In this case, however, the chains are chemically bound to each other. This often means a broader interphase and good compatibility between the phases. It usually also means that the domain sizes are limited by the size of the blocks or grafts, since all such chains must have one or more mers in the interphase.

As opposed to the rubber-toughened plastics of Chapter 8, useful block and graft copolymers are often elastomeric in nature, forming tough elastomers or elastomeric fibers. In such cases, the soft blocks are amorphous polymers with low glass transition temperatures.

Since the advent of anionic, living polymerization methods, some kinds of block copolymers can be made with very narrow molecular weight distributions and almost perfect, reproducible architecture. The original materials of this nature were the diblock and triblock materials composed of polystyrene and polybutadiene, first synthesized in the early 1960s (4–6). The history and development of block and graft copolymers is discussed in some detail by Legge et al. (7). The synthesis of true block copolymers of styrene and butadiene and related monomers was made possible by the discovery of Szwarc and co-workers in 1956 of homogeneous "living" anionic polymerization (8,9).

While block and graft copolymers exhibit similar morphologies and behavior patterns for similar compositions, the uses of these materials has tended to be far different. Graft copolymers are often used as compatibilizers in polymer blends; see Chapter 6 and Section 8.6. As such, they may have irregular structures and are rarely separated out in the pure state. By contrast, the block copolymers are often highly regular in structure, used in the pure state, and form a class of highly distinctive materials. This chapter will briefly discuss the synthesis, morphology, and behavior of block and graft copolymers, with emphasis on the former.

9.2. BLOCK AND GRAFT COPOLYMER SYNTHESIS

9.2.1. Some Irregular Structures

The synthetic methods for preparing block and graft copolymers vary from the most haphazard to the most regular and exquisite. A simple method of obtaining some limited block and graft copolymer formation involves dissolving a certain amount of monomer B or sometimes polymer B in polymer A, and masticating the mass (10–12). The extensive shearing of the mass causes side chain moieties to be peeled off and backbone chains to scission. A series of free radicals, anions, and cations are generated on such chains, leading to graft and block copolymer formation. If monomer B is present, usually significant amounts of homopolymer B are also formed.

Another method, especially for graft copolymers, involves polymerization of monomer B in the presence of polymer A, especially when polymer A contains unsaturation and monomer B is being polymerized via free radical chemistry. An important example was discussed in Section 8.2, where polymer A was polybutadiene and monomer B was styrene. The relatively small amount of graft copolymer produced in this way contributes greatly to the interfacial adhesion between the polystyrene and the polybutadiene.

Similarly, the introduction of reactive groups such as hydroxyl or carboxyl groups along a chain may lead to various ester or amide formation. Chapter 6 and Section 8.6 describe such methods for graft copolymer formation.

Another method involves the use of azo or peroxide groups in a polymer chain, such as (13,14):

$$P-N=N-P' \quad \text{or} \quad P-O-O-P' \tag{9.1}$$

Upon application of heat or ultraviolet light, these groups break down, creating free radicals. In the presence of the appropriate monomers, a series of block copolymers will be formed. If the groups form a side chain, as in a hydroperoxide, for example, a graft copolymer plus homopolymer will be formed.

9.2.2. Poly(ether ester) Block Copolymers

This class of linear $(AB)_n$ block copolymers is composed of polyether segments such as the poly(alkylene oxides) or poly(arylene oxides) and polyester segments such as the poly(alkyene terephthalates). For example, hydroxyl-terminated poly(tetramethylene glycol), molecular weight of 1000–3000 g/mol, is reacted with dimethyl terephthalate and 1,4-butanediol,

$$HO[(CH_2)_4O]_nH + mCH_3O-\overset{\overset{\displaystyle O}{\|}}{C}-\underset{}{\bigcirc}-\overset{\overset{\displaystyle O}{\|}}{C}-OCH_3$$

$$+mHO(CH_2)_4OH \rightarrow CH_3O[\overset{\overset{\displaystyle O}{\|}}{C}-\underset{}{\bigcirc}-\overset{\overset{\displaystyle O}{\|}}{C}-O(CH_2)_4O]_m[(CH_2)_4O]_nH \tag{9.2}$$

Hard block   Soft block

where methanol is given off in an ester exchange side reaction. This particular poly(ether ester) block copolymer is known as Hytrel™ (15). In this material, the polyether block serves as the "soft block" and the poly(butylene terephthalate) block serves as the "hard block." The soft block is usually elastomeric, having a low glass transition temperature, while the hard block may be either glassy or crystalline, the latter in this case. Such alternating hard and soft blocks produce excellent elastomeric fibers, for example.

9.2.3. Polyurethanes

Similarly, polyethers or polyesters can be reacted with diisocyanates to form polyurethanes. Most polyurethane multiblock copolymers are formed in such a way that the chemical reaction employs both a monomer and a preformed functionally terminated oligomer. The polyester or polyether soft segments typically have molecular weights in the range of 600–6000 g/mol. Some of the basic reactions involving isocyanate groups are as follows (16):

$$R-NCO + R'OH \rightarrow R-\underset{\underset{H}{|}}{N}-\underset{\underset{O}{\|}}{C}-O-R'$$

Urethane linkage

$$R-NCO + R'NH \rightarrow R-\underset{\underset{H}{|}}{N}-\underset{\underset{O}{\|}}{C}-\underset{\underset{H}{|}}{N}-R' \tag{9.3}$$

Urea linkage

$$R-NCO + R'\overset{\overset{O}{\|}}{C}-OH \rightarrow R-\underset{\underset{H}{|}}{N}-\underset{\underset{O}{\|}}{C}-R' + CO_2$$

Amide linkage

Basically, the polyurethane elastomers consist of three components: a polyester or polyether diol, a chain extender such as a short-chain diol, and a diisocyanate. Thus, a polyester-based polyurethane known as Estane^{TM} uses diphenylmethane-p,p'-diisocyanate (MDI) as the hard block:

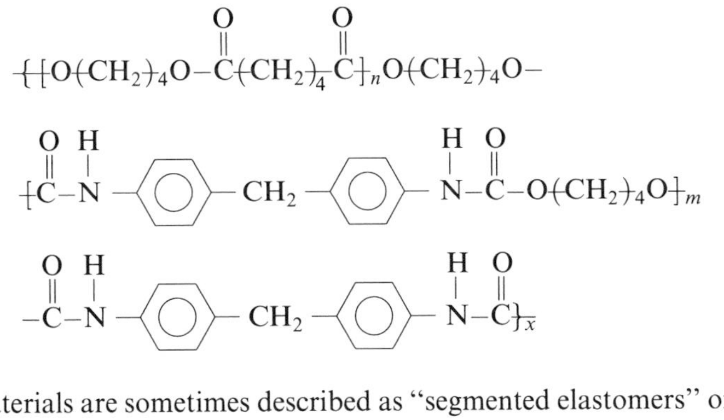

$$\begin{aligned}
&\{\!\{O\{CH_2\}_4O-\overset{\overset{O}{\|}}{C}\{CH_2\}_4\overset{\overset{O}{\|}}{C}\}_nO\{CH_2\}_4O- \\
&\{\overset{\overset{O}{\|}}{C}-\underset{\underset{H}{|}}{N}-\!\!\bigcirc\!\!-CH_2-\!\!\bigcirc\!\!-\underset{\underset{H}{|}}{N}-\overset{\overset{O}{\|}}{C}-O\{CH_2\}_4O\}_m \tag{9.4} \\
&-\overset{\overset{O}{\|}}{C}-\underset{\underset{H}{|}}{N}-\!\!\bigcirc\!\!-CH_2-\!\!\bigcirc\!\!-\underset{\underset{H}{|}}{N}-\overset{\overset{O}{\|}}{C}\}_x
\end{aligned}$$

These materials are sometimes described as "segmented elastomers" or thermoplastic urethanes (TPU).

9.2.4. Anionic Block Copolymer Synthesis

Anionic polymerizations require a great deal of purity, that is, freedom from such impurities as carbon dioxide or water. In such cases, there is no termination reaction. Only the initiation and propagation steps are entailed. Thus, after all of the monomer has been consumed, the anionic end of the chain remains "living." This feature allows the synthesis of predetermined and well-controlled block copolymer structures, almost unique in the polymer world.

Anionic polymerizations utilize alkyllithium initiators, frequently butyl lithium. The most important monomers are styrene and butadiene. The synthesis of a diblock copolymer is carried out as follows (15):

Initiation

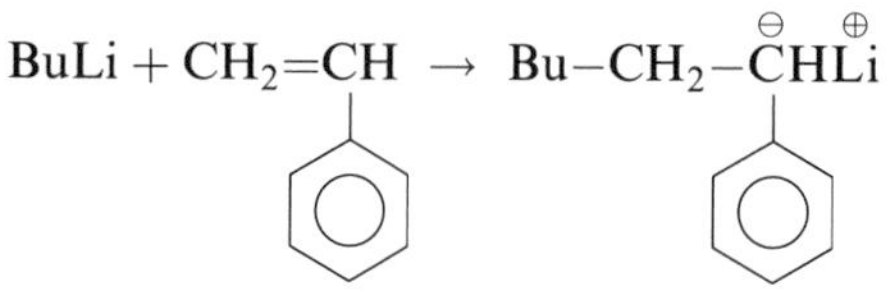

$$BuLi + CH_2{=}CH \longrightarrow Bu{-}CH_2{-}\overset{\ominus}{C}H\overset{\oplus}{Li}$$

First Propagation

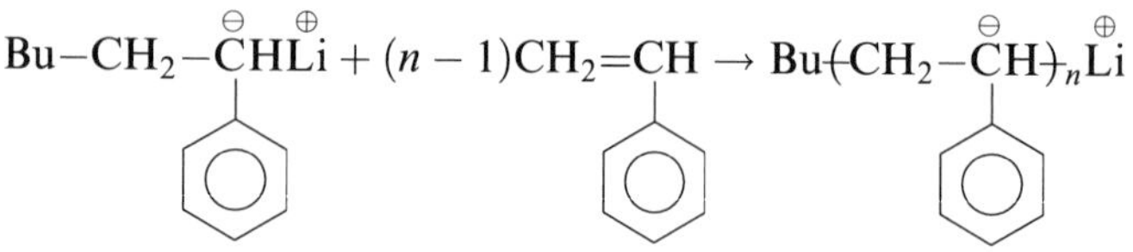

$$Bu{-}CH_2{-}\overset{\ominus}{C}H\overset{\oplus}{Li} + (n-1)CH_2{=}CH \longrightarrow Bu{\left(CH_2{-}\overset{\ominus}{C}H\right)}_n\overset{\oplus}{Li}$$

Cross-Initiation

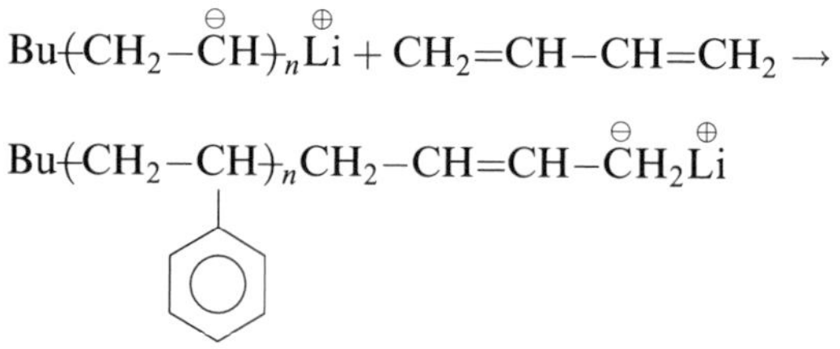

$$Bu{\left(CH_2{-}\overset{\ominus}{C}H\right)}_n\overset{\oplus}{Li} + CH_2{=}CH{-}CH{=}CH_2 \longrightarrow$$

$$Bu{\left(CH_2{-}CH\right)}_nCH_2{-}CH{=}CH{-}\overset{\ominus}{C}H_2\overset{\oplus}{Li} \tag{9.5}$$

Second Propagation

$$Bu{\left(CH_2{-}CH{-}CH_2{-}CH{=}CH{-}\overset{\ominus}{C}H_2\overset{\ominus}{Li} \right)} + (m-1)CH_2{=}CH{-}CH{=}CH_2 \longrightarrow$$

$$R{\left(CH_2{-}CH\right)}_n{\left(CH_2{-}CH{=}CH{-}CH_2\right)}_m{-}CH_2{-}CH{=}CH{-}\overset{\ominus}{C}H_2\overset{\oplus}{Li}$$

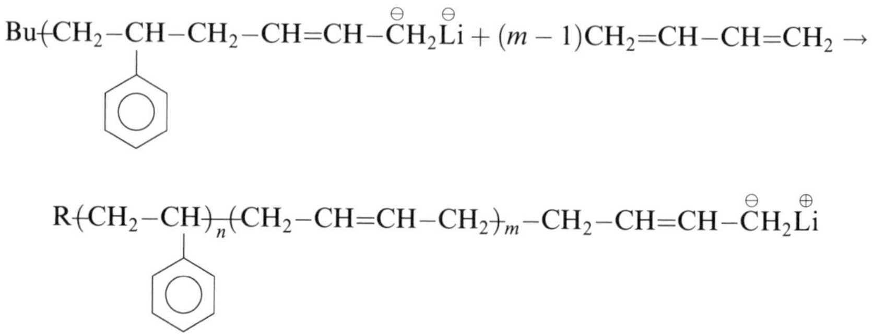

Other monomers include isoprene, α-methyl styrene, and methyl methacrylate. The reactions are usually carried out in a hydrocarbon solvent such as tetrahydrofuran. While the reactions can be terminated by exposure to air or water, specific end groups may be added. These include the addition of ethylene oxide to produce hydroxyl end groups, or carbon dioxide to form carboxyl end groups

(11). This technique proves especially useful in the synthesis of star block copolymers.

Triblock copolymer syntheses often utilize dianions in their synthesis. In this method, the chain is started from the middle and grown in both directions simultaneously (17,18).

A major advantage of anionic polymerization is the control provided over the molecular weight distribution. Normally, all of the chains are initiated simultaneously and grow at approximately the same rate. An analogy would be a horse race, where all the horses start at the bell and finish at nearly the same time. This results in a Poisson statistical molecular weight distribution, which is normally very narrow. Polydispersity indices, M_w/M_n, may be in the range of 1.01–1.05, as compared with free radical or condensation polymerizations, often having polydispersity indices in the range of 2.0.

A significant problem with the use of diene-type polymers involves oxidative and ultraviolet degradation. A solution to this problem, widely practiced in the block copolymer field, is the use of selective hydrogenation. Here, the polybutadiene or polyisoprene is hydrogenated. Since usually the original polymerization gives a copolymer of the 1,2- and 1,4-adducts, hydrogenation of the polybutadiene results in a copolymer of ethylene and butylene. This latter tends to stay amorphous. If the ethylene content is too high, however, crystallization may set in. For 100% 1,4-polyisoprene, an interesting alternating copolymer of ethylene and propylene results.

A very significant limitation of anionic synthesis of block copolymers is the small number of monomers that are amenable to such syntheses. Monomers that can be used include styrene, butadiene, isoprene, and α-methylstyrene. Acrylic monomers such as methyl methacrylate, methyl acrylate, and acrylonitrile undergo anionic polymerization, but the pendant ester and nitrile groups constitute sources of complicating side reactions. In unsubstituted polyacrylates, the labile tertiary hydrogen presents further opportunities for unwanted side reactions. Other block copolymers synthesized include polysiloxanes, polylactones, and poly(alkylene sulfides).

9.2.5. Use of Telechelic and Macromer Polymers for Synthesis

The terms *telechelic polymers* and *macromers* refer to low-molecular-weight polymers that have reactive groups on their ends. While usage differs, the term macromers generally refers to materials having one reactive group at one end only, such as a double bond, while the term telechelic refers to materials having two reactive groups (19). If the functional groups are different, the polymer is said to be heterotelechelic. Branched or star-shaped polymers having more than two chain ends that are reactive are called *polytelechelic* (20). The diols used in polyurethane polymerization are examples of telechelics; see Section 9.2.3.

The telechelic and macromer polymers are important building blocks for graft and block copolymers, and also for AB-crosslinked polymers; see Figure 9.2 (19). They are also the preferred starting materials for the synthesis of model

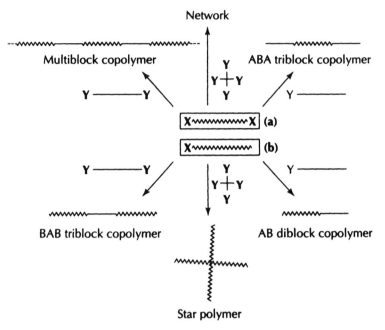

Figure 9.2 Block copolymers and other polymer structures can be synthesized through coupling reactions. (a) a telechelic and (b) a monotelechelic polymer.

polymer networks, that is, networks in which the chain length between two crosslinks as well as the crosslink functionality are well-known. Besides double-bond and hydroxyl functionalities, various amines, halogens, carboxyl, and other groups have been used in telechelics.

9.3. MORPHOLOGY OF BLOCK COPOLYMERS

Similar to polymer blends, most block copolymers are phase separated. A major difference between polymer blends and block copolymers, however, is that the chains in the latter are chemically bound to each other. This restricts the size of the domains and produces domain morphologies that depend primarily on the relative block lengths. As shall be shown below, however, the shapes of the domains depend primarily on the length of one block relative to the other.

9.3.1. Electron Microscopy

Most block copolymers phase separate as in blends. However, the block junctions limit the size of the domains to the molecular length of the blocks. Otherwise, lower than natural densities will occur in the center of the domains. If the blocks are of uniform size, i.e., monodisperse in block length, the phase domains can be surprisingly uniform in appearance, and regularly spaced.

9.3.1.1. SBS Triblock Copolymers Typical morphologies for block copolymers are illustrated in Figure 9.3 (21), obtained by transmission electron microscopy. Each of these figures are of polystyrene-*block*-polybutadiene-*block*-polystyrene (SBS), where the diene portion has been stained with osmium tetroxide. The samples were prepared by casting the block copolymers from toluene solution. Three morphologies are seen: spheres, cylinders, and lamellar structures, depending on the relative block lengths. The absolute DP sequences are 720–350–720, 1840–2450–1840, and 740–2240–740, respectively.

With SBS-1 (20% butadiene), similar patterns were observed in both normal and parallel sections to the film direction. This indicates spherical domains (Figure 9.3a). The normal and parallel sections are shown for SBS-3 (40% butadiene) in Figures 9.3b and 9.3c. As shown in Figure 9.3b, what appears to be both cylinders and spheres appears. Figure 9.3c, from the same material, shows only cylinders. It was concluded that this is a cylindrical morphology, oriented in the plane of the film. With SBS-5 (60% butadiene), the normal and parallel sections show the formation of alternating lamellae. Similar electron micrographs have been obtained many times (7).

These morphologies are modeled by Molau (22), see Figure 9.4. The highly regular appearance of the domains is stressed by this figure, and also the inversion of the morphology with composition past the 50/50 composition line. The organization of the domains increases into almost crystal-like structures with increasing number of blocks in the same polymer. Thus, an ABA triblock copolymer tends to be more regular than the equivalent AB diblock copolymer, and so on.

9.3.1.2. OBDD Morphology In 1986, Alward and co-workers (23,24) discovered a new morphology, which they called the ordered, bicontinuous double-diamond (OBDD). The OBDD morphology appears most often with multiarmed star block copolymers. Literally, the morphology consists of a tetrahedral arrangement of short rods. Such units are interconnected on a cubic lattice, such that the resultant structure consists of two translationally displaced, mutually interwoven, but unconnected three-dimensional networks of rods embedded in a matrix to form a double diamond morphology.

Figure 9.5 (25) shows the OBDD morphology for a star block copolymer of eight arms. Also shown in Figure 9.5a is a three-dimensional triply periodic surface of constant mean curvature. Such surfaces belong to the family of surfaces known as the Schwarz-D minimal surface. An obvious other requirement is that the double diamonds be placed in such a manner that the spacing of the interfaces are easily reached by the block copolymer junctions. Figure 9.5c shows the corresponding two-dimensional simulated projection of the structure in Figure 9.5b. The relationship between Figure 9.5b and c is striking. Sometimes these morphologies have been called "wagon wheels" because of their obvious similarity to such.

Thus, there are three major domain morphologies, spheres, cylinders, and lamella, and the OBDD morphology, which appears under special circumstances.

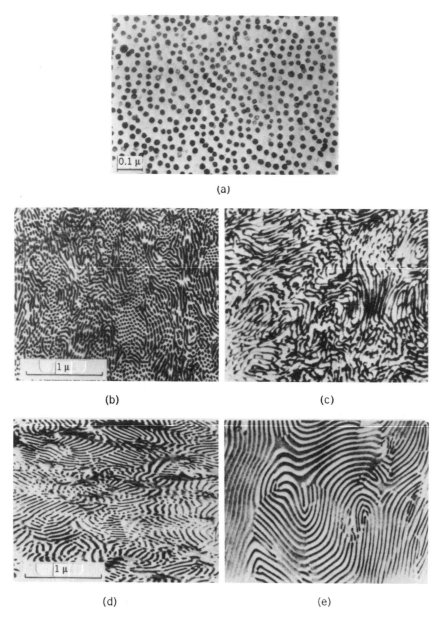

Figure 9.3 Polystyrene-*block*-polybutadiene-*block*-polystyrene (SBS), triblock copolymer morphology dependence on composition. (a) SBS-1, 20% B, (b,c) SBS-3, 40% B, vertical and horizontal cuts, respectively, (d,e) SBS-5, 60% B, vertical and horizontal cuts, respectively. Polymers cast from toluene solution, cut to about 600 Å, and stained with osmium tetroxide. Interpretation: SBS-1 forms B spherical domains, SBS-3 forms B cylindrical domains, and SBS-5 forms alternating lamellae.

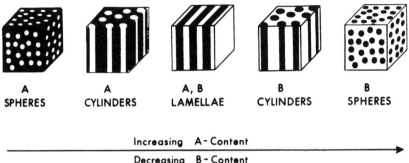

A	A	A, B	B	B
SPHERES	**CYLINDERS**	**LAMELLAE**	**CYLINDERS**	**SPHERES**

Increasing   A - Content

Decreasing   B - Content

Figure 9.4 Cartoon of block copolymer morphologies, illustrating the transformation from spheres to cylinders to lamellae and back to cylinders and spheres of the inverse morphology as the ratio of the two blocks is changed.

These are, together with the general composition range in which they appear for block copolymers (26), shown in Table 9.1.

The above morphologies are for block copolymers containing only two different kinds of blocks. More recently, Mogi et al. (27) investigated the morphologies of ABC triblock copolymers and some ABCB tetrablock copolymers, see Figure 9.6. In this figure, the osmium tetroxide staining produced black, white, and gray images that denote polyisoprene, polystyrene, and poly(2-vinyl pyridine) domains, respectively. Since the morphologies did not

Figure 9.5 Understanding the OBDD morphology. (a) Computer-generated double diamond structure of constant mean curvature. (b) Bright-field TEM of PS/PI block copolymer (SI 8/30/10, 8-armed, 30% PS in the outer arms, and a PS block molecular weight of 10×10^3 g/mol). A (111) crystallographic projection of the morphology via computer projection.

Table 9.1 Block copolymer morphologies

Morphology	Experimental Range[a]	Theoretical Prediction
Spherical	0 to 0.20–0.25	$0–0.20^{b,c}$
Cylindrical	0.20–0.25 to 0.35–0.40	$0.20–0.33^{b,c}$
OBDD	0.27–0.38 and 0.62–0.66	Above $0.262^{d–f}$
Lamellar	0.35–0.40 to 0.60–0.65	$0.33–0.67^{b,c}$
Ripple or perforated-lamellar	Various	Not yet well defined[g]

[a] Assumed symmetrical in composition, unless indicated.
[b] T. Ohta and K. Kawasaki, *Macromolecules*, **19**, 2621 (1986).
[c] D. J. Meier, in *Block and Graft Copolymers*, J. I. Burke and V. Weiss, eds., Syracuse University Press, Syracuse, 1973.
[d] H. Hasagawa, H. Tanaka, Y. Yamasaki, and T. Hastimoto, *Macromolecules*, **20**, 1651 (1987).
[e] E. L. Thomas, D. M. Anderson, C. S. Henkee, and D. Hoffman, *Nature*, **334**, 598 (1988).
[f] E. L. Thomas, D. B. Alward, D. L. Kinning, D. L. Martin, D. L. Handlin, Jr., and L. J. Fetters, *Macromolecules*, **19**, 2197 (1986).
[g] K. Almdal, K. A. Koppi, F. S. Bates, and K. Mortensen, *Macromolecules*, **25**, 1743 (1992).

change on annealing, they were considered to be near thermodynamic equilibrium. Figure 9.6a shows a lamellar structure of an ABCB polymer of polyisoprene-*block*-polystyrene-*block*-poly(2-vinyl pyridine)-*block*-polystyrene. Hogi and co-workers (27) called this the three-phase four-layer lamellar structure. Figure 9.6b shows a morphology that is considered to consist of two kinds of mutually interpenetrated frameworks formed by the two end block polymers, embedded in a matrix composed of the middle block polymer. They called this structure an ordered tricontinuous double-diamond (OTDD), after the OBDD terminology. Figure 9.6c shows a cylindrical morphology of a triblock copolymer with a composition of 1:4:1. This structure consists of two kinds of cylindrical domains formed by two end block polymers. Figure

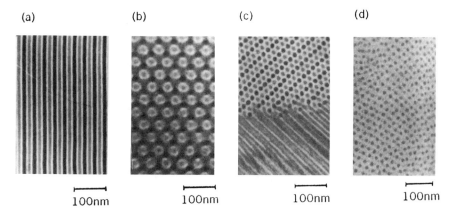

(a)　　　　　(b)　　　　　(c)　　　　　(d)

100nm　　100nm　　100nm　　100nm

Figure 9.6 Morphologies of ABC and ABCD block copolymers. Samples: (a) ISP-4, (b) ISP-3, (c) ISP-18, and (d) ISP-12. See text for exact block sequences.

9.6d shows a micrograph of spherical domains of two end block polymers in a polystyrene matrix with a composition of $1:8:1$. Sakurai (28) recently reviewed the morphology of block copolymers.

9.3.2. Order–Disorder Transition

Classical composition–temperature phase diagrams cannot be obtained for block copolymers because the shape of the diagram depends on the volume fractions of the two polymers. In this case, the chains are attached, and hence the volume fraction and the molecular weights of the blocks cannot be independently varied. However, people talk about the order–disorder transition (ODT), sometimes called the microphase separation transition (MST); to describe where given block copolymer phase domain structures become thermodynamically stable, showing the conditions under which a one-phased melt becomes two phased.

There have been two sets of theories from which people have calculated such diagrams. One of these focuses on the strong segregation regime, that is, far from the MST, while the other considers the weak segregation regime, that is, in some sense near the MST. For the strong segregation regime, Helfand and Wasserman used a self-consistent mean-field theory (29–35) and a narrow interphase approximation (NIA) to calculate phase diagrams. They found that the phase boundaries separating the different morphologies were almost independent of χZ; rather they depended almost solely on the weight fractions of the different blocks. Here, Z represents the copolymer degree of polymerization, and χ is the Flory interaction parameter.

The weak segregation regime was treated for model systems by Leibler (36), Fredrickson and Helfand (37), and Mayes and Olvera de la Cruz (38–40). Such approaches all predict phase diagrams near the MST that were qualitatively different from those predicted for the strong segregation regime, with curved microphase boundaries. Thus, for asymmetric copolymers, as a function of increasing χZ, the first equilibrium microphase would consist of spheres, followed by transitions to cylinders and then lamellae.

A fuller specification of a system requires at least seven independent quantities: the copolymer degree of polymerization Z, the volume fraction of either block f_A or f_B ($f_A = 1 - f_B$), the two Kuhn statistical lengths b_A and b_B, the pure component densities, and the Flory χ parameter (41). The most important of these is the volume fraction of the blocks. Figure 9.7 (42) shows the appearance of the several morphologies, where H is homogeneous, S is spheres, C is cylindrical, and L is lamellar. The solid curves are based on the self-consistent mean-field theory. In the left-hand panel, the dashed curves were calculated using the theory of Leibler. For the right-hand panel, the theory of Helfand and Wasserman was used. The OBDD structure is not shown. All phase boundaries converge for symmetrical block copolymers, $f_A = 0.5$, and $\chi Z = 10.5$. At smaller values of χZ, miscible materials are predicted. Thus, the Y axis increases with total molecular weight (43). Similar figures are reported

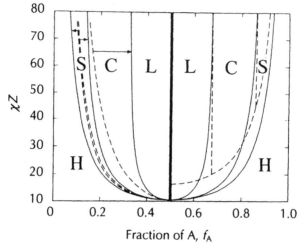

Figure 9.7 Phase diagram for a block copolymer. For $f_A < 0.5$, the dashed curves were calculated using Leibler's theory, arrows indicating the corresponding actual phase boundaries determined experimentally (see text). For $f_A > 0.5$, the dashed curves were calculated via the theory of Helfand and co-workers (29,37).

elsewhere (41). Physically, χkT represents the free energy change per segment of polymer A in taking a chain of A from pure A surroundings and transferring it to pure B surroundings, while χZ represents the heat of mixing per chain.

The temperature dependence of χ is usually given by

$$\chi = A + B/T \tag{9.6}$$

Thus, the value of χ decreases with increasing temperature. Therefore, Figure 9.7 shows that upper critical solution temperature behavior should be expected. The upper critical solution temperature finding results from an increase in the interaction energy between the segments with decreasing temperature, as quantified by the Flory–Huggins segmental interaction parameter, χ. Thus, for most polymer pairs, where χ is positive, an increase in the temperature weakens the repulsion between unlike mers. The shape of the curves in Figure 9.7 shows that both the experimental and theoretical ranges of the various morphologies shown in Table 9.1 are only approximate and must be expected to vary with temperature.

There are a number of experiments, the results of which tend to support the idea of an upper critical solution temperature; see Table 9.2. Here, T_s is the phase separation temperature (or ODT or MST). The polymer in the experiment cited in footnote b of Table 9.2, with a higher molecular weight, is seen to phase separate at a higher temperature than those cited in footnotes c and d, all three being SBS triblock copolymers. The polymer in the experiment cited in footnote e is a hydrogenated cis-1,4-polyisoprene-*block*-1,2-polybutadiene. More recently, a symmetric block copolymer of poly(perdeuterated

Table 9.2   Research supporting an ODT for block copolymers

Block Copolymer[a]	Exp'tl Method	T_s (°C)	Ref.
SBS (12.5–75–12.5)	DSC + laser light transmission	217	b
SBS (7–43–7)	Melt rheological behavior	150–175	c
SBS (7–43–7)	Dynamic shear moduli	142	d
Poly(ethylenepropylene)-*block*-poly(ethylethylene) (61–33)	SANS	175	e

[a] Molecular weights in thousands.
[b] D. F. Leary and M. C. Williams, *J. Polym. Sci., Polym. Phys. Ed.,* 12, 265 (1974).
[c] C. I. Chung, and J. C. Gale, *J. Polym. Sci., Polym. Phys. Ed.,* 14, 1149 (1976).
[d] E. V. Gouinlock and R. S. Porter, *Polym. Eng. Sci.,* 17, 534 (1977).
[e] K. Almdal, K. A. Koppi, F. S. Bates, and K. Mortensen, *Macromolecules,* **25**, 1743 (1992).

polystyrene) and poly(*n*-butyl methacrylate) of molecular weight 99,000 g/mol was found to have an upper critical ordering temperature (UCOT) at 60–80°C and a lower critical ordering temperature (LCOT) at a temperature just below the decomposition temperature (44), as studied by X-ray scattering. The LCOT is entropy driven, similar to that found in polymer blends. (The notation here varies among authors pending development of a definitive nomenclature.)

Hasegawa et al. (45) investigated the various morphologies found in poly-styrene–polyisoprene P(S–PI) diblock copolymers as a function of composition and molecular weight. The results are summarized in Figure 9.8. Hasegawa and co-workers found spheres, cylinders, lamella, and the OBDD morphology, which they called tetrapod. Substantially no dependence of the morphology on the total molecular weights of the block copolymers was observed. The transition from one morphology to another was found to be markedly sharp.

Hasegawa et al. (45) remarked that cast films (and also molded films) are not necessarily in their thermodynamically equilibrium state. The morphology tends to be frozen in during solvent evaporation or cooling. Thus, the morphology may be kinetically controlled to a significant extent.

It must be noted that a UCOT implies a nucleation and growth mechanism for the kinetics of phase separation. This was found using time-resolved small-angle X-ray scattering when the undercooling, that is, the difference between the measurement and microphase separation transition temperatures, was small (46–48).

9.3.3. Free Energies of Domain Formation

9.3.3.1. Molecular Basis   On a molecular scale, the morphology can be depicted as chains wandering from domain to domain, each kind of domain containing primarily one kind of chain; see Figure 9.9. Here, a triblock copolymer, perhaps SBS, is illustrated with the spherical domains containing

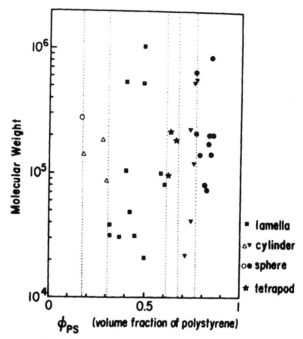

Figure 9.8 Microdomain morphology changes in toluene-cast films of polystyrene–polyisoprene diblock copolymers as a function of total molecular weight and volume fraction of polystyrene component. Designations: PI spheres, filled circles; PI cylinders, filled triangles; PI tetrapod network, stars; PS-PI lamellae, squares; PS cylinders, open triangles; PS spheres, open circles.

Figure 9.9 Molecular architecture of a phase-separated ABA triblock copolymer. Each kind of block resides in a different phase. Spherical domains in a matrix are illustrated.

the polystyrene blocks, and the matrix containing the polybutadiene center block. Of course, in addition to a chain wandering from domain to domain, it may also circle back, and the two polystyrene blocks be in the same domain. As drawn, the reader may imagine that the "narrow interphase approximation" (see below) holds. There are two critical questions yet to be addressed:

1. What are the free energies for each of the important kinds of domain shapes, spheres, cylinders, and lamellae?
2. What are the thicknesses of the interphases between the domains?

The first of these questions will be addressed in this section; the second question will be addressed in Section 9.3.4.

There are three main contributions to the free energy changes associated with the spatial constraints and the interactions energies between the segments, as follows (26):

1. There is a heat of mixing, ΔH_m, between the various mers when a random mixture transforms into a microphase-separated system. For finite interphase thicknesses, the local heat of mixing across the interphase thickness, calculated separately, is ΔH_{int}.
2. The restriction on the placement of the A–B junction to the interphase in the microphase-separated state results in a loss of entropy, ΔS_p.
3. In the microphase-separated state, the A and B segments are segregated into their respective domains, that is, into restricted regions of space, resulting in a loss of entropy, ΔSc.

Additionally, the size of the domains, and hence their interphase surface areas, are controlled by the lengths (molecular weights) of the individual blocks. This last results from the space-filling requirement of near constant density in each domain of the material. Thus, if the domains are much larger than the blocks, there will be a hole in the middle. If the domains are too small, the chains may have reduced conformational dimensions, resulting in another kind of entropy loss. While the chain conformations in phase separated block copolymers are not exactly random coils, an important part of the calculation must include the conformational entropies of the chains.

The relative contributions to the free energy of domain formation, per chain, is given in Figure 9.10 (26). The domain repeat thickness is $T_{AB} = (T_A + T_B)/2$. While each contribution to the free energy behaves differently, a distinct minimum in the free energy occurs at a particular domain repeat thickness.

Each of the domain morphologies has a free energy associated with it throughout composition space. However, only that domain morphology having the lowest free energy will appear under equilibrium conditions. Ohta and Kawasaki (49) derived the free energies for each of the domain shapes; see Figure 9.11. The quantity Q is a function of the volume fraction, the chain radius of gyration, and the interphase thickness. The predicted transitions in

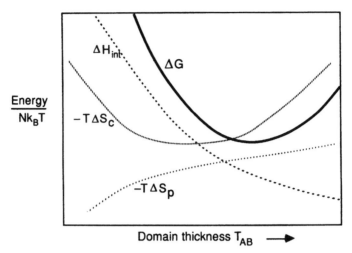

Figure 9.10 The relative contributions per chain to the free energy of domain formation as a function of domain size. The minimum in the free energy involves the interfacial energy, ΔH_{int}, which decreases monotonically with an increase of the domain thickness since the relative fraction of material in the interface then decreases. The placement entropy, ΔS_p, increases monotonically with increasing domain size. However, the constraint entropy contribution, ΔS_c, shows a minimum, resulting from the interplay of the boundary constrains that favor an increase in the space available to the chains, and the density constraint, which forces the chains to fill space uniformly.

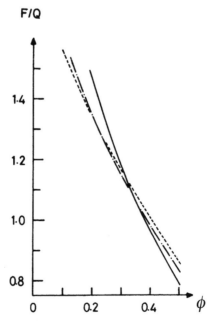

Figure 9.11 The free energy of formation of different domain shapes depends on the volume fraction, ϕ, of the domain, the lowest free energy being the domain shape that appears. (—) lamellar, (–·–) cylindrical, and (– – –) spherical structures.

morphology are volume fractions equal to 0.215 for the spherical-to-cylindrical domain shapes and 0.355 for the cylindrical-to-lamellar morphological transition.

9.3.3.2. Domain Sizes The domain size clearly depends on the molecular weight of the blocks in the domains, as described above. Today, we know that the domain size also depends on the dimensions of the interphase. However, approximate relations of the size of the domains to the polymer chain molecular weight was given by Meier (50,51):

$$\text{Sphere:} \qquad R = 1.33\alpha K M^{1/2} \qquad\qquad (9.7)$$

$$\text{Cylinder:} \qquad R = 1.0\alpha K M^{1/2} \qquad\qquad (9.8)$$

$$\text{Lamella:} \qquad R = 1.4\alpha K M^{1/2} \qquad\qquad (9.9)$$

where M is the molecular weight of the block in the domain, and R is a characteristic dimension: the radius for spheres and cylinders, and the half-thickness for lamellae. The quantity α measures the chain perturbation existing in the block copolymer morphology, absent in a bulk homopolymer. Values of α vary between 1.0 and 1.5 for most cases of interest, with a value of 1.25 being sufficient for simple calculations.

The quantity K represents the ratio of the unperturbed root-mean-square end-to-end distance to the square root of the molecular weight. For some common polymers, K for R in Ångstroms is given by

Polystyrene: 670×10^{-3}
Polyisobutylene: 700×10^{-3}
Polybutadiene: 880×10^{-3}
Poly(methyl methacrylate): 565×10^{-3}

Using polystyrene-*block*-polyisoprene diblock copolymers, Hashimoto et al. (52) determined the domain sizes experimentally via small-angle X-ray scattering (SAXS); see Figure 9.12. The average spacing, D, was calculated from each maximum on the basis of Bragg's equation. They found that the results agreed quite well with those of transmission electron microscopy.

The straight-line relationship in Figure 9.12 was found to follow the empirical relationship,

$$D = 0.024 M^{2/3} \qquad\qquad (9.10)$$

The exponent power of $\frac{2}{3}$ is substantially greater than the power of $\frac{1}{2}$ expected for the simple case of unperturbed molecular dimensions. Since the extent of perturbation increases with molecular weight, a higher power dependence is found. This is equivalent to saying that the quantity α in equations (9.7)–(9.9)

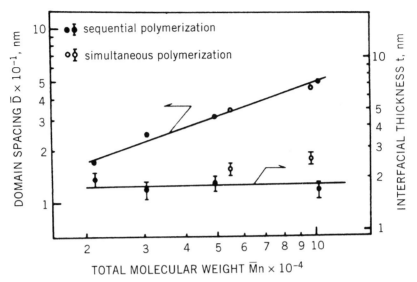

Figure 9.12 Domain sizes D and interfacial thickness t of polystyrene-*block*-polybutadiene diblock copolymers as a function of molecular weight.

may depend on M to the $\frac{1}{6}$th power. In fact, the confined chain model predicts a 0.65 power dependence of the domain spacing on the molecular weights (53).

More recently, work has centered on the effect of shear deformation (54) and the influence of shear (55). One of the results is that the order–disorder transition increases in temperature during mechanical work on the system. Following the cessation of shearing, a poly(ethylene-*alt*-propylene)-*block*-poly(ethylethylene) material was found to have a transition from hexagonally ordered cylinders to disorder at 157°C.

9.3.3.3. Example Calculation of Domain Shapes and Sizes Suppose a polystyrene-*block*-polybutadiene-*block*-polystyrene triblock copolymer is synthesized, with molecular weights in thousands of 10–80–10. What is the morphology of the material?

Since its total molecular weight is 100,000 g/mol, it is 20% polystyrene. Therefore, it is likely to have spherical domains of polystyrene embedded in a matrix of polybutadiene.

What are the sizes of the domains?
Equation (9.7) can be used:

$$R = 1.33 \times 1.25 \times 670 \times 10^{-3}(10,000)^{1/2} \text{ in Ångstroms}$$

$$R = 111\,\text{Å or } 11.1\,\text{nm}$$

The radius of gyration of a 10,000-g/mol polystyrene chain in its relaxed state is 2.75 nm, and its end-to-end distance is 6.7 nm. These dimensions are comparable to the domain size. However, this represents only one block; the chain wanders on. An interesting conclusion from this calculation is that the block copolymer chains as a whole are larger than the domains they occupy.

9.3.3.4. Example Calculation of Domain Contents   How many chains exist in one polystyrene domain in Example 9.3.3.3?

The density of polystyrene is 1.05 g/cm^3, and the molecular weight of one mer is 104 g/mol. The number of mers per cm^3 is

$$(1.05 \, \text{g/cm}^3 / 104 \, \text{g/mol}) \times 6.02 \times 10^{23} \, \text{mers/mol} = 6.07 \times 10^{21} \, \text{mers/cm}^3$$

yielding $1.64 \times 10^2 \, \text{Å}^3/\text{mer}$.

The volume of one domain is

$$V = \tfrac{4}{3}\pi R^3 = 5.72 \times 10^6 \, \text{Å}^3$$

which contains 3.49×10^4 mers.

A 10,000-g/mol chain contains 96.1 mers of polystyrene, so there are 363 chains per domain. The area of one sphere is given by $A = 4\pi R^2$, yielding $1.54 \times 10^5 \, \text{Å}^2/\text{domain}$. The approximate cross-sectional area of one mer is given by the $\tfrac{2}{3}$ power of the volume of one mer, or $(1.64 \times 10^2 \, \text{Å}^3)^{2/3} = 30 \, \text{Å}^2$. The 363 junctions occupy $1.09 \times 10^4 \, \text{Å}^2$, so that approximately 7% of the 5.13×10^3 possible junctions in a domain interphase are occupied. Thus, taking the "narrow interphase approximation" (below) to the limit, 93% of the interphase is occupied by non-junction portions of the chains.

9.3.4. Domain Interphase Thickness

Because of the junctions tying the various blocks in a block copolymer together, these materials are significantly more miscible than their polymer blend counterparts of the same molecular weights, per block. This results in a broader interphase region than in the corresponding polymer blend. Since the domains in block copolymers are usually smaller than the domains in most polymer blends, a correspondingly larger volume fraction of the material is occupied by interphase material.

There are two immediate effects:

1. The mechanical behavior, particularly $\tan \delta$ vs. temperature, is changed; see Section 9.5.
2. The thermodynamics and, in particular, the entropy calculations will be altered by the large interphase volumes. These effects will be examined here.

There have been two basic models developed to treat the interphase thickness. One is called the "confined-chain" model, first presented by Meier (50) and then modified and extended by him and his co-workers to its current form (56).

The mean-field theory was first presented by Helfand and co-workers (29–35), and continued by Noolandi and Hong (57–59). Most of this work focused on the strong segregation limit (SSL), predicting microstructural dimensions in reasonably good agreement with experimental data.The classical interphase distance divided by the actual domain width is given by:

$$a/d_k = (\tfrac{2}{6})^{1/2} b/\chi \gg 1 \qquad (9.11)$$

where b is the segment length. A number of simplifying assumptions can be made, known as the "narrow interphase approximation" (NIA). The NIA works in the range of $\chi Z_k \geq 20$, where Z_k is the degree of polymerization of a block. Leibler (36) proposed a mean-field formalism that identified the weak segregation limit (WSL), and the conditions responsible for the order–disorder transition. More recently, Melenkevitz and Muthukumar (60) and Shull (61) proposed mean-field formalisms that predict microstructural dimensions in both segregation regimes. Spontak and Zielinski (62) examined the dependence of the interphase thickness on chain length in block copolymers, finding that as χN increases, the domain interphase thickness gradually decreased to a limiting value.

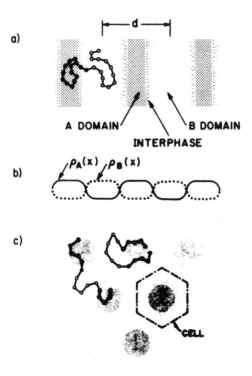

Figure 9.13 Schematic illustration of how block copolymer chains and their domains fit into the polymeric mass. (a) Lamellar microdomains, (b) density profiles indicate that the interphase is of finite width, and (c) cross-section of cylindrical microdomains and their unit cell.

Thus, the theoretical approaches to the problem depend on the assumptions: whether one is far from or near the ODT, and whether one has a narrow or wide interphase. The problem is depicted in Figure 9.13 (63). Note that in Figure 9.13b the density of block A falls off as the density of block B builds up. This occurs over a distance a; see equation (9.11). While the distance d is the entire periodicity distance, d^k is the width of the individual domains. These various theories have been used to good effect in calculating the free energy of domain formation (see Figure 9.11), and phase diagrams (see Figure 9.7). Although the assumptions seem to be quite different, the results are in quite good agreement.

9.4. MORPHOLOGY OF CRYSTALLIZABLE BLOCK COPOLYMERS

In the case of SBS and SIS block copolymers discussed above, the various blocks are amorphous. At room temperature, typically, there is a hard, glassy block, and a soft, elastomeric block. Quite different morphologies and property profiles can be attained if the hard block is crystalline, rather than glassy. While the purely amorphous block copolymers phase separate primarily because of the immiscibility of the two blocks, semicrystalline block copolymers can microphase separate by a second mechanism: crystallization of one or more blocks. Thus, crystallization may take place from a homogeneous melt. If it takes place from a microphase-separated state, the final morphology may be expected to be quite different. The latter may be true if the blocks have high molecular weights, so that they phase separate in the melt state.

It has long been known that the polyester block in poly(ether ester) block copolymers is crystalline (64). For example, when the hard block is poly(tetra-methylene terephthalate) of molecular weight near 1000 g/mol, and the soft block is poly(tetramethylene ether), with molecular weight near 1000 g/mol, the hard block melts at 189°C (65). Similar results have been found for poly(ester ester) block copolymers (66).

Many of the thermoplastic polyurethane elastomer hard blocks also crystallize (67). Typical melting temperatures are in the range of 200°C or slightly higher. Ordinarily, the soft blocks are not crystalline in the relaxed state but may crystallize on extension. It must be remembered that both the poly(ether ester) and the polyurethane types of multiblock copolymers have much shorter blocks than the SBS and SIS types, and thus have significantly greater miscibility in the melt state. Of course, both the polyester and the polyurethane blocks are highly polar, and usually hydrogen bond to each other.

9.4.1. Crystallizable Short Block Copolymers: Model Materials

Rangarajan et al. (68,69) investigated the dynamics of crystallization and concomitant phase separation in a model material, polyethylene-*block*-poly(ethylene-*alt*-propylene), synthesized by hydrogenating a block copolymer based on 1,4-polybutadiene and 1,4-polyisoprene. The molecular weights were

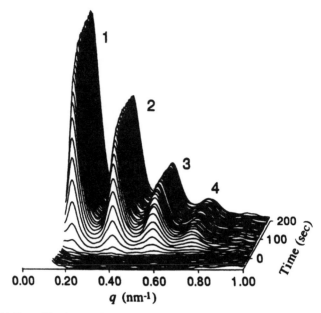

Figure 9.14 Crystallization and phase separation in polyethylene-*block*-poly(ethylene-*alt*-propylene) model copolymer, studied by SAXS at 98°C.

low, in the range of 11,000–50,000 g/mol. These materials were homogenous in the melt state, so that crystallization occurred from the one-phased state.

They used time-resolved SAXS to follow the crystallization; see Figure 9.14 (68). The Y axis, not shown, is q^2I, which magnifies small reflections. Bragg's law relations yields an overall lamellar repeat distance of approximately 28.5 nm. Since the molecular weight of the crystallizable block in this particular sample was 11,130 g/mol, with 0.44 weight fraction of crystallizable block. This corresponds to a 12.5-nm crystalline lamellar thickness with a maximum stem distance of 100 nm, the individual crystallizing chains can make a maximum of eight folds. In the actual experiment, only 20% of the polyethylene actually crystallized, so the number of folds and stems per crystallizable block was less.

In Figure 9.14, the higher order peaks appear at integral multiples of the position of the first-order peak, indicating a rather well-ordered alternating lamellar morphology. At very short times, the sample is observed to be amorphous. The resulting morphology is modeled in Figure 9.15 (69). Here, the crystalline lamellae are shown to be significantly thinner than the domain diameter. The authors also note that in addition to the lamellar morphology, the materials also exhibited a spherulitic superstructure. Interestingly, then, the dimensions of the lamellae do not define the dimensions of the crystalline domains.

When the molecular weights of the polymers are higher, the block copolymer melt may phase separate before crystallization sets in. Douzinas and Cohen (70)

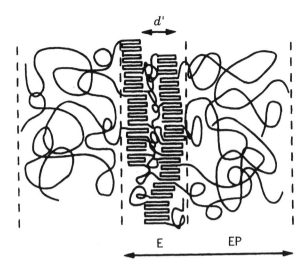

Figure 9.15 Proposed morphology of a polyethylene-*block*-poly(ethylene-*alt*-propylene) block copolymer, with polyethylene segments long enough to phase separate before crystallization.

investigated such a case. They found that their crystallizing block's chain axis runs parallel to the domain interface, rather than perpendicular as shown in Figure 9.15.

9.4.2. Poly(ether ester) Block Copolymer Crystallization

These multiblock copolymers are most often used as fibers, which are highly oriented. Using X-ray orientation studies, Cella and co-workers (71,72) found that the molecular axes of oriented materials were aligned parallel to the stress direction.

Using low-angle laser light scattering and polarized light microscopy, Seymour et al. (73) found spherulitic morphologies for unoriented poly(ether esters). Since these materials crystallized similar to homopolymer materials, the assumption was that the materials crystallized from a single phase. The molecular weights of each block of poly(tetramethylene oxide) and poly(tetramethylene terphthalate) were in the range of a few thousand. Via chemical etching and molecular weight determination, the average fold period was found to be 7 to 8 mer units of the 8,000 g/mol hard block.

Commercial polyester elastomers include Hytrel^R, by du Pont; see equation (9.2), and Pelprene by Toyobo (74). The resulting linear polymers are random segmented copolymers containing (in some cases) approximately 58 wt% tetramethylene terephthalate (4GT) hard block of approximately 1500 g/mol (seven mers). The soft block of poly(tetramethylene oxide) with a terphthalate mer attached (PTMOT), has a number-average molecular weight near 1000 g/mol.

9.4.2.1. SAXS Studies   The theoretical basis of scattering by visible light, X-rays, and neutrons has been reviewed many times. In this section, and several sections to follow, both SAXS and small-angle neutron scattering (SANS) will be utilized. Most of the data of interest here involve one or more peaks in the angular profile of the scattering data. The methods of analyzing SAXS (and SANS) data are summarized in Appendix 9.1, with special reference to data that exhibit peaks in the angular pattern.

While there has been some overlap in the use of SANS and SAXS, since SAXS is a lot cheaper, it has been used wherever possible. Measurement of chain conformation via Zimm or Guinier plots, using deuterated polymers requires SANS. Domain sizes, shapes, and spacing can be done via SAXS.

The crystalline morphology of the poly(ether esters) was examined by Perego et al. (75) and others (76,77) using SAXS. Perego and co-workers found the lamellar crystalline domain thickness to be of the order of 40 Å, depending on the exact nature of the corrections to the results. Noting that the length of the 4GT unit in the crystalline state, projected perpendicular to the lamella surface, is 9.7 Å (d_{001}), this corresponds to around 4 mers. With a mer molecular weight of 220 g/mol, and a block length of around a 1000 g/mol, one may conclude that the average crystalline block traverses the crystalline lamella just once, forming a single stem. For these materials, it must be emphasized that the molecular weight distribution is broad, and hence the crystal structure, especially at the interface, must be somewhat irregular.

The values obtained via SAXS were obtained independently via electron microscopy. Cella (71) stained a cast poly(ether ester) film with phosphotungstic acid, to obtain the morphology shown in Figure 9.16. Lamellar structures of approximately 100 Å thickness and several thousand Ångstroms in length are seen. The morphology is modeled in Figure 9.17 (71). The morphology is

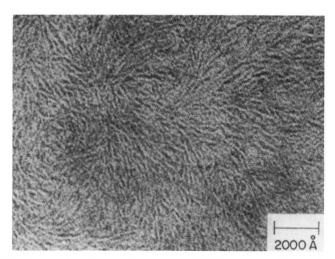

Figure 9.16 Morphology of a polyester film cast from 1,1,2,2-tetrachloroethane solution, stained with phosphotungstic acid, and examined via transmission electron microscopy.

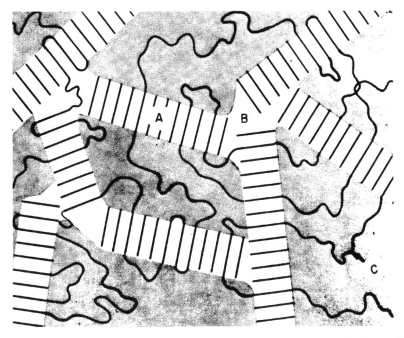

Figure 9.17 Schematic diagram of multiblock copolymer morphology, with one block crystalline. On the macroscopic scale, the system is isotropic.

depicted as containing three different structures: the crystalline domains, the junctions between the crystalline domains, and the amorphous material. Cella drew the crystalline lamellar regions with very few chain folds, it must be noted.

More recently, Bandara and Droscher (78) used SAXS to investigate the effect of annealing on isotropic poly(ether esters). They deduced two quantities of special interest:

1. The diffuse boundary zone between the crystalline and amorphous blocks is approximately 9 Å wide.
2. Samples were annealed at various degrees of undercooling from their nominal melting temperature. The specific internal surface of their samples, O^s ($= S_{sp}$; see Section 4.6.2), increases as the extent of undercooling is increased; see Figure 9.18 (78), meaning smaller or less organized domains. Specific internal surface areas of the order of 100–200 m^2/cm^3 are noted. This is in the true colloid range. As O_s increased, of course, the average chord lengths decreased. For materials having about 6.3 4GT mers per hard block and about 14 poly(tetramethylene oxide) (PTMO) mers per soft block, the chord lengths of the hard block domains were of the order of 50 Å, while those of the soft block domains were of the order of 200 Å in many cases.

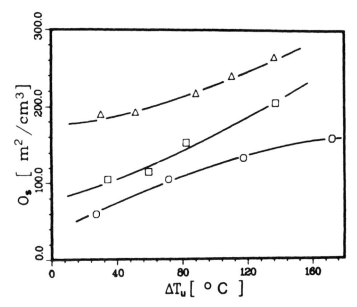

Figure 9.18 Annealing studies of isotropic poly(ether esters) as a function of undercooling. At greater undercoolings, the specific internal surface area increases.

9.4.2.2. SANS Studies While the SAXS experiments reveal the dimensions of the domains, SANS is capable of examining the individual chain conformations through deuteration of probe molecules (79). Miller and co-workers (80,81) examined both the soft and hard segment molecular conformations via SANS. At room temperature, they found that the average soft segment was slightly expanded over that of the random coil conformation, although some of the segments were probably more extended.

The radius of gyration of the hard segments was found to increase with increasing temperature (81). Their notation 5-HS and 10-HS refers to the number of terephthalate units (or 4GT mers) per hard block. For the 5-HS sample, the radius of gyration increased from 12.5 to 15.7 Å. This compares with a value of 18.0 Å calculated for a five mer hard block in an extended crystal conformation. While a 5-mer hard block seems typical of the industry, Miller et al. (80,81) also examined the 10-mer hard block, finding dimensions of 12.5–25.8 Å as the temperature was increased from 25 to 190°C.

Since the radii of gyration for the actual chains were smaller than those expected for the extended chain radii of gyration, Miller et al. (80,81) concluded that chain folding was occurring in the hard segments of these materials. Obviously, more chain folding was occurring in the 10-mer block than in the 5-mer block. At room temperature, in the latter case, one fold associated with every second stem approximately fits the data. Noting the polydisperse nature of the blocks, it is probably the longer blocks that are doing most of the folding. Incidently, this is almost exactly what is drawn into the model in Figure 9.17.

9.4.3. Phase Separation in Segmented Polyurethanes

The polyurethanes present a structure somewhat similar to that already examined in the poly(ether esters); see Section 9.4.1. Typical block molecular weights are in the range of 500–4000, with the hard blocks being either of roughly equal length or somewhat shorter for thermoplastic elastomers. The range of interest for commercial materials seems to be broader than that of the poly(ether esters). The segmented polyurethanes differ from their poly(ether ester) counterparts in another way: Phase separation may precede crystallization or occur simultaneously with crystallization. Also, some types of polyurethane hard blocks have irregular structures, which do not ordinarily crystallize.

9.4.3.1. SAXS Studies   Koberstein and Stein (82,83) used SAXS to investigate the domain size of some linear and crosslinked segmented polyurethane elastomers. For a toluene diisocyanate/ethylene glycol (TDI/EG) material, domain spacings of 12–14 nm were found with a material having 21% of hard segment of molecular weight of 1700 g/mol formed from toluene diisocyanate and ethylene glycol. The authors comment that the asymmetry of the TDI residue in the hard block (based on an 80:20 commercial mixture of the asymmetric 2,4-isomer and the symmetric 2,6 isomer) prevents crystallization. The methyl group on the benzene ring also leads to atacticity. These irregular structures inhibit ordering, and subsequently these hard blocks do not produce significant chain alignment perpendicular to the domain. The diffuse phase boundary thickness for this sample was 0.6 nm.

On the other hand, diphenylmethane diisocyanate/butane diol (MDI/BD) based on the symmetric mer of 4,4'-diphenyl methane diisocyanate chain extended with butane diol can crystallize. Three MDI units correspond to the thickness of a hard-segment domain as determined from the SAXS measurements. Since the average hard domain was about seven mers, a significant amount of chain folding is probable. Since butanediol residues are highly flexible, it is probable that folding occurs in this portion of the mers.

Using single-crystal X-ray methods, model MDI-butanediol hard segment crystal structures were determined; see Figure 9.19 (84). Three mers are depicted. This structure allows for the formation of an optimum number of hydrogen bonds. It is this kind of organization that is likely in the polyurethane thermoplastic elastomers.

A number of recent studies have been carried out using SAXS as well as other methods (85). Koberstein and co-workers (86–89) investigated the kinetics and thermodynamics of crystallization and phase separation using SAXS and differential scanning calorimetry (DSC). While the SAXS provided information about the crystal size and structure, the DSC provided information about the extent of crystallization and multiple types of crystals present. The samples were melted and then annealed at various temperatures. The polyurethanes studied were prepared from MDI/BD and poly(ethylene glycol) soft blocks.

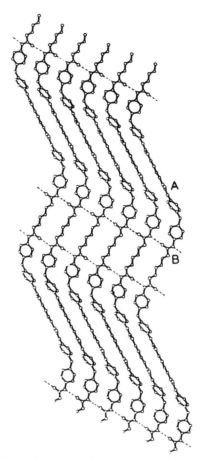

Figure 9.19 Packing of three mers of MDI–butanediol hard segments.

Basic properties of typical specimens were (88) T_m of the hard block phase was 240°C, microphase separation temperature at 150–190°C, T_g of the hard phase 70–110°C, and T_g of the soft phase −40 to +30°C, increasing with increasing hard block content.

At high enough temperatures, MDI/BD hard blocks dissolved in the poly (ethylene glycol) of 2000 g/mol. As in the case of the poly(ether esters), there is a microphase separation transition temperature (TMST). Upon quenching to a temperature (TA) below the TMST, microphase separation and/or crystallization can occur. For TA close to or below the hard phase glass transition temperature, the sample is substantially amorphous, and the DSC thermograms at 120 and 130°C [Figure 9.20 (86)] are independent of TA. This is known as region I. The primary endotherm in this region occurs at the TMST, see peak I, and is associated with dissolution of the noncrystalline microphase structure.

In region II, between the glass transition temperature of the hard phase, 110°C, and TMST, crystallization occurs that leads to two endotherms. At TA

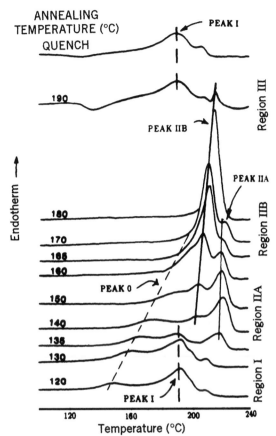

Figure 9.20 Differential scanning calorimetry at 40°C/min of MDI/BDO/PPG polyurethane after melt crystallization at the temperature indicated. The specimen contains 60% by weight of hard segments.

just above the glass transition of the hard phase, the higher melting temperature forms; see region IIA. As TA is raised, the amount of the high melting temperature crystals diminishes while the amount of the lower melting temperature crystals increases, region IIB.

When the annealing temperature was above the TMST, region III, the behavior was, surprisingly, similar to that of region I. The DSC thermograms show an endotherm at the TMST and also exhibit a phase separation exotherm near 185°C.

Wide-angle X-ray diffraction studies of the two crystalline melting peaks shown in region II of Figure 9.20 show that the crystalline structure is identical. However, they belong to two distinct crystal populations that apparently depend on the exact conditions of formation. During each endothermic process, the hard segment crystals not only melt but also spontaneously mix with the soft microphase. Apparently, the exact position of the melting peaks depends very much on the thermal history of the polymer. The lower temperature melting peak is able to reorganize, perhaps from a folded to a more elongated structure, on annealing. One might speculate that one of the crystalline peaks forms via crystallization from the homogeneous state, while the other peak forms from crystallization from an already phase-separated (liquid–liquid) state. Under some conditions, only one melting peak is observed.

The effects of soft segment length and structure (90) and hard segment flexibility (91) were investigated by Chu and co-workers via SAXS. The microphase separation kinetics from a sample quenched from the melt state to some lower annealing temperature could be described by a single relaxation process in most cases. By increasing the PTMO soft segment molecular weight from 1000 to 2000 g/mol, the relaxation time was decreased from about 1000 to 64 s. This is because the shorter soft blocks are dominated to some extent by the immediate presence of the hard block. The hard segment interdomain spacing was found to be 12.6 nm for the 2000 g/mol PTMO soft block, and a material containing 39% hard block. The interdomain spacings remained substantially constant throughout the kinetics of crystallization, due to the effects of connectivity.

9.4.3.2. SANS Studies Miller and co-workers (92–94) examined the conformation of PTMO chains in segmented polyurethanes via SANS. Special deuterated chains were synthesized, having number-average molecular weights of near 1000 g/mol. After correcting for polydispersity, etc., an average value for the number-average radius of gyration of the soft block chains in the polyurethane was 15.6 Å. This value compared to a value of 11.7 Å found by intrinsic viscosity methods in Flory θ solvents (95–97). This represents an expansion of about 35% in the soft block copolymer chains. Therefore, the chains are somewhat extended. Primarily, this is the result of having the chain ends having to be placed at the outer edge of the soft block, at the interface, rather than having the opportunity of being placed randomly. Figure 9.21 (92) provides a model of the morphology of the segmented polyurethane. Note that substantially no chain folding is included, with the soft segments being illustrated as being fairly extended.

The radius of gyration of the PTMO soft segments was also studied as a function of temperature; see Table 9.3 (93). As the temperature increases, the radius of gyration of the soft segments initially decreases. This is thought to be associated with the increased chain mobility at higher temperatures. As the temperature is raised above 160°C, a slight upturn in the soft segment radius of gyration is noted. This upturn may be due to substantial phase mixing at higher temperatures. Small-angle neutron scattering studies on the soft blocks by themselves, in the melt state at room temperature, gave a radius of gyration of 12.1 Å (94). This tends to confirm the more extended chain conformation in the block copolymer.

Chee and Farris (98) proposed a model that may explain in more detail the initial increase in the soft segment radius of gyration. Their model for the

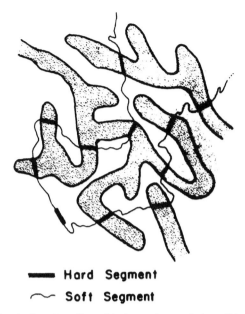

━━ **Hard Segment**

⌒ **Soft Segment**

Figure 9.21 Model of a single polyurethane block copolymer chain as it traverses the hard and soft domains of the system. Note apparent dual-phase continuity in the model.

kinetics of phase separation states that as the temperature rises, the entropic retractive force arising from the extended soft segment chains increases. This may be especially true for the taut tie molecules present in the system. The increased force facilitates the pulling of hard segments out of the hard

Table 9.3 Polyurethane soft segment radii of gyration vs. temperature

Sample	T (°C)	R_g (Å)	Error[a] ($\pm$Å)
No. 1	25	15.6	0.27
	65	15.5	0.42
	95	15.2	0.32
	125	14.4	0.34
	160	14.6	0.43
	190	15.1	0.42
	215	18.4	0.39
No. 2	25	14.9	0.21
	50	14.8	0.47
	80	14.1	0.44
	110	13.8	0.50
	140	13.6	0.50
	170	14.2	0.49

[a] Error estimates listed are 95% confidence limits established by the non-linear regression routine.

phase into the soft phase, allowing the radius of gyration of the soft phase to decrease.

The thermodyamics of pulling crystalline stems out of a lamellar structure via single chain rubber elasticity forces is treated in Section 9.4.3.3. The conclusion is that the more highly extended soft chains may indeed pull single stems out, since the energies holding a relatively short segment are limited.

9.4.3.3. Example Problem: Block Copolymer Stem Pull-Out   "Taut tie molecules" can sometimes pull a crystalline block copolymer out of the lamella. What thermodynamic forces are involved?

A poly(ether ester) block copolymer (see Section 9.4.2) is assumed. It is composed of five 4GT mers and 1000 g/mol of PTMO soft block. The heat of fusion of the 4GT is 7600 cal/mol (of mers).

Calculations   The energy to pull one stem out of its lamella, on a mer basis, after converting to ergs, is five times the energy per mer, or 2.64×10^{-12} ergs. The rubber elasticity force per chain is given by

$$f = 3kTr/r_0^2 \tag{9.12}$$

The work is obtained by integrating the force over the distance,

$$W = \int_0^r = \tfrac{3}{2}kT(r^2/r_0^2) = \tfrac{3}{2}kT\alpha^2 \tag{9.13}$$

$$2.64 \times 10^{-12} \text{ ergs} = 1.38 \times 10^{-16} \text{ ergs/}^\circ\text{K}(298\,^\circ\text{K})(\alpha^2)$$

$$\alpha^2 = 42.8 \qquad \alpha = 6.5$$

The chain must be extended about 6.5 times from its random coil conformation to pull the stem out. This is about the extension of most rubber bands. It must be noted that for very short PTMO chains, such as those specified above, rubber elasticity theory is only very approximately applied.

9.4.3.4. Other Kinds of Blocks   Of course, there are many other hard blocks that have been used or examined for thermoplastic polyurethanes. Materials based on piperazine rings separated by butanediol chain extenders have received considerable attention (99,100). The piperazine rings have made some good model materials because the transurethanization reaction, which may destroy any tailored primary structure, cannot easily take place. This is because the =NCOO– urethane group has no active hydrogens. As mentioned earlier (82,83), the TDI-based polyurethanes are difficult to crystallize, if at all, due to irregularities in their structure. However, they are strongly held together via hydrogen bonding (101). Other kinds of multiblock copolymers include ether–ether, ether–vinyl, ester–ester, and polymers containing polycarbonate, polyamide, polysulphone, and polysiloxane blocks (102).

In summary, both the poly(ether esters) and the polyurethanes are multi-block materials that phase separate primarily because of the onset of crystallinity, rather than liquid–liquid phase separation. The domains tend to be very thin and lamellar in nature.

9.5. IONOMERS

Ionomers are polymers that contain ionic groups in either their backbone or side chains. Because ionic groups are so vastly different from hydrocarbon groups in nature, even a relatively few such groups cause phase separation. Thus, singly spaced ionic groups in concentrations of only 5 mol% may phase separate. Since the ionic groups are tightly bonded to each other, they form a type of physical crosslink similar to that existing in the block copolymers. Also, similar to the block copolymers, the bonds become weaker at elevated temperatures, permitting flow. Therefore, a brief treatment of the materials will be given here. Important evidence for the phase-separated nature of the ionomers was summarized by Eisenberg and King (103) as one-mer blocks.

9.5.1. Types of Ionomers

There are several types of these materials. Elastomers can be physically cross-linked if they contain carboxyl groups. These can be neutralized with zinc oxide to form Zn^{++} ions that hold the chains together. Surlyn[R], a du Pont product, is made by incorporating 5% of sodium carboxylate groups in polyethylene (104). In this case, methacrylic acid is copolymerized with ethylene to make a statistical copolymer. The copolymer is subsequently neutralized, usually with sodium hydroxide. This product is especially useful as a packaging film in the meat-packaging industry, because it is heat sealable in the presence of aqueous fluids. Nafion[R], also a du Pont product, is an ionomer based on polytetrafluoroethylene with sulfonic acid or salt groups placed at the ends of side chains. This material is used as separators in electrochemical applications.

In general, the physical bonding will be tighter as the valence of the cations increases from 1 to 3. Thus, aluminum ions usually produce intractable materials, while ions such as sodium or calcium allow flow at elevated temperatures.

These materials phase separate on several levels. At the mer level, there is multiplet formation, consisting of at most eight ion pairs. These are gathered together in clusters, forming domains of the order of 5–7 nm in diameter. A model has been proposed by MacKnight et al. (105) that combines the results of SAXS, which defines the domain size by the angular position of its peak, by wide-angle X-ray diffraction, and other methods; see Figure 9.22. Obviously, the ionic portions of the molecules are far separated from the aliphatic portions.

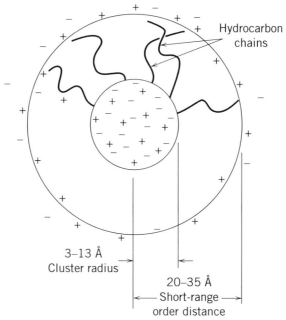

Figure 9.22 In ionomers, the ionic groups attract one another, causing phase separation. A model of the ionic cluster.

9.5.2. SANS Studies

One of the more interesting current debates in the literature concerns the question of whether the presence and subsequent aggregation of the ionic groups causes expansion of the chain to accommodate the selective motion of the ionic groups. Here, SANS experiments come to the fore as one of the best ways to establish chain radii of gyration (106–108). Included in these SANS experiments are two recent ones on polyurethane ionomers (109,110). The results are contradictory, with references 106–110 finding no chain expansion, while references 107 and 109 found modest expansion, of the order of 25%. Apparently, the result depends on the level of ionomer concentration, as well as other chain factors. Theoretically, a modest expansion of chain dimensions should be expected.

More recent work has emphasized the role of ionic groups in increasing polymer blend miscibility. Materials such as carboxylated or sulfonated polystyrene ionomers can be made miscible with polyamides through ion–dipole interactions, which have larger negative heats of interaction than dipole–dipole interactions (111,112). Lithium ions are reported to be more effective at increasing miscibility than sodium ions. The blending of poly-amides with polyethylene ionomers containing ionic carboxylate groups such as sodium methacrylate can lead to significant increases in impact properties (113–115).

9.6. PHYSICAL AND MECHANICAL BEHAVIOR OF BLOCK COPOLYMERS

Since most block copolymers are phase separated, they exhibit either two glass transitions or a glass transition and a melting transition. If the crystalline phase is incompletely crystallized, the usual case, and is present in significant volume fraction, there may be two glass transitions and a crystalline melting transition. While the materials are phase separated, usually phase separation remains incomplete, significantly more so than the corresponding blends. This arises as a direct result of the chain portions being chemically bound to each other. Thus, the glass transition in the polystyrene blocks in an SBS triblock copolymer is usually around 60–95°C rather than near the polystyrene homopolymer glass transition temperature of 100°C, a result of two factors: First, some polybutadiene is dissolved in the polystyrene-rich phase, and second, the blocks have rather low molecular weights. By contrast, the glass transition of polystyrene in styrene–butylene–styrene (SEBS) triblock copolymers is 80–95°C, because of the reduced miscibility (116). These results hold for the standard glass transition conditions of 0.1 Hz, and for polystyrene blocks of 6000–9000 g/mol, the range of many commercial materials.

If the molecular weight of the polystyrene blocks is higher, the glass transition behavior will be more normal. Figure 9.23 (117) illustrates the

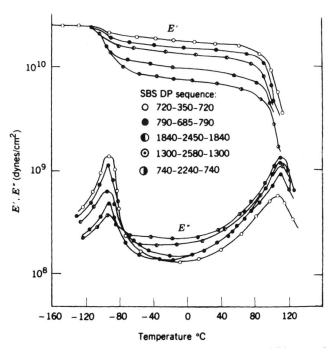

Figure 9.23 Long block copolymers often phase separate to exhibit two glass transition temperatures. Dynamic mechanical behavior of polystyrene-*block*-polybutadiene-*block*-polystyrene for several such sequences.

modulus–temperature behavior of SBS triblock copolymers of various styrene–butadiene ratios (118). All of these materials have enough polystyrene so that it forms a continuous phase; hence the modulus remains high in the plateau region between the two glass transitions. If the polystyrene domains are spherical, then the modulus between the transitions is much lower, around $5 \times 10^7 \, \text{dyn/cm}^2$ (119).

9.6.1. Example Calculation: Young's Modulus

As a first approximation, the modulus in the plateau region can be calculated through the use of the appropriate Takayanagi model. Thus, pure polystyrene has a modulus of $3 \times 10^{10} \, \text{dyn/cm}^2$. The 1300–2580–1300 SBS sample in Figure 9.23 is approximately 50% rubber, having a lamellar morphology. Since the polystyrene is a continuous phase, an upper bound Takayanagi model is needed. As shown in Section 2.1, for 50/50 compositions, the modulus should be about half that of the pure hard block, or about $1.5 \times 10^{10} \, \text{dyn/cm}^2$. Examination of Figure 9.23 shows a modulus only slightly lower than predicted.

9.6.2. Stress–Strain Studies

The stress–strain data on these materials, see [Figure 9.24 (120)], shows increasing moduli with increasing polystyrene content. At 13% polystyrene, there is substantially no phase separation, and the material is very soft, behaving like an undercured conventional vulcanizate. At 28% polystyrene, in the range of commercial thermoplastic elastomers, the material elongates to about 600%, then stiffens greatly. At 80% polystyrene, the material exhibits a stress–strain behavior typical of polystyrene. At intermediate compositions, the materials exhibit a leathery behavior, exhibiting a yield region followed by a drawing region, after which the modulus appears to increase.

Of course, the glassy block provides a temporary type of crosslinking. It also provides a significant amount of reinforcement. In this manner, the hard block behaves like carbon black in rubber, raising the modulus, and increasing the toughness.

The transition behavior of the multiblock copolymers varies with the composition. Table 9.4 (121) shows the glass and melting temperature ranges of the soft blocks in polyurethanes (122). Despite the fact that the melting points of many of the soft blocks are above room temperature, they are normally in the amorphous state. However, elongation induces crystallinity, leading to a self-reinforcing effect. The glass transition temperature of most useful soft blocks must be below $-30°C$, so as to exhibit good elasticity.

MacKnight et al. (123) examined the glass transitions of the hard blocks in segmented polyurethanes (Table 9.5), as well as their melting temperatures (Table 9.6) as functions of composition (124). Here, the type A block is based on an aliphatic hexamethylene diisocyanate, the type B block is based on MDI, while the Type C block is based on 2,4- and 2,6-TDI mixtures. Only types A and

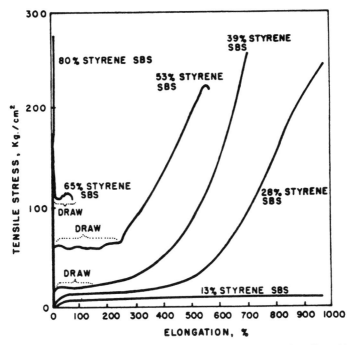

Figure 9.24 Stress–strain curves at 2 in./min for polystyrene-*block*-polybutadiene-*block*-polystyrene block copolymers at various polystyrene contents. Note that decreasing polystyrene contents result in decreasing moduli and increasing strain, as can be predicted by the Takayanagi and other models.

Table 9.4   Important polyols[a] and corresponding thermoplastics polyurethane elastomers[b]

	Polyols		Elastomers
Polyol Nomenclature	T_g, °C	T_m, °C	T_g, °C
Poly(ethylene adipate)glycol	−46	52	−25
Poly(butylene-1,4 adipate) glycol	−71	56	−40
Poly(ethylene butylene-1,4 adipate) glycol	−60	17	−30
Poly(hexamethylene 2,2- dimethylpropylene adipate) glycol	−57	37	−30
Polycaprolactone glycol	−72	59	−40
Poly(diethylene glycol adipate) glycol	−53	/	−30
Poly(hexanediol-1,6 carbonate) glycol	−62	49	−30
Poly(oxytetramethylene) glycol	−100	32	−80

[a] Molecular weight 2000 g/mol.

[b] ca. 85 Shore hardness.

Table 9.5 Glass transition temperatures for urethane homopolymers

X	T_g (°C)		
	Type A	Type B	Type C
2	56	139	52
3	55	119	72
4	59	109	42
5	58	95	52
6	59	91	32
7	55	84	61
8	58	79	64
9	58	72	62
10	55	72	18

B crystallized, while type C, being irregular, remained amorphous. The quantity X represents the number of $-CH_2-$ units making up the polyether soft block. Except for type A blocks, both the glass transition and the melting temperature are seen to decrease as the aliphatic portion of the soft block gets longer.

9.6.3. Dynamic Mechanical Behavior

The dynamic mechanical behavior of some 4GT-PTMO polyurethanes is illustrated in Figure 9.25 (125). The storage modulus, E', is higher in the rubbery plateau region for the case of Q6 because of its higher polyurethane content. The slight rise in E' in the glass transition region for Q5 in Figure 9.25 may be due to some additional crystallization on heating. Above 395 K, the polyurethane domains begin to disorder. This temperature also corresponds to

Table 9.6 Melting points of urethane homopolymers

X	T_m (°C)	
	Type A	Type B
2	166	265
3	163	241
4	183	248
5	157	192
6	171	200
7	151	198
8	162	201
9	154	194
10	161	194

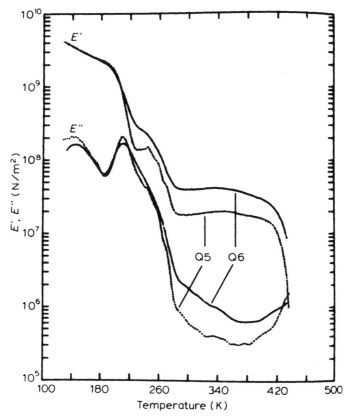

Figure 9.25 Dynamic mechanical behavior of some 4GT-PTMO polyurethanes at 110 Hz, with a heating rate of 2 K/min.

the onset of melting of the hard block, as determined separately by DSC. The modulus drops off quickly above this temperature.

9.6.4. Melt Viscosities

The melt viscosity of block copolymers is usually much higher than those of the equivalent homopolymers or blends. Figure 9.26 (126) shows SBS and poly-butadiene polymers, both of about 75,000 g/mol. The much higher viscosity of the triblock copolymer is attributed to the persistence of a two-phase structure in the melt similar to that existing at lower temperatures. Even though both polymers are above their glass transition temperatures, the polybutadiene block must pull a polystyrene block out of its domain in order for there to be viscous flow; this action costs energy.

As a consequence of their high melt viscosities, triblock copolymers are difficult to process. Formulations often include 50–80 phr of plasticizing oils to reduce their viscosity. Such oils also reduce the cost of the materials. Of course,

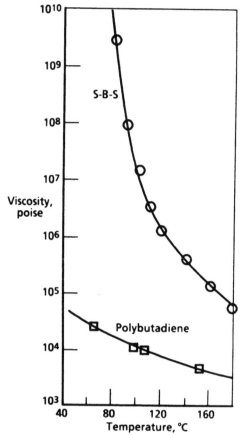

Figure 9.26 At the same total molecular weight and shear level, phase separation causes triblock copolymers to have melt viscosities orders of magnitude higher than the equivalent homopolymer.

the oils also serve to soften the final product, especially those materials in which the polystyrene is in the spherical domain form. Oils low in aromatics are preferred because they do not plasticize the polystyrene block as much. Polystyrene is often used as a compounding ingredient for SBS block copolymers because it acts as a processing aid and makes the final products stiffer.

9.7. APPLICATIONS OF BLOCK COPOLYMERS

Major applications of SBS and SEBS triblock copolymers include shoe soles, sound-deadening materials, wire insulation, sealants, lubricating oil additives, and pressure sensitive adhesives (127). Applications in athletic footwear such as sneakers have been especially important.

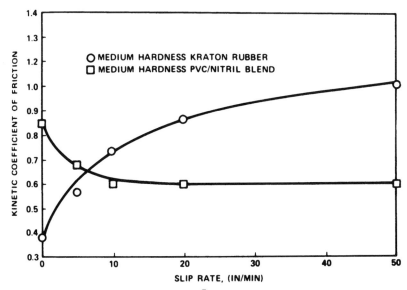

Figure 9.27 Coefficients of friction for a Kraton® triblock copolymer vs. a PVC blend. At high slip rates, the triblock copolymer exhibits the higher kinetic coefficient of friction, highly desired for shoe and sneaker soles. Friction is measured against wet vinyl tile.

9.7.1. Shoe and Sneaker Soles

A particular advantage of the triblock copolymer shoe soles apparently lies in its high coefficient of friction, especially under high slide rates [Figure 9.27 (128)]. One may imagine that under high slide rates, frictional work heats the surface of the shoe soles, up to the glass transition temperature of the hard block. Only the surface 40–50 nm or so need to be heated. Above the glass transition temperature of the hard block, the elastomer becomes an adhesive. (Note the definitions of an elastomer and an adhesive: An elastomer is a crosslinked amorphous polymer above its glass transition, while an adhesive is a linear amorphous polymer above its glass transition temperature.) Above the glass transition temperature of the hard block, under stress the hard block chains may be pulled out of the domains under the strains imposed. By contrast, chemically cross-linked homopolymer elastomers may shred at high slide rates, reducing their coefficients of friction.

9.7.2. Pressure-Sensitive Adhesives

Pressure-sensitive adhesives make use of triblock copolymers with a tackifier added (129). The triblock copolymers must have the elastomer as the only continuous phase, so that the compliance is high, and hence the modulus is low enough. Unlike a plasticizer, the tackifier must both soften the material and increase its glass transition temperature. Noting the Fox equation, the tackifier must have a higher glass transition than the elastomer. It must be miscible only with the elastomer phase, and not with the hard phase. The very soft resulting product allows the adhesive to conform to the surface of the microscopically rough substrate. This is the bonding stage of adhesion, and is relatively slow, of the order of one second. Removal of the adhesive is a much faster process. Here, the function of the tackifier is to adjust the glass transition temperature of the elastomer phase such that the adhesive stiffens up during (attempted) removal from the substrate, increasing the stresses necessary for its removal. Some commercial tackifiers include coumarone–indene polymer, and phenol–form-aldehyde materials. Thus, the concepts of creep compliance and the Williams–Landel–Ferry (WLF) equation are brought into this viscoelastic phenomenon.

For sealant applications, the block copolymer may be added as a hot melt. On cooling, they self-crosslink reversibly, without the need of vulcanization (hence the name, thermoplastic elastomers).

9.7.3. Lubricating Oil Additives

Diblock copolymers have found extensive use as lubricating oil additives (130). At low temperatures, a diblock copolymer such as styrene–ethylene–butylene (S-EB) forms a micellar structure in the oil, contributing little to the fluid's viscosity. Usually, one block remains miscible with the oil, while the other phase separates, creating a stable colloidal system of particles called micelles (131). In the case of S-EB diblocks, it is the polystyrene that forms the micelles, while the EB block remains as a type of cilia in the oil. At higher temperatures, the entire diblock copolymer dissolves in the oil. The larger hydrodynamic volume of the individual random coils increases the oil's viscosity. Of course, without the additive, the viscosity of the oil decreases with increasing temperature. With the diblock copolymer additive, it can be made to remain substantially constant. These products are sometimes called viscosity improvers or multiviscosity or multigraded lubricating oils. Thus, multigraded crankcase oils containing viscosity improvers meet the Society of Automotive Engineers (SAE) require-ments of two or more viscosity grades simultaneously.

Besides the S-EB diblock copolymers, several other types of viscosity improvers exist (132). These include ethylene–propylene copolymers, poly(alkyl methacrylate) copolymers, maleic anhydride–styrene copolymers, polyisobutyl-ene, and poly(vinyl acetate-alkyl fumarate) copolymers. While some of these increase their hydrodynamic volume on increasing temperature as the mechan-ism of viscosity control, others have ionic groups, mixtures of straight-chain and branched alkyl groups, or "runs" of one mer in the chain, all intended to increase the heterogeneity of the system, and hence provide mechanisms for viscosity improvement.

9.7.4. Elastomeric Fibers

The multiblock copolymers also have a variety of uses. The polyurethanes form excellent fibers, sometimes known as Spandex® fibers (124). The term

"Spandex"® applies to elastomeric fibers with a composition of at least 85% segmented polyurethane. These fibers are characterized by a very high elongation at break, very low modulus, and high recovery from large deformations, compared to ordinary crystalline fibers. Thus, the plain elastomeric fiber may have extensions up to 600% (133). Spandex® fibers are usually processed into fabrics as covered yarns. The covering material may be either staple or continuous filament hard fibers, such as polyesters, polyamides, and the like. The fiber is wrapped while it is under stretch, and thus prevented from returning to its original length when the stretching forces are removed. As a result, the fiber operates farther out on the stress–strain curve, with concomitant higher elastic power. Obviously, the total remaining extension is reduced. The cover also protects the fiber. The final fabric retains its high stretch, and most importantly, its high recoverability. Specific applications include swim wear, sock tops, and the elastic threads in foundation garments (134).

9.7.5. Surfactants

There are, of course, many other applications of block copolymers. One that must be mentioned is that of surfactants. One way to make these materials is by using block copolymers of poly(ethylene oxide) and poly(propylene oxide). The former is hydrophilic, while the latter is hydrophobic. Usually the blocks are short. Thus, oily materials are held by the oil-soluble block, while the other pulls it out into the aqueous phase. An important trade name is Pluronics of Wyandotte (124).

9.7.6. Reaction Injection Molding: Polyurethanes and Other Polymers

Reaction injection molding (RIM), is used for producing large polymeric moldings very rapidly (135–137). The reactant monomers, prepolymers, and blocks are intimately mixed by a high-pressure feed impingement technique. A schematic representation of the RIM process is illustrated in Figure 9.28 (137). The components are initially stored in separate tanks under nitrogen or dry air blankets. Precise amounts of each monomer mix is drawn from each tank with a metering cyclinder or pump. The components meet in the mixing head under high pressure. The high-pressure streams meet at 180° angles, providing the mixing energy. Then, a plunger pushes the already reacting fluid into the mold.

Since many RIM parts are foams, the molds are incompletely filled. The vacant area of the mold cavity is filled by the expansion of the foam as the flowing agent is vaporized by the heat of the reaction. The final density may be in the range of 0.7–0.9 g/cm³. The polymer forming reaction takes place in a very short time, usually only 5 s to 5 min. Thus, the "dry" molding can be removed from the mold within a few minutes.

While RIM materials are principally polyurethanes in structure, epoxy and vinyl materials can be used. A major requirement is that the monomer, prepolymer, or block copolymer system have an initially high ratio of molecular

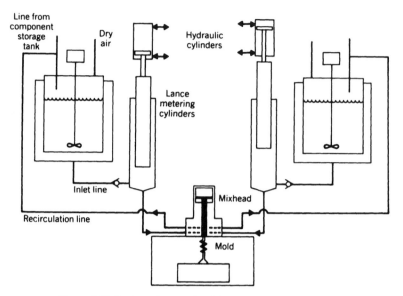

Figure 9.28 Design of a reaction injection molding machine.

weight to reacting sites, so that the temperature rise due to the reaction(s) will be within managable ranges. The reactions themselves should proceed vigorously to ultimate completion.

Chopped glass fiber and/or other fibers and fillers can be incorporated as well. These are called reinforced reaction injection molding, or RRIM.

The RIM and RRIM materials see important applications in the automotive industry, from door panels to front-end moldings to bumpers. Some predictions indicate that the RIM process will soon see application in business machine housings, home construction, and structural parts.

9.7.7. Smart Materials

Block copolymers may also have characteristics of *smart materials*. In one example, polyurethanes (PU) may have shape memories (138). The shape-memory PUs have rather high glass transition temperatures, in the range of 10–50°C. When set in a particular shape above T_g and cooled to a temperature below T_g, they hold this particular shape until heated above T_g at some future time. Then, they return to their original shape. While the process may be repeated, the cyclic action requires the shape to be reset each time. The shape-memory property was ascribed to the molecular motion of the amorphous soft segments, thus making it subject to rubber elasticity theory when heated.

Polyurethanes may also be made to undergo reversible gelation (139). The particular compositions used were based on polyoxyethylene and polyoxypropylene prepolymer chains reacted with hexamethylene diisocyanate. In a simple multiblock copolymer state, the basic compositons resemble the Pluronics described in Section 9.7.4. However, the aqueous solutions of these materials undergo an ODT or an LCST at 30–86°C, depending on the exact compostion. Thus, at low temperatures they are miscible and flow, when the temperature is raised to the critical temperature, the viscosity increases rapidly, with the onset of turbidity.

The thermosensitive polyether polyurethanes according to the present invention (139) contain polyoxyethylene chain(s) and polyoxypropylene chain(s) having average molecular weights of 700–20,000 g/mol in specific proportions. According toTsukanome et al. (139):

> The polyether polyurethane according to the present invention is hydrated at its ether linkage sites and dissolved in water when its aqueous solution is prepared. When this aqueous solution is heated, the polyether polyurethane first undergoes dehydration at its polyoxypropylene chain segment and turns hydrophobic, and so adjacent polyoxypropylene chains associate with each other by hydrophobic bonding, resulting in gelation. When further heated, the polyurethane also undergoes dehydration at polyoxyethylene chains bonded to the polyoxypropylene chains, and so the whole molecule turns hydrophobic to separate from water. That is, the same phenomenon as the clouding point phenomenon characteristic of nonionic surfactants occurs, and the aqueous solution becomes cloudy. This allows the polyether polyurethane according to the present invention to optionally preset the phase transition temperature by appropriately changing the weight ratio between the polyoxyethylene chain(s) and the polyoxypropylene chain(s), which make up the molecule of the polyether polyurethane according to the present invention, and the average molecular weight of the polyoxypropylene chain(s).

Applications mentioned for this system include thermosensitive compositions in applications such as thickeners, gelatinizers, adsorbents, resisting agents, resist printing pastes, printing pastes, release-controlled agents, cosmetic compositions, light-screening materials, adhesives or pressure-sensitive adhesives, dispersion stabilizers, fragrances, heat-regenerating media for cooling, discharge printing pastes and printing adhesives, and production process for these materials. The concept of light screening as a smart material application is also discussed in Section 10.12.3.

It must be pointed out that gelatin in water undergoes a similar transition, gelling when cooled to 40°C. In this latter case, the proteins form triple helices. Edible compositions are usually consumed cold, when half-percent concentrations of protein gel sufficiently to be cut and handled with silverware.

9.7.8. Selected Patents

A computer-driven search of the patent literature revealed in excess of 140,000 patents on block copolymers since 1967! In the spirit of providing a peek at the future, the following is a summary of selected recent patents on block copolymers.

Biomedical applications have received much attention lately. Red blood cells are damaged when exposed to high modulus surfaces. In this area, important advances have been made in developing gradient modulus materials that are strong enough on the interior to be used for mechanical purposes but have very low modulus surfaces, usually in the form of a soft gel. Natural blood vessels have this property. Onishi et al. (140) developed an antithromogenic medical device having outstanding surface lubricity. The surface layer forms a hydrogel when wetted by water. The composition is based on a crosslinked polyurethane, coated with heparin. Heparin helps prevent thrombus formation. Applications include catheters, which then have slimy, low-friction surfaces. Ward and White (141) used surface active end group-containing polyurethanes based on poly-tetramethylene glycol soft segment and a diamine-MDI hard segment. The surface-active end group serves to lower the surface tension to body fluids. Applications include catheters and artificial hearts. Nakabayashi et al. (142) improved hemocompatibility of their materials by using a phosphate salt copolymer coated onto a polyurethane membrane. Presumably, the salt-containing polymer swells in contact with blood fluids.

Wang and Chen (143) made angioplasty catheter balloons from block copolymer thermoplastic elastomers in which the hard segments were based on a polyester or polyamide and the soft segments based on a polyether. The block copolymer has a low flexural modulus, and high strength, the latter attributed to the hard segments, which compose over 50% of the composition.

Coatings, adhesives, printing inks, sealants, and binders have long been important polyurethane and block copolymer applications. St. Clair (144) prepared coatings and adhesives containing polyurethane monohydroxylated diene polymers and epoxidized derivatives. The diene polymers were hydrogenated. After the reactions were complete, the coating film had a 2H pencil hardness. Leir et al. (145) used diamino polysiloxanes that were polymerized with diisocyanates to give block copolymers having repeating units comprised of polysiloxane and urea segments. These block copolymer polysiloxanes can be used to prepare pressure-sensitive adhesive compositions. Noting that these workers assigned their patent to 3M, one will note the relation to Scotch Tape[R]. Jacob (146) made a low-viscosity hot-melt pressure-sensitive adhesive composition from a polystyrene–polyisoprene triblock copolymer composition.

Fibers and nonwovens are another area of interest. Masami and Yamada (147) prepared novel polyether–polyester elastomers for fibers, with polyethylene glycol for the hard segment and a poly(alkylene oxide) soft segment. Umezawa et al. (148) made durable polyurethane fibers suitable for undergarments. These were based on etheylene glycol-MDI and polytetraethylene glycol block copolymers. Ido et al. (149) used an ε-caprolactone-MDI-based polyurethane to prepare Spandex-type fibers with improved heat resistance and elastic recovery. Murayama and Saito (150) used a nonwoven fabric made of a blend of polystyrene-*block*-polyisoprene-*block*-polystyrene, and polypropylene, laminated with a polyurethane film and an acrylic adhesive to prepare porous adhesive bandages.

Thermoplastic elastomers constitute one of the oldest applications of block copolymers, yet it is still a very active area for invention and development. Takahasi et al. (151) developed polyurethane thermoplastic elastomers based on adipic acid and tetramethylene glycol, reacted with 4,4′-diphenyl methane diisocyanate to produce shoe soles and nonskid tires with good abrasion resistance. Naritomi (152) formed a composite body comprising a rigid synthetic resin and a thermoplastic elastomer based on polystyrene–polybutadiene blocks and thermoplastic polyurethane block copolymer elastomers. The thermoplastic block copolymer urethane rubber was fusion bonded to an injection-molded polycarbonate body to prepare goggles.

9.7.9. Trade Names

Lastly, there are very many trade names used in the field of block copolymers; see Table 9.7 (153). Obviously each product is engineered to highly specific specifications. RIM materials are not included because they are sold as monomers, prepolymers, and individual blocks.

9.8. RELATION OF THEORY TO PRACTICE

Some types of block copolymers, particularly the SBS, SIS, and the SEBS types, are among the most regular polymers made up to this time. Hence, they have constituted wonderful model materials, leading to numerous polymer physics experiments and detailed theories of behavior. Clearly, they have also found numerous applications in modern society. Other materials are quite irregular, those of the poly(ether ester) and segmented polyurethane types having polydispersity indices of approximately two for the individual blocks as well as the whole polymer chain. Being irregular in structure, however, has not stopped people from utilizing these materials.

If the blocks are long enough, and/or the thermodynamic parameters different enough, these materials exist in the phase-separated state, where one block is in one domain, and the second block in a second domain. At higher temperatures, they may become miscible, only to phase separate via nucleation and growth kinetics as the temperature is lowered enough.

For shorter blocks, especially those that crystallize, the blocks may remain thermodynamically miscible, but one component crystallized. At a higher temperature, the crystalline portion melts, leading immediately to a homogeneous material. While the exact mechanics of phase separation mimics those of the longer blocks, the details are seen to be quite different.

This chapter has emphasized the thermoplastic elastomers. The idea of a thermoplastic elastomer is seen where the hard block, be it glassy or crystalline, provides a physical type of crosslinking. On heating, the hard blocks soften or

Table 9.7 Trade names and manufacturers of thermoplastic rubbers[a]

Trade Name	Manufacturer	Type[b]	Hard Segment	Soft Segment
a. *General-purpose soluble*				
Kraton D®	Shell Chemical	Triblock (S-B-S or S-I-S)	S	B or I
Solprene 400®	Phillips Petroleum	Branched (S-B)$_n$	S	B or I
Stereon®	Firestone Co.	Triblock (S-B-S)	S	B
Tufprene®	Asahi	Triblock (S-B-S)	S	B
Europrene SOL T®	EniChem	Triblock (S-B-S or S-I-S)	S	B or I
K-Resin®	Phillips	Triblock (S-B-S)	S	B
Styrolux®	BASF	Triblock (S-B-S)	S	B
b. *Improved stability*[d]				
Kraton G®	Shell Chemical	Triblock (S-EB-S)	S	EB
c. *Wire and cable*				
Elexar®	Shell Chemical	Triblock (S-EB-S and S-B-S)	S	EB or B
d. *Hard, abrasion and oil resistant*				
Estane®	BF Goodrich	Multiblock	U	Polyether polyester or polycapro-lactone
Texin®	Mobay Chemical Corp	Multiblock	U	
Pellethane®	Dow Chemical	Multiblock	U	
Cyanoprene®	Amer. Cyanamid Co.	Multiblock	U	
Rucothane®	Hooker Chemical	Multiblock	U	
Hytrel[e]®	E.I. duPont de Nemours & Co.	Multiblock	Polyester	
Lycra (Spandex)®	GAF	Multiblock	U	
Gaflex®	GAF		Polyester	
e. *Hard, low density blend*[f]				
TPR®	Reichhold-Cook		PP or PE	Cross-linked EPDM
Santoprene®	Monsanto Co.		PP	Cross-linked EPDM
Ren-Flex®	Research Polymers		PP	EPDM
Polytrope®	Schulman		PP	EPDM
Somel®	Colonial		PP	EPDM
Telcar®	Teknor-Apex		PP	EPDM
Vistaflex®	Reichhold-Cook		PP	EPDM
Ferroflex®	Ferro		PP	EPDM
ETA®	Republic Plastics		PP	EPDM

[a] Courtesy of the American Chemical Society.

[b] S, polystyrene; B, polybutadiene; I, polyisoprene; EB, poly(ethylene-*co*-butadiene); u, urethane.

[c] Now withdrawn from the market.

[d] Soluble when not compounded.

[e] Similar to polyurethane, but better at low temperature.

[f] Not highly filled.

melt, allowing chain pull-out of the domains, and providing a mechanism for flow. Usually, these materials are quite reversible, capable of going through several cycles with only slow degradation of mechanical properties due to shear and thermal treatments. This latter phenomenon has made for another kind of utility, that of reuse of scrap material. Once the polymer is vulcanized, as in a rubber tire, the chemistry is changed forever. Recovery of materials becomes quite complicated and is today still an incompletely solved problem. Not so with the thermoplastic elastomers!

APPENDIX 9.1

ON THE USE OF SAXS AND OTHER SCATTERING METHODS, ESPECIALLY WHEN SCATTERING INTENSITY MAXIMA APPEAR

Any kind of electromagnetic radiation can be used in scattering experiments to determine information about the size and shape of objects in space. If the scattering objects are uniform enough in size, then one or more maxima will appear in an intensity–angle scattering plot. While all kinds of scattering have much in common, their origin and methods of analysis do differ. What follows expands on Section 1.4.2 and Figure 1.6.

Small-angle X-ray scattering (SAXS) can be employed to measure any structural feature that is of the order of 2 to 100 nm size, if sufficient contrast in scattering power exists, that is, the electron density of the two phases or structures differs significantly. SANS scattering, by difference, utilizes different atomic isotopes rather than electrons. Light scattering uses differences in refractive index. Much of what follows can, in general, be used for all three types of scattering studies.

In wide-angle X-ray scattering (WAXS), the various intensity maxima are usually related via Bragg's law to spacings in crystalline lattices, or in the case of amorphous halos, to the average distances between chains. While peaks in SAXS may be characterized in a simple sense using the Bragg equation to define interdomain spacing and the like, the interpretation of the data today is often very much more sophisticated (154–156). Some definitions of terms as they often appear in the literature:

Bragg's law: For domain spacing estimation, Bragg's law has the form $d = 2\pi/q_{max}$ where d is the interdomain spacing and q_{max} is the value of $q = 2\pi \sin\theta/\lambda$ at the first intensity maximum, and θ is the angle of scattering and λ is the wavelength. Thus, the first maximum in the scattering yields the domain spacing as $d = \lambda/\sin\theta$. Sometimes the scattering angle is defined as 2θ; then the domain spacing is given by $\lambda/(2\sin\theta)$. There are various correction factors added; these usually are not large but are sometimes applied in cases where particular shapes are assumed. The reader should note that different symbols are used by different authors.

Desmearing: This often refers to the correction of the scattering error when slit collimation is used, and the data is reduced to the equivalent pinhole collimation geometry. This term sometimes also refers to edge effect correction, etc.

Methods of plotting data: There are three classical methods of presenting data, each of which assume an isotropic system:

1. A plot of the function Is^2 vs. s, where I is the scattering intensity, and $s = 2(\sin\theta)/\lambda$, where θ is half of the full scattering angle and λ is the wavelength. This is referred to as a Lorentz-corrected plot. One value of using Is^2 vs. I alone is that small peaks at large s are magnified.

2. A plot of Is^4 vs. s is referred to as a Porod law plot, and $\ln(Is^4)$ vs. s, referred to as a modified Porod plot. For an ideal two-phase system with sharp phase boundaries, the product Is^4 should reach a constant value at high enough s values. This approach can be used to determine the nonideal nature of real two-phased systems. Positive deviations from Porod's law, that is, a positive slope in a plot of Is^4 vs. s or s^2, can be attributed to thermal density fluctuations or to phase mixing. On the other hand, the presence of interphase material between the two phases results in negative deviations.

3. Fourier-inverted data, where a one-dimensional correlation function, $\gamma(x)$, is determined from

$$\gamma(x) = \int_0^\infty I(s)s^2 \cos(2\pi xs)\mathrm{d}s \qquad (A9.1)$$

which contains the Lorentz correction. The quantity x is the distance within the material from the origin or zero position. Equation (A9.1) represents the probability of finding the same or a different phase or morphology, starting from an origin or zero position, which is the center of a given phase domain or crystalline domain. A three-dimensional correlation function is also used in the literature. Correlation distances can also be determined; see Section 4.6.2, and related discussion. Today, equation (A9.1) is used to treat the experimental data via computer analysis. Some of the results that can be obtained are illustrated in Figure A9.1. Quantities that can be determined include the interlamellar spacing or long period, denoted by L, the lamella thickness, denoted by C, and the overall (bulk) volume fraction of crystallinity, denoted by V_c.

To determine the radius of gyration, R_g, of these domains, a Guinier plot, $\ln(I)$ vs. s^2 can be used if there are no maxima or minima in the scattering intensity. Here, the angle usually must be very small. Guinier's law reads:

$$\lim_{q\to0} I(q) = I(0)\, e^{(R_g^2 q^2/3)} \qquad (A9.2)$$

where the Guinier plot yields R_g from the slope.

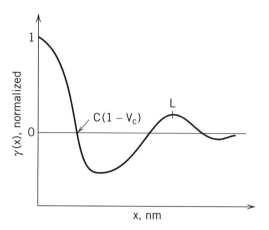

Figure A9.1 Schematic of a correlation function plot, showing the determination of the expressions $C(1 - V_c)$ and L. The quantity C lies between $C(1 - V_c)$ and the first minimum and can be determined analytically. Note the *ringing* of the data, a measure of regularity in the system.

Table A9.1   Information provided by scattering methods

SAXS Relationship	Characteristic Information
Initial slope	Radius of gyration
Overall intensity	Degree of phase separation
Breadth of intensity	Size distribution of the phase
Angular position of maxima	Interdomain spacing
Shape of tail region	Thickness of diffuse boundary between phases (Interphase)

Musselman and Sperling (157) summarize much of the above, as shown in Table A9.1.

REFERENCES

1. H. Iatrou, A. Avgeropoulos, and N. Hadjichristidis, *Macromolecules*, **28**, 6232 (1994).

2. H. Iatrou and N. Hadjichristidis, *Macromolecules*, **27**, 2479 (1993).

3. N. Hadjichristidis, presented at the Polymers for Advanced Technologies meeting, Pisa, Italy, June, 1995.

4. J. T. Bailey, presented at the ACS Rubber Division, October 22, 1965.

5. J. T. Bailey, *Rubber Age*, October, 1966, p. 69.

6. G. Holden, E. T. Bishop, and N. R. Legge, *J. Polym. Sci*, **26C**, 37 (1969).

7. N. R. Legge, G. Holden, and H. E. Schroeder, in *Thermoplastic Elastomers: A Comprehensive Review*, N. R. Legge, G. Holden, and H. E. Schroeder, Eds., Hanser, Munich, 1987, Chapter 1.

8. M. Szwarc, M. Levy, and R. Milkovich, *J. Am. Chem. Soc.*, **78**, 2656, (1956).

9. M. Szwarc, *Nature (London)*, **178**, 1168 (1956).

10. A. Casale and R. S. Porter, *Polymer Stress Reactions*, Academic, New York, Vol. 1, 1978, Vol. 2, 1979.

11. R. S. Porter and Á. Casale, *Polym. Eng. Sci.*, **25**, 129 (1985).

12. R. S. Porter and A. Casale, in *Encyclopedia of Polymer Science and Engineering*, Vol. 9, 2nd ed., Wiley, New York, 1987, p. 467.

13. G. Hurtrez, D. J. Wilson, and G. Riess, in *Polymer Blends and Mixtures*, D. J. Walsh, J. S. Higgins, and A. Maconnachie, Eds., Martinus Nijhoff, Dordrect, The Netherlands, 1985.

14. J. P. Kennedy, in Recent Advances in *Polymer Blends, Grafts, and Blocks*, L. H. Sperling, Ed., Plenum, New York, 1974.

15. J. E. McGrath, in *Block Copolymers: Science and Technology*, D. J. Meier, Ed., Harwood Academic, Chur, Switzerland, 1983.

16. S. Abouzahr and G. L. Wilkes, in *Processing, Structure and Properties of Block Copolymers*, M. J. Folkes, Ed., Elsevier Applied Science, Essex, England, 1985.

17. L. J. Fetters and M. Morton, *Macromolecules*, **2**, 453 (1969).

18. M. Morton and L. J. Fetters, German Offen. 2,042,624 (1971).

19. P. Van Caeter and E. J. Goethals, *TRIP (Trends in Polym. Sci.)*, **3**, 227 (1995).

20. E. J. Goethals, Telechelic Polymers: Synthesis, *Properties and Applications*, CRC Press, Boca Raton, FL, 1989.

21. M. Matsuo, *Jpn. Plastics*, **2**(July), 6 (1968).

22. G. E. Molau, in *Block Copolymers*, S. L. Aggarwal, Ed., Plenum, New York, 1970.

23. D. B. Alward, D. J. Kinning, E. L. Thomas, and L. J. Fetters, *Macromolecules*, **19**, 215 (1986).

24. E. L. Thomas, D. B. Alward, D. J. Kinning, D. L. Handlin, Jr., and L. J. Fetters, *Macromolecules*, **19**, 2197 (1986).

25. E. L. Thomas, D. M. Anderson, C. S. Henkee, and D. Hoffman, *Nature*, **334**, 598 (1988).

26. D. J. Meier, in *Thermoplastic Elastomers: A Comprehensive Review*, N. R. Legge, G. Holden, and H. E. Schroeder, Eds., Hanser, Munich, 1987.

27. Y. Mogi, H. Kotsuji, Y. Kaneko, K. Mori, Y. Matsushita, and I. Noda, *Macromolecules*, **25**, 5408 (1992).

28. S. Sakurai, *TRIP (Trends in Polymer Science)*, **3**, 90 (1995).

29. E. Helfand and Z. R. Wasserman, in *Developments in Block Copolymers*, I. Goodman, Ed., Elsevier, New York, 1982, Vol. 1.

30. E. Helfand, in *Recent Advances in Polymer Blends, Grafts and Blocks*, L. H. Sperling, Ed., Plenum, New York, 1974.

31. E. Helfand, *Macromolecules*, **8**, 552 (1975).

32. E. Helfand, *J. Chem. Phys.*, **62**, 999 (1975).

33. E. Helfand and Z. R. Wasserman, *Macromolecules*, **9**, 879 (1976).

34. E. Helfand and Z. R. Wasserman, *Macromolecules*, **11**, 960 (1978).

35. E. Helfand and Z. R. Wasserman, *Macromolecules*, **13**, 994 (1980).

36. L. Leibler, *Macromolecules*, **13**, 1602 (1980).

37. G. H. Fredrickson and E. Helfand, *J. Chem. Phys.*, **87**, 697 (1987).

38. A. M. Mayes and M. Olvera de la Cruz, *Macromolecules*, **24**, 3975 (1991).

39. M. Olvera de la Cruz, *Phys. Rev. Lett.*, **67**, 85 (1991).

40. A. M. Mayes and M. Olvera de la Cruz, *J. Chem. Phys.*, **91**, 4670 (1991).

41. J. D. Vavasour and M. D. Whitmore, *Macromolecules*, **26**, 7070 (1993).

42. J. D. Vavasour and M. D. Whitmore, *Macromolecules*, **25**, 5477 (1992).

43. E. Helfand and Z. R. Wasserman, in *Developments in Block Copolymers*, Vol. 1, I. Goodman, Ed., Applied Science, Essex, England, 1982.

44. T. P. Russell, T. E. Karis, Y. Gallot, and A. M. Mayes, *Nature*, **368**, 729 (1994).

45. H. Hasegawa, H. Tanaka, K. Yamasaki, and T. Hashimoto, *Macromolecules*, **20**, 1651 (1987).

46. G. Floudas, T. Pakula, E. W. Fischer, N. Hadjichristidis, and S. Pispas, *Acta Polym.*, **45**, 176 (1994).

47. J. Rosedale and F. S. Bates, *Macromolecules*, **23**, 2329 (1990).

48. M. Schuler and B. Stuhn, *Macromolecules*, **26**, 112 (1994).

49. T. Ohta and K. Kawasaki, *Macromolecules*, **19**, 2621 (1986).

50. D. J. Meier, *J. Polym. Sci.*, **26C**, 81 (1969).

51. D. J. Meier, *Polym. Prepr.*, **11**, 400 (1970).

52. T. Hashimoto, M. Shibayama, and H. Kawai, *Macromolecules*, **13**, 1237 (1980).

53. T. Hashimoto, M. Shibayama, H. Kawai, and D. J. Meier, *Macromolecules*, **18**, 1855 (1985).

54. S. Okamoto, K. Saijo, and T. Hashimoto, *Macromolecules*, **27**, 5547 (1994).

55. F. S. Bates, K. A. Koppi, M. Tirrell, K. Almdal, and K. Mortensen, *Macromolecules*, **27**, 5934 (1994).

56. T. Hashimoto, M. Shibayama, H. Kawai, and D. J. Meier, *Macromolecules*, **18**, 1855 (1985).

57. J. Noolandi and K. M. Hong, *Ferroelectrics*, **30**, 117 (1980).

58. K. M. Hong and J. Noolandi, *Macromolecules*, **13**, 727 (1981).

59. J. Noolandi, *Ber. Bunsenges, Phys. Chem.*, **89**, 1147 (1985).

60. J. Melenkevitz and M. Muthukumar, *Macromolecules*, **24**, 4199 (1991).

61. K. R. Shull, *Macromolecules*, **25**, 2122 (1992).

62. R. J. Spontak and J. M. Zielinski, *Macromolecules*, **26**, 396 (1993).

63. E. Helfand and Z. R. Wasserman, in *Developments in Block Copolymers*, Vol. 1, I. Goodman, Ed., Applied Science, Essex, England, 1982.

64. J. R. Wolfe, Jr., in *Block Copolymers: Science and Technology*, D. J. Meier, Ed., Harwood Academic, Chur, Switzerland, 1983.

65. J. R. Wolfe, Jr., *Adv. Chem. Ser.*, **176**, 129 (1979).

66. R. W. M. van Berkel, S. A. G. de Graaf, F. J. Huntjens, and C. M. F. Vrouenraets, in *Developments in Block Copolymers*, Vol. 1, I. Goodman, Ed., Applied Science, Essex, England, 1982.

67. W. Meckel, W. Goyert, and W. Wieder, in *Thermoplastic Elastomers: A Comprehensive Review*, N. R. Legge, G. Holden, and H. E. Schroeder, Eds., Hanser, Munich, 1987.

68. P. Rangarajan, R. A. Register, D. A. Adamson, L. J. Fetters, and W. Bras, *Macromolecules*, **28**, 1422 (1995).

69. P. Rangarajan, R. A. Register, and L. J. Fetters, *Macromolecules*, **26**, 4640 (1993).

70. K. C. Douzinas and R. E. Cohen, *Macromolecules*, **25**, 5030 (1992).

71. R. J. Cella, *J. Polym. Sci., Polym. Symp.*, **42**, 727 (1973).

72. W. H. Buck and R. J. Cella, *Polym. Prepr.*, **14**(1), 98 (1973).

73. R. W. Seymour, J. R. Overton, and L. S. Corley, *Macromolecules*, **8**, 331 (1975).

74. R. K. Adams and G. K. Hoeschele, in *Thermoplastic Elastomers: A Comprehensive Review*, N. R. Legge, G. Holden, and H. E. Schroeder, Eds., Hanser, Munich, 1987.

75. G. Perego, M. Cesari, and R. Vitali, *J. Appl. Polym. Sci.*, **29**, 1157 (1984).

76. W. H. Buck, R. J. Cella, E. K. Gladding, and J. R. Wolfe, Jr., *J. Polym. Sci., Symp.*, **48**, 47 (1974).

77. G. Wegner, T. Fujii, W. Meyer, and G. Lieser, *Angew. Makromol. Chem.*, **74**, 295 (1978).

78. U. Bandara and M. Droscher, *Colloid Polym. Sci.*, **261**, 26 (1983).

79. J. A. Miller and S. L. Cooper, in *Thermoplastic Elastomers: A Comprehensive Review*, N. R. Legge, G. Holden, and H. E. Schroeder, Eds., Hanser, Munich, 1987.

80. J. A. Miller and S. L. Cooper, *Makromol. Chem.*, **185**, 2429 (1984).

81. J. A. Miller, J. M. McKenna, G. Pruckmeyer, J. E. Epperson, and S. L. Cooper, *Macromolecules*, **18**, 1727 (1985).

82. J. T. Koberstein and R. S. Stein, *J. Polym. Sci., Polym. Phys. Ed.*, **21**, 1439 (1983).

83. J. T. Koberstein and R. S. Stein, *J. Polym. Sci., Polym. Phys. Ed.*, **21**, 2181 (1983).

84. J. Blackwell and K. H. Gardner, *Polymer*, **20**, 13 (1979).

85. A. J. Ryan, S. Naylor, B. Komanschek, W. Bras, G. R. Mant, and G. E. Derbyshire, in *Hyphenated Techniques in Polymer Characterization*, T. Provder, M. W. Urban, and H. G. Barth, Eds., ACS Symposium Series No. 581, American Chemical Society, Washington, DC, 1994.

86. J. T. Koberstein, C. C. Yu, A. F. Galambos, T. P. Russell, and A. J. Ryan, *Polym. Prepr.*, **31**(2), 110 (1990).

87. J. T. Koberstein and A. F. Galambos, *Macromolecules*, **25**, 5618 (1992).

88. L. M. Leung and J. T. Koberstein, *Macromolecules*, **19**, 706 (1986).

89. J. T. Koberstein and T. P. Russell, *Macromolecules*, **19**, 714 (1986).

90. B. Chu, T. Gao, Y. Li, J. Wang, C. R. Desper, and C. A. Byrne, *Macromolecules*, **25**, 5724 (1992).

91. Y. Li, Z. Ren, M. Zhao, H. Yang, and B. Chu, *Macromolecules*, **26**, 612 (1993).

92. J. A. Miller, S. L. Cooper, C. C. Han, and G. Pruckmeyer, *Macromolecules*, **17**, 1063 (1984).

93. J. A. Miller, G. Pruckmeyer, J. E. Epperson, and S. L. Cooper, *Polymer*, **26**, 1915 (1985).

94. J. A. Miller and S. L. Cooper, *Makromol. Chem.*, **185**, 2429 (1984).

95. M. Kurata, H. Utiyama, and K. Kamada, *Makromol. Chem.*, **88**, 281 (1965).

96. K. Bak, E. Elefante, and J. E. Mark, *J. Phys. Chem.*, **71**, 4007 (1967).

97. J. M. Evans and M. B. Huglin, *Makromol. Chem.*, **127**, 141 (1969).

98. K. K. Chee and R. J. Farris, *J. Appl. Polym. Sci.*, **29**, 2529 (1984).

99. D. Meltzer, H. W. Spiess, C. D. Eisenbach, and H. Hayen, *Macromolecules*, **25**, 993 (1992).

100. J. A. Kornfield, H. W. Spiess, H. Nefzger, H. Hyen, and C. D. Eisenback, *Macromolecules*, **24**, 4787 (1991).

101. C. S. Paik Sung and N. S. Schneider, *Macromolecules*, **10**, 452 (1977).

102. I. Goodman, in *Developments in Block Copolymer*, Vol. 1, I. Goodman, Ed., Applied Science, Essex, England, 1982.

103. A. Eisenberg and M. King, *Ion-Containing Polymers: Physical Properties and Structure*, Academic, New York, 1977.

104. R. W. Rees, U.S. Pat. 3,264,272 (1966) and 3,404,134 (1968).

105. W. J. MacKnight, W. P. Taggart, and R. S. Stein, *J. Polym. Sci., Polym. Symp.*, **45**, 113 (1974).

106. M. Pineri, R. Duplessix, S. Gauthier, and A. Eisenberg, *Adv. Chem. Ser.*, **187**, 283 (1980).

107. T. R. Earnest, J. S. Handlin, W. J. MacKnight, *Macromolecules*, **14**, 192 (1981).

108. R. A. Register, S. L. Cooper, P. Thiyagarajan, S. Chakrapani, and R. Jerome, *Macromolecules*, **23**, 2978 (1990).

109. R. A. Register G. Pruckmayr, and S. L. Cooper, *Macromolecules*, **23**, 2978 (1990).

110. S. A. Visser, G. Pruckmayr, and S. L. Cooper, *Macromolecules*, **24**, 6769 (1991).

111. A. Molnar and A. Eisenberg, *Polymer*, **34**, 1918 (1993).

112. A. Molnar and A. Eisenberg, *Macromolecules*, **25**, 5774 (1992).

113. Asahi Chemical Industries, Co. Ltd., U.S. Pat. 4,429,076 (1977).

114. E. I. Du Pont de Nemours and Co., Eur. Pat. 34,704 (1981).

115. Allied Corp. U.S. Pat. 4,404,325 (1983).

116. W. P. Gergen, R. G. Lutz, and S. Davison, in *Thermoplastic Elastomers: A Comprehensive Review*, Hanser, Munich, 1987.

117. M. Matsuo, *Jpn. Plastics*, **2**, 6(July) (1968).

118. M. Matsuo, T. Ueno, H. Horino, S. Chujya, and H. Asai, *Polymer*, **9**, 425 (1968).

119. R. Angelo, R. M. Ikeda, and M. L. Wallach, *Polymer*, **6**, 141 (1965).

120. G. Holden, E. Bishop, and N. R. Legge, *J. Polym. Sci.*, **26C**, 37 (1969).

121. W. Goyert and H. Hespe, *Kunstoffe*, **68**, 819 (1978).

122. W. Meckel, W. Goyert, and W. Wieder, in *Thermoplastic Elastomers: A Comprehensive Review*, Hanser, Munich, 1987.

123. W. J. MacKnight, M. Yang, and T. Kajiyama, in *Analytical Calorimetry*, R. S. Porter and J. F. Johnson, Eds., Plenum, New York, 1968.

124. A. Noshay and J. E. McGrath, *Block Copolymers: Overview and Critical Survey*, Academic, New York, 1977, pp. 392–393.

125. M. A. Vallance, J. L. Castles, and S. L. Cooper, *Polymer*, **25**, 1734 (1984).

126. G. Holden and N. R. Legge, in *Thermoplastic Elastomers: A Comprehensive Review*, N. R. Legge, G. Holden, and H. E. Schroeder, Eds., Hanser, Munich, 1987.

127. G. Holden in *Thermoplastic Elastomers: A Comprehensive Review*, N. R. Legge, G. Holden, and H. E. Schroeder, Eds., Hanser, Munich, 1987.

128. G. Holden, in *Recent Advances in Polymer Blends, Grafts, and Blocks*, L. H. Sperling, Ed., Plenum, New York, 1974.

129. G. Kraus, K. W. Rollmann, and R. A. Gray, *J. Adhesion*, **10**, 221 (1979).

130. G. Kraus and D. S. Hall, in *Block Copolymers: Science and Technology*, MMI Press, Chur, Switzerland, 1983.

131. H. Watanabe and T. Kotaka, *Macromolecules*, **16**, 1783(1983).

132. S. K. Baczek and W. B. Chamberlin, in *Encyclopedia of Polymer Science and Engineering*, Vol. 11, 2nd ed., J. I. Kroschwitz, Ed., Wiley, New York, 1988.

133. L. Rebenfeld, in *Encyclopedia of Polymer Science and Engineering*, Vol. 6, 2nd ed., J. I. Kroschwitz, Ed., Wiley, New York, 1986.

134. A. J. Utlee, in *Encyclopedia of Polymer Science and Engineering*, Vol. 6, 2nd ed., J. I. Kroschwitz, Ed., Wiley, New York, 1986.

135. D. C. Allport, C. Barker, and J. F. Chapman, in *Developments in Block Copolymers I*, I. Goodman, Ed., Applied Science, London, 1982.

136. C. W. Macosko, *RIM: Fundamentals of Reaction Injection Molding*, Hanser Gardner, Cincinnati, 1988.

137. L. T. Manzione, in *Encyclopedia of Polymer Science and Engineering*, Vol. 14, 2nd ed., J. I. Kroschwitz, Ed., Wiley, New York, 1988, p. 72.

138. T. Takahasi, N. Hayashi, and S. Hayashi, *J. Appl. Polym. Sci.*, **60**, 1061 (1996).

139. M. Tsukanome, I. Kuroda, K. Kamio, M. Kawashima, and M. Moriya, Eur. Pat. Appl. EP 692,506 (1996).

140. M. Onishi, K. Shimura, and N. Ishii, Can. Pat. Appl. CA 2,153,466 (1996).

141. R. S. Ward and K. A. White, PCT Int. Appl. WO 95 26,993 (1995).

142. N. Nakabayashi, K. Ishihara, S. A. Jones, and P. W. Stratford, PCT Int. Appl. WO 95 05,408 (1995).

143. L. Wang and J. Chen, PCT Int. Appl. WO 95 23,619 (1995).

144. D. J. St. Clair, U.S. Pat. 5,459,200 (1995).

145. C. M. Leir, J. J. Hoffman, L. A. Tushaus, G. T. Widerholt, M. H. Mazurek, and A. A. Sherman, PCT Int. Appl. WO 95 03,354 (1995).

146. L. E. Jacob, PCT Int. Appl. WO 95 16,755 (1995).

147. M. Masami and H. Yamada, JP 07,316,277 (1995); CA 124:204814g.

148. M. Umezawa, T. Watanabe, and H. Nakanishi, PCT Int. Appl. WO 95 23,883 (1995).

149. Y. Ido, S. Nakamura, and H. Suzuki, JP 07,292,063 (1995); CA 124:204807g.

150. E. Murayama and T. Saito, Eur. Pat. Appl. EP 673,657 (1995).

151. N. Takahashi, W. Norihiro, and K. Wakabayashi, JP 07,330,857 (1995); CA 124:204685r.

152. M. Naritomi, U.S. Pat. 5,472,782 (1995).

153. G. Holden, in *Encyclopedia of Polymer Science and Engineering*, Vol. 5, 2nd ed., J. I. Kroschwitz, Ed., Wiley, New York, 1986.

154. M. E. Myers, A. M. Wims, T. S. Ellis, and J. Barnes, *macromolecules*, **23**, 2807 (1990).

155. D. Tyagi, J. E. McGrath, and G. L. Wilkes, *Polym. Eng. Sci.*, **26**, 1371 (1986).

156. R. J. Goddard and S. L. Cooper, *J. Polym. Sci., Part B, Polym. Phys.*, **32**, 1557 (1994).

157. S. G. Musselman and L. H. Sperling, to be published.

10

INTERPENETRATING POLYMER NETWORKS

10.1. DEFINITION OF TERMS

Interpenetrating polymer networks (IPNs) are defined as combinations of two or more polymers in network form, at least one such polymer being polymerized and/or crosslinked in the immediate presence of the other(s) (1–4). The original concept of IPNs suggested materials that had extensive molecular interpenetration, that is, were miscible. In practice, most IPNs form immiscible compositions, usually phase separating during some stage of the polymerization. Figure 1.1 compares the molecular topology of an IPN with those of the common polymer blends, blocks, and grafts.

Many different kinds of IPNs exist. The following definitions identify some of the more important types. While these definitions imply two-polymer IPNs, the concepts are perfectly general, and many three- and four-polymer IPNs have been prepared. Additional variations on the theme will be developed later in the chapter.

Sequential IPN. Polymer network I is synthesized first. Then, monomer II plus crosslinker and activator are swollen into network I and polymerized in situ, see Figure 10.1a (5).

Simultaneous interpenetrating network (SIN). The monomers and/or prepolymers plus crosslinkers and activators of both networks are mixed, followed by simultaneous polymerizations via noninterfering reactions; see Figure 10.1b. Interference is minimized if one polymerization involves chain polymerization while the other involves step polymerization kinetics. While both reactions proceed simultaneously, the rates of polymerization are rarely identical.

Latex IPN. In this case, the polymers are synthesized in the form of latexes, each particle constituting a micro-IPN. Frequently, a core–shell structure develops to a greater or lesser extent. Of course, after film formation, further

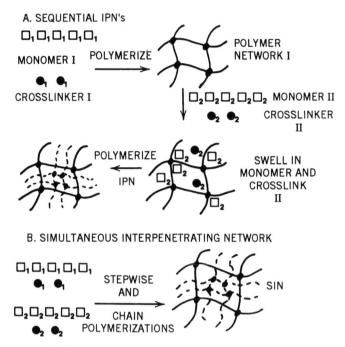

Figure 10.1 Synthesis of sequential and simultaneous types of IPNs.

crosslinking between the particles may take place. A related material utilizes the mixing of two kinds of latex particles, followed by film formation and crosslinking of both polymers usually in the form of a three-dimensional mosaic structure. This latter material is known as an interpenetrating elastomeric network (IEN), especially when both polymers are above T_g.

Gradient IPN. The overall composition or crosslink density of gradient IPNs varies from location to location on the macroscopic level. One way of preparing these materials involves partial, nonequilibrium swelling of polymer network I by the monomer II mixture with crosslinker, etc., followed by rapid polymerization before diffusional equilibrium takes place. Thus, a film can be made with polymer network I predominantly on one surface and polymer network II on the other surface, with a gradient composition existing throughout the interior.

Thermoplastic IPNs. These materials contain physical crosslinks rather than chemical crosslinks, and as such are hybrids between polymer blends and IPNs. Such crosslinks may utilize block copolymers, ionomers, and/or semicrystallinity. As thermoplastics, they flow at elevated temperatures.

Semi-IPNs. Assuming a combination of only two polymers, on selected introduction of crosslinks, there are four possibilities: If there are no crosslinks, the composition is actually a polymer blend. If both polymers contain crosslinks, it is an IPN, sometimes distinguished as a full IPN; see sequential and simultaneous IPNs above. If only one of the two polymers is crosslinked, then two possibilities exist, if the synthesis of the two polymers was sequential in nature: If polymer I contains the crosslinks, it is called a semi-I IPN. If polymer II contains the crosslinks, it is called a semi-II IPN. At this time, no generic notation has been developed for SIN syntheses where only one polymer is crosslinked. These must be distinguished by naming the polymers and crosslinkers.

Hundreds of topological variations of the above materials exist in the scientific and patent literature. Of course, each topological variation can be made with a variety of polymers. Since these materials are important to industry, new IPNs are appearing in the literature at a rate difficult for any one person to follow completely.

10.2. PHASE SEPARATION AND MORPHOLOGY IN IPNS

The development of phase separation was investigated for the sequential IPN *net*-polybutadiene-*inter-net*-polystyrene (6). First, the polybutadiene network was swollen with styrene plus divinyl benzene and peroxide. The styrene was partly polymerized and the remaining monomer evaporated. The polybutadiene portion was stained with osmium tetroxide, and salami slices of some 60 nm thickness were cut for transmission electron microscopy. The resulting morphology is shown in Figure 10.2 (6). At 7% conversion of the styrene, Figure 10.2J, spherical domains appear. At 46% conversion (Figure 10.2K), the

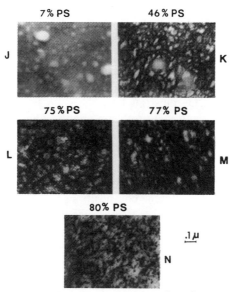

Figure 10.2 Evolution of morphology in a *net*-polybutadiene-*inter-net*-polystyrene sequential IPN during polymerization of the polystyrene. See text for description.

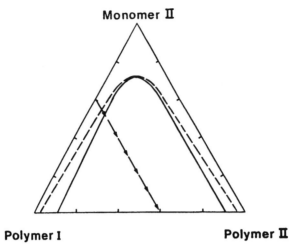

Figure 10.3 Phase diagram for the polymerization of monomer II in the presence of polymer I.

spheres are still present, but a new, ellipsoidal morphology appears. The latter was interpreted as the cutting of cylindrical domains at various angles. At higher conversions, the morphology gives the appearance of a snowstorm, being significantly disordered. This last was interpreted as being caused by the important presence of crosslinks and some grafts, which reduced the mobility of the forming polystyrene to phase separate properly.

In addition, small-angle neutron scattering (SANS) and other experiments were performed on this system (7,8). The resulting phase diagram is shown in Figure 10.3 (7). Below about 3% conversion of the polystyrene, the system is clear, the system scattering only a low intensity of neutrons, presumably being one phased. The dashed line in Figure 10.3 is the binodal, where nucleation and growth begins in the range of approximately 3–6% conversion. At 7% conversion, the system is significantly phase separated, as already indicated in Figure 10.2. The elliptical figures, interpreted as cuts through cylinders, is indicative of spinodal decomposition, as illustrated by the solid line in Figure 10.3. The spinodal decomposition is thought to begin at approximately 10–25% conversion of the styrene, according to the SANS experiments. When the polymer I–polymer II line is reached, polymerization is complete, in this case a sample of some 80% polystyrene.

Donatelli et al. (9) compared the morphology of completely polymerized *net*-polybutadiene-*inter-net*-polystyrene compositions with those of the semi-IPNs and polymer blends in Figure 10.4. If there are no crosslinks, then a solution graft copolymer develops. If the solution is *stirred* during the polymerization, then a phase inversion takes place; see Figure 10.4, upper left. This material is the well-known high-impact polystyrene (HIPS); see Chapter 8. If the material is polymerized quiescently, then it does not undergo phase inversion and polybutadiene remains the continuous phase; see Figure 10.4, upper right. This last material was invented in 1927 by Ostromislensky (10). (One might be

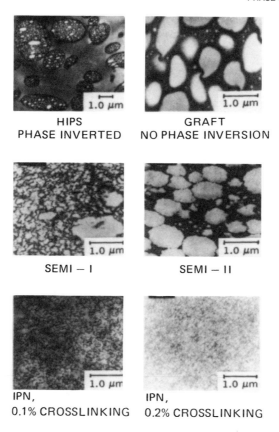

HIPS
PHASE INVERTED

GRAFT
NO PHASE INVERSION

SEMI — I

SEMI — II

IPN,
0.1% CROSSLINKING

IPN,
0.2% CROSSLINKING

Figure 10.4 Several morphologies of polybutadiene–polystyrene compositions via TEM of OsO$_4$ stained salami slices.

tempted to speculate that if Ostromislensky had proper stirring equipment in his laboratory, HIPS might have been invented a quarter century earlier than described in Chapter 8!) The middle left and middle right compositions are, respectively, the semi-I and semi-II compositions. The bottom left and right are the full sequential IPNs, at lower and higher crosslink levels in the polybutadiene, as indicated.

As shall be shown, the mechanical behavior follows the morphology. Thus, the upper left has an Izod impact resistance of approximately 1.5 ft-lb per inch of notch, the middle left about 3 ft-lb per inch of notch, and the bottom left composition about 5 ft-lb per inch of notch.

10.3. PHASE DIAGRAMS

Sophiea et al. (11) worked out a classical composition–temperature phase diagram for the semi-IPN *net*-polyurethane-*inter*-poly(vinyl chloride); see

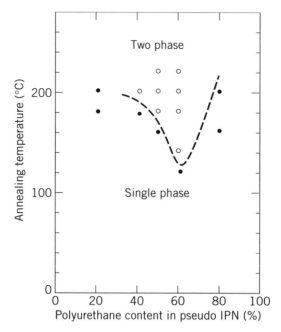

Figure 10.5 Phase diagram for a poly(vinyl chloride)–polyurethane semi-IPN. Note lower critical solution temperature.

Figure 10.5. In this case, the polyurethane was based on a mixture of two polyols: poly(ε-caprolactone diol) (PCL) and poly(oxypropylene diol) (PPG). The PCL portion is miscible in all proportions with the poly(vinyl chloride) (PVC) while the PPG portion is immiscible with the PVC.

The lower critical solution temperature for this system appears at approximately 120°C, when the polyurethane (PU) is 80/20 PCL/PPG based. Below this temperature, the system is one-phased, and above, two-phased. This type of phase diagram is characteristic of polymer blends.

10.4. APPLICATION OF THE TTT CURE DIAGRAM

Kim et al. (12) developed a new time–temperature–transformation (TTT) cure diagram for the system poly(ether sulfone)-*inter-net*-epoxy semi-II IPN. They mixed tetraglycidyl 4,4′-diaminediphenylmethane with dicyandiamide curing agent for the epoxy component and poly(ether sulfone) (PES) of $M_n = 1.77 \times 10^3$ g/mol. Without the curing agent, a lower critical solution temperature (LCST) of 265°C was found. The main instruments used were light scattering, torsional braid analysis, differential scanning calorimetry (DSC), and scanning electron microscopy (SEM).

For the complete system, the TTT cure diagram is shown in Figure 10.6 (12). The transformation involves five steps for semi-IPN formation: onset of phase

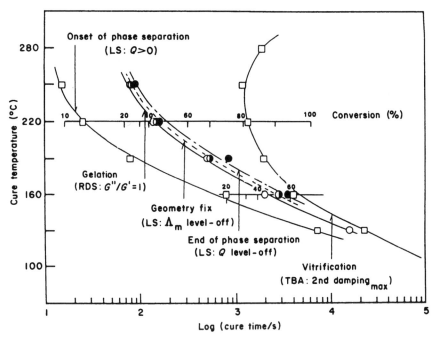

Figure 10.6 Time–temperature–transformation (TTT) cure diagram for the system poly(ether sulfone)-*inter-net*-epoxy semi-II IPN. The instruments employed are LS, light-scattering; RDS, dynamic mechanical spectroscopy via a Rheometrics; and TBA, torsional braid analysis.

separation, gelation of the epoxy component, fixation of the phase-separated morphology (domain size becomes fixed), end of phase separation, and vitrification of the epoxy component. For the gelation step, the method of Winter and Chambon (13,14) was used. In this method, the dynamic loss and storage moduli, G'' and G', respectively, are plotted vs. cure time. At first, the G'' term is larger than the G' term. The gel point is taken as the crossover point of these two quantities.

Figure 10.6 shows that the onset of phase separation precedes gelation for this system. After gelation of the epoxy, the domains are still able to increase in size slightly, then become fixed in size before the end of the phase separation phenomenon. In the last stages of phase separation, there is an increase in the compositional difference between the poly(ether sulfone) (PES)-rich and the epoxy-rich domain regions, as indicated by the light-scattering intensities.

10.5. THE METASTABLE PHASE DIAGRAM

10.5.1. Three Significant SIN Events

In Figure 10.6, it was noted that phase separation took place *before* gelation. For an SIN, the system actually has three such significant events: gelation of polymer I, gelation of polymer II, and phase separation of polymer I from polymer II. In general, these three events may occur in any time order, the number of independent ways of carrying such a polymerization being three factorial or six such possible ways. Each of these six possible ways is expected to produce a specific morphology, with concomitant specific sets of physical and mechanical properties. By contrast, in the case of sequential IPNs gelation of polymer I always is the first significant event.

As an example, let us consider polyurethane-based SINs, which are very popular materials. Most frequently the other polymer is poly(methyl methacrylate) (PMMA), followed by polystyrene. In fact, there are some 50 such studies treating these systems (15). The morphologies, where studied, vary significantly. In dynamic mechanical behavior studies, some of these SINs show two well-defined glass transitions (16,17), one broad glass transition (18,19), or glass transitions shifted inward toward each other (20,21), all from substantially the same starting materials. One possible explanation is that the time order of the three events mentioned above might be different for these otherwise very similar chemistries.

An important problem in characterizing these morphologies and glass transition behavior patterns has been the lack of approach to true equilibrium. The presence of two interlocking networks prevents complete phase separation with two completely relaxed networks. Also, since one or both of the networks may vitrify, the kinetic approach to equilibrium may be further slowed. This aspect of metastability has been experimentally observed by Zhou and Frisch (22,23) for the SIN system PU–PMMA. The term *metastable phase diagram* was developed in an effort to portray the lack of equilibrium, while delineating possible approaches to study the time order of the three events (24–26).

10.5.2. Tetrahedron Construction

Figures 10.7 (24) and 10.8 (24) describe the PU–PMMA SIN system in the form of tetrahedrons. While all of the information for the metastable phase diagram can be represented in only one figure, such a drawing has too many points and lines for easy analysis. The four vertices of both figures represent the four pure components: the two monomers and the two polymers, represented in these figures as MMA, "U," PMMA, and PU. The four triangular faces of both figures represent the ternary systems delineated by the compositions indicated in the vertices. The inset in Figure 10.7 indicates experimental detail as to how the curvilinear line "U"-A-D-C was constructed, see below.

In the specific system studied, the PMMA contained 0.5% tetraethylene glycol dimethacrylate, and gelled after about 8% conversion, indicated by the G_1-U-PU plane (Figure 10.7). The phase separation curve for the ternary system MMA-PMMA-"U" on polymerization of only the MMA is indicated by the points C-D-A-E. Similarly, the phase separation curve for the MMA-PMMA-PU system (rear triangle, Figure 10.7), is indicated by the points J-B-K-L. The entire tetrahedron volume is divided into two regions: one phase separated and

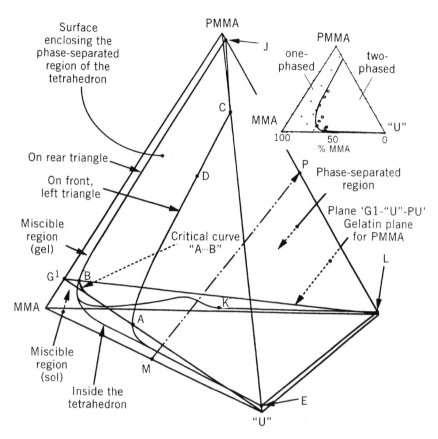

Figure 10.7 Tetrahedron metastable phase diagram for the SIN polymerization of poly(methyl methacrylate) and polyurethane.

one single phased, separated by *C-D-A-E-L-K-B-J*. The characteristic sail-like shape of this surface is of interest, and tracing it aids in following the separation of the two regions of space.

The inset in Figure 10.7 shows the actual experimentally determined curve (*C-D-A-E* of the tetrahedron), with data points for near the front left face of the tetrahedron. The filled points indicate a one-phased determination; the open points a two-phased determination. The monomer conversion was determined as a function of time via Fourier transform infrared (FTIR), while the actual phase separation times were determined visually with the onset of turbidity. The gelation times were also determined visually, by noting the point where a vial of reacting material, upon inversion, failed to flow significantly.

The curvilinear line *A-B*, indicating the intersection of the PMMA gelation plane with that of the phase separation sail-like surface, represents the critical line along which there is simultaneous gelation of PMMA and phase separation of the PU from the PMMA. Reactions moving to the left of this curve will have

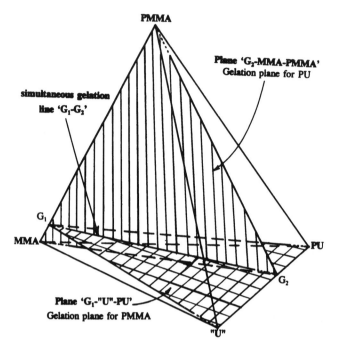

Figure 10.8 Continuation of Figure 10.7, illustrating the gelation planes of both poly(methyl methacrylate), G_1, and polyurethane, G_2. The intersection of these two planes delineates a line of simultaneous gelation.

the PMMA gel before phase separation, while reactions to the right of A-B will phase separate before gelation.

Figure 10.8 illustrates the gelation plane of the polyurethane, G_2-MMA-PMMA, as well as the G_1 gelation plane of the PMMA, described above. The intersection of the two planes, G_1-G_2 describes the line of simultaneous gelation of both polymers. Reactions passing to one side or the other of this line will have one polymer or the other gel first. The line G_1-G_2 of Figure 10.8 also intersects the line A-B of Figure 10.7, not shown. The intersection represents a triple critical point, where both polymers simultaneously gel and phase separate. Thus, this last intersection might be used to define the ideal SIN synthesis.

All SIN (and IPN polymerizations) must begin somewhere along the edge line MMA-"U", and end along the edge line PMMA-PU, noting the line M-P. However, the polymerization lines need not be straight or even single lines. Starting at M, the polymerization will intersect the sail-like phase separation surface. After that, the line divides into two portions, in this case ending at points J and L. (Ending at the point P would imply a miscible SIN composition.)

A straight line M-P along even the initial portion of the line would indicate SIN polymerization with equal polymerization rates. Usually, the polymerization rates are quite unequal.

A sequential IPN synthesis can also be represented by Figures 10.7 and 10.8 type of tetrahedrons. Taking the sequential IPN system PU-PMMA, with the PU polymerized first, leads to a polymerization along the "U"-PU line. On swelling in the MMA component, the locus is moved to a point on the MMA-PU line. (Note that so far, phase separation has not taken place!) On polymerization of the MMA in the triangle making up the back face, phase separation takes place, and the final compositions again emerge at *J* and *L*. However, one would expect the morphologies and properties to be different from the SIN synthesis described above.

Most interesting is that Figures 10.7 and 10.8 can be generalized to any and all SIN and IPN materials, except perhaps the thermoplastic IPNs and the gradient IPNs. The semi-IPNs will clearly lack the appropriate gelation plane. The corresponding polymer blend polymerization of monomer II in the presence of polymer I (see Chapter 8), will contain only the phase separation curvilinear plane, without either gelation plane.

10.5.3. Temperature-Dependent Ternary Phase Diagrams

Ternary phase diagrams individually provide a wealth of information about phase separation kinetics and thermodynamics; see Figure 10.3. If the experiment is repeated as a function of temperature, much more information can be obtained.

In very recent work, Du Prez et al. (27) investigated the behavior of ternary phase diagrams of poly(ethylene oxide)-*inter-net*-poly(methyl methacrylate) semi-II IPNs as a function of temperature; see Figure 10.9. The cloud points were determined visually. At the onset of turbidity, the samples were cooled and precipitated in chloroform, a solvent for the poly(ethylene oxide) (PEO), and remaining monomer. Thus, the insoluble fraction was the crosslinked PMMA. While the temperature range was narrow, the available evidence (especially by comparison to the corresponding blend figure, not shown) points to an upper critical solution temperature (UCST). This is in contrast with the results of Sophiea et al. described in Section 10.3, where an LCST phase diagram was found.

10.6. PHASE DOMAIN SIZE IN IPNS

10.6.1. Donatelli and Yeo Equations

In amorphous block copolymers such as the polystyrene-*block*-polybutadiene-*block-polystyrene* materials (Chapter 9), the phase domain size is governed primarily by the molecular weight of the individual blocks. In the case of sequential IPNs, the phase domain size is controlled primarily by the molecular weight between crosslinks of polymer I. There have been two derivations of equations for determining the phase domain size of polymer II in sequential

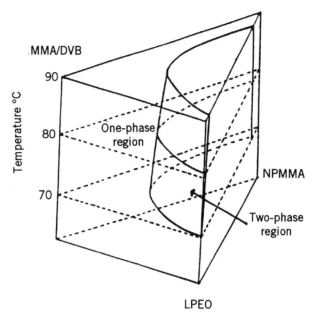

Figure 10.9 Ternary phase diagram illustrating the temperature dependence of phase separation curves of MMA/NPMMA/LPEO semi-IPNs at different temperatures during synthesis. M_n(TEST-PEO) = 2500 g/mol; BPO and DVB: 3 wt% with regard to MMA. Notation: TEST, α,ω-bis(triethoxysilane) terminated; L, linear; N, network.

IPNs after complete polymerization has taken place. The first of these was by Donatelli et al. (28), followed by the work of Yeo et al. (29). Both of these derivations assumed that the polymer II disperse phase diameter, D_2, addressed spherical domains. More recent studies (see Section 10.2) suggest that the domains of interest were formed via spinodal decomposition kinetics and are cylindrical. While the Donatelli equation was semi-empirical in nature, the Yeo equation made fewer assumptions. The Donatelli equation may be written

$$D_2 = 2\gamma v_2/RT\nu_1[1/v_1^{2/3} - \tfrac{1}{2}] \tag{10.1}$$

and the Yeo equation may be written

$$D_2 = 4\gamma/RT(A\nu_1 + B\nu_2) \tag{10.2}$$

where

$$A = \tfrac{1}{2}(1/v_2)(3v_1^{1/3} - 3v_1^{4/3} - v_1 \ln v_1) \tag{10.3}$$

and

$$B = \tfrac{1}{2}(\ln v_2 - 3v_2^{2/3} + 3) \tag{10.4}$$

where γ is the interfacial tension, RT the gas constant times the absolute temperature, ν_1 and ν_2 represent the concentration of effective network chains of polymers I and II, respectively, and v_1 and v_2 represent the volume fraction of polymers I and II, respectively.

10.6.2. Example Calculations

In the following, a solution to the Yeo equations [(10.2), (10.3), and (10.4)] will be examined. The full IPN of polybutadiene and polystyrene will be taken as a basis; see lower left and lower right figures of Figure 10.4. An important value to be determined is the interfacial tension. This value is approximated as 3 ergs/cm^2 at 100°C; see Table 4.2 and also by the calculational scheme by Bianchi et al. (30). According to the data by Donatelli et al. (9), the molecular weights between crosslinks for the lower left and lower right compositions, respectively, were 5.0×10^4 g/mol and 2.7×10^4 for the polybutadiene, and 8.6×10^3 g/mol for the polystyrene. The compositions of the final IPN were approximately 20% polybutadiene and 80% polystyrene in both cases. Then, the value for A in equation (10.3) is given by:

$$A = \tfrac{1}{2}(1/0.8)[3(0.2)^{1/3} - 3(0.2)^{4/3} - 0.2\ln 0.2]$$
$$A = 1.078$$

The value for B in equation (10.4) is given by:

$$B = \tfrac{1}{2}[\ln 0.8 - 3(0.8)^{2/3} + 3]$$
$$B = 0.096$$

Substituting these values into equation (10.2) yields

$$D_2 = 4 \times 3\,\text{ergs/cm}^2/8.31 \times 10^7\,(\text{ergs/mol K}) \times 373\,\text{K}$$
$$\times [1.078 \times 2 \times 10^{-5}\,(\text{mol/cm}^3) + 0.096 \times 1.16 \times 10^{-4}\,(\text{mol/cm}^3)]$$

$$D_2 = 1.184 \times 10^{-5}\,\text{cm} \quad \text{or} \quad 1184\,\text{Å}$$

for the lower left composition, assuming that $\nu = \rho/M_c$, where ρ, the density of polybutadiene is approximately 1.0 g/cm^3 and that of polystyrene is approximately 1.05 g/cm^3. A corresponding calculation for the lower right composition gives 616 Å. The smaller diameter is caused by the greater crosslink density in the polybutadiene.

Examination of Figure 10.4 shows domain sizes of approximately 650 Å for the lower left, and approximately 450 Å for the lower right, although the morphology of the latter is somewhat less distinct. The important point is that the Yeo equation provides a valuable estimate of the domain sizes, useful in estimating in advance what crosslink levels will produce a particular set of desired morphologies, and hence properties.

10.7. DUAL-PHASE CONTINUITY

In the previous sections, IPNs were shown to be phase separated, a trait in common with most polymer blends, blocks, and grafts. In Chapter 9, block copolymers with cylindrical or lamellar morphologies were already shown to possess dual-phase continuity. It is relatively easy to make IPNs with dual-phase continuity, albeit the morphologies are quite different from those of the block copolymers, and among the IPNs themselves, are also quite different from each other.

10.7.1. Sequential IPNs

The evidence suggests that sequential IPNs tend to phase separate via spinodal kinetic mechanisms, forming cylinders of polymer network II in polymer network I. Classical theory suggests that these cylinders should be interconnected. This should produce dual-phase continuous materials.

Widmaier and Sperling (31) studied the system *net*-poly(*n*-butyl acrylate)-*inter-net*-polystyrene sequential IPNs. First, a network of poly(*n*-butyl acrylate) was formed, using acrylic acid anhydride (AAA) as the crosslinker, and a small amount of dodecane thiol as a chain transfer agent to reduce the branching/gelation effect of the α-hydrogen on the *n*-butyl acrylate. The value of the AAA is that it is easily hydrolyzed by dilute, warm ammonia water. Then, styrene and divinyl benzene were swelled in and polymerized to create the second network.

Samples of the IPNs with the labile AAA crosslinker were soaked in a 10% aqueous ammonium hydroxide solution for about 12 h. Decrosslinking was effective, as shown by the complete dissolution of a poly(*n*-butyl acrylate) homopolymer network in organic solvents while the original network only swelled in the same solvents.

After decrosslinking of the AAA groups, the resulting linear polymer was extracted from the former IPN in a Soxhlet extractor, using either toluene or acetone. Both the soluble and insoluble portions were dried and characterized.

The sol fraction amount of poly(*n*-butyl acrylate) more or less followed the fraction inserted in the IPN, plus a small amount of solubles from the polystyrene. The molecular weights of the soluble polymer approximated those of the homopolymers, prepared without crosslinker.

Most interestingly, above about 20% of polymer network II, its phase domain structure was continuous after the extraction. A brittle porous solid of polystyrene was obtained, with densities in the 0.7–0.8 g/cm^3 range. Thus, there was an apparent partial collapse of the morphology during the hot extraction.

In summary, polymer network I was made in a continuous film. Polymer network II was formed inside of network I, with the normal volume increase. Polymer network I was decrosslinked and extracted, to leave a continuous, albeit porous, brittle polymer network II behind. Thus, the evidence points to dual-phase continuity for the original IPN, since polymer network I was continuous initially and totally extractable, leaving behind a continuous, porous solid.

More recently, a related idea was used by Gankema et al. (32) to produce nanoporous membranes. They used a crosslinkable methyl methacrylate mix as the solvent for two classes of entirely different compounds. One was based on (perfluoroalkyl) alkanes, $F(CF_2)_m(CH_2)_nH$, and the other was based on benzo-15-crown-5 benzoate [6,7,9,10,12,13,15,16-octahydro-1,4,7,10,13-pentaoxaben-zocyclo-pentadecen-2-ylmethyl   3,4,5-tris(p-dodecyloxy-benzyloxy)benzoate], DOBOB-CE. The $F(CF_2)_m(CH_2)_nH$ undergoes thermoreversible gelation in methyl methacrylate, tending to form long needle crystals from bundles of cylindrical aggregates. The DOBOB-CE undergoes self-assembly in methyl methacrylate solution to form cylindrical moieties upon cooling from dilute solution.

The following procedure was used by Gankema et al. (32):

1. A homogeneous solution was formed by dissolution of one of the two special components in methyl methacrylate.

2. The solution was cooled below the gelation temperature to induce thermoreversible gelation. This forms a physical, phase-separated network.

3. Ultraviolet (UV) polymerization and crosslinking of the methyl methacrylate solvent–monomer was carried out. This leads to a semi-IPN.

4. The special component was removed from the PMMA network by leaching or by sublimation, leading to a microporous membrane. Both components must be continuous in space after initial polymerization for such a membrane to function.

10.7.2. Thermoplastic IPNs

Physical crosslinks can be introduced into polymers via multiblock copolymers, ionomers, or semicrystallinity. In order to form a thermoplastic IPN, such polymers must behave as if crosslinked at ambient temperatures, yet flow at a higher temperature. Any combination of two or more such polymers may be used to form a thermoplastic IPN. Another class of thermoplastic IPN involves the (partial) chemical crosslinking of one of the polymers during a shearing action on the system, to produce a cylindrical or fiber-like morphology of the crosslinked component.

By mixing two immiscible polymers following equation (2.16), dual-phase continuity can be achieved, as illustrated in Figure 10.10 (33). In Figure 10.10a,

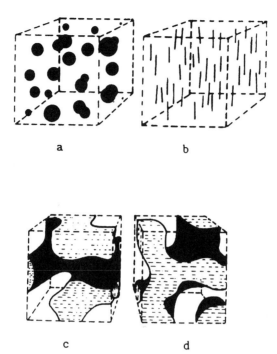

Figure 10.10 Various morphologies illustrating the development of dual-phase continuity. (a) Droplets dispersed in a continuous medium; (b) dispersed fibers or cylinders; (c) and (d) together illustrate dual-phase continuity, where both phases are locked together.

droplets of material are dispersed in a continuous matrix, a morphology observed in rubber-toughened plastics. Figure 10.10b shows dispersed fibers, observed in liquid crystal polymers and some thermoplastic elastomers. Figures 10.10c and 10.10d together depict a co-continuous morphology, where the two phases are interlocked or are interpenetrating. Thus, the morphology depends upon the volume fraction of the components, the shear viscosity of the phases, the amount of interfacial compatibility, and the relative rates of crosslinking and phase separation.

In the original patents by Davison and Gergen (34a,34b) the thermoplastic IPNs were based on polystyrene-*block*-polybutadiene-*block*-polystyrene (or the equivalent saturated center block), semicrystalline polymers such as polyamides or thermoplastic saturated polyesters. Table 10.1 provides a typical composition (34b). The reader should note the presence of a third polymer, polypropylene. This latter is used as a compatibilizer, appearing at the interfaces between Figures 10.10c and 10.10d. Thus, three co-continuous phases are formed in this material!

Davison and Gergen (34a) point out that their high-melting semicrystalline polymers are quite incompatible with the SEBS block copolymers, and ordinary blends produce grossly heterogeneous mixtures with few useful properties. Their improvements arise from blending in such a way as to form interpenetrating

Table 10.1   Composition of co-continuous interlocking network phases

Component	Composition by Parts	
	A	B
25,000–100,000		
25,000 SEBS	100	100
Poly(butylene terephthalate)	100	70
Shellflex 790 extending oil	100	—
Tuffto 6050 oil	—	50
Polypropylene	10	10
Irganox 1010 antioxidant	0.2	0.2
Dilaurothiodipropionate antioxidant	0.5	0.5
TiO$_2$	5	5

phases; see Figure 10.10c and 10.10d. Each component behaves as a thermoset at use temperatures, but flows at more elevated temperatures. Crosslinked interpenetrating phases are important in the formation of rubbery and leathery materials, or else the composition is likely to behave as an adhesive rather than as an elastomer. Davison and Gergen (34a) continue with the requirements for forming a thermoplastic IPN:

> Without wishing to be bound to any particular theory, it is considered that there are two general requirements for the formation of an interpenetrating network. First, there must be a primary network formed or in the process of forming in the shearing field. This requirement is fulfilled by employing the block copolymers of the instant invent on having sufficiently high molecular weight to retain domain structure in processing. Second, the other polymer employed must be capable of some kind of chemical or kinetic reaction to form an infinite network from a disassociated melt. The polymer must possess sufficient fluidity to penetrate the interstices of the primary network. This second requirement is fulfilled by employing the instant polyamides.

Below, the Davison and Gergen thermoplastic IPNs will be shown to be useful as under-the-hood electrical insulators because they are substantially constant modulus between the glass transition of the ethylene–butylene (EB) center block and the melting temperature of the poly(butylene terephthalate) (PBT) or polyamide (PA).

In 1981, Siegfried et al. (35) investigated thermoplastic IPNs based on SEBS triblock copolymers with a polystyrene ionomer. Two subclasses of the thermoplastic IPNs were identified: (*a*) prepared by a sequential polymerization method, and (*b*) prepared by mechanically blending separately synthesized polymers. The first subclass exhibited lower melt viscosities, but more nearly equal dual-phase continuity was achieved with the second subclass.

Ohlsson et al. (36) investigated the properties of thermoplastic IPNs based on SEBS triblock copolymers and polypropylene (PP). In the composition range of

10–55% polypropylene, these materials were found to have a co-continuous morphology. Processing oil, Witco Sonneborn Plastic Oil 260, was used to improve the processing behavior. Many of these materials showed stress–strain behavior similar to rubber, with no signs of necking, and extending to 700–1000%. Interestingly, these workers defined a distribution coefficient for the oil, k, as the ratio of the weight fraction of oil in the SEBS and polypropylene components, respectively. Values of k ranged from 0.47 at 90% PP to 0.33 at 10% PP. No corrections were made for the structure of the material, noting that this material actually has four phases, crystalline and amorphous PP, polystyrene, and the ethylene–butylene copolymer. After correction for the nonswellable crystalline portions, the results are comparable to those of Mishra et al. (37), where the partition coefficient was found to be approximately unity. A theoretical discussion is found in Section 8.2.2.

Wei et al. (38) investigated the interfacial areas of thermoplastic elastomers based on SBS/poly(styrene-*stat*-methyl methacrylate)-ionomer via SAXS. They found areas in the range of 145–360 m²/g, indicating phase dispersions in the range of true colloids. In the Ohlsson et al. (36) compositions above, the

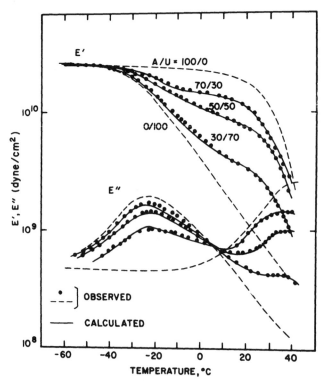

Figure 10.11 Dynamic mechanical behavior of acrylic/urethane interpenetrating elastomeric networks. Solid circles: experimental values; solid lines, homopolymer values. Note two glass transitions, with a plateau in between that exhibits leathery behavior.

physical crosslinks were developed through the styrene hard blocks and the polypropylene semicrystallinity. In the Wei et al. (38) case, the triblock hard domains are the same, but the ionomers form another kind of physical cross-link.

10.7.3. Interpenetrating Elastomer Networks

Two kinds of latex particles, such as acrylics and urethanes can be mixed and subsequently crosslinked to form IENs (39). The effect of composition on the storage and loss moduli are illustrated in Figure 10.11 (39). The appearance of two glass transitions, of course, indicates the phase-separated nature of the materials. To quantify the range of dual-phase continuity, the storage moduli values were plotted against composition; see Figure 10.12 (39). The Takayanagi models (see Section 2.1) were used to indicate the upper and lower bound

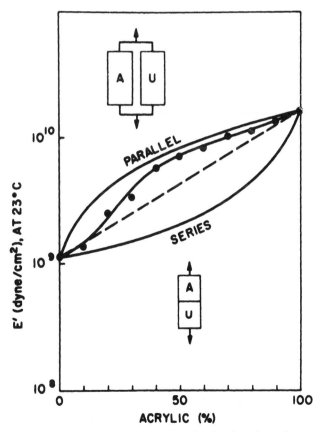

Figure 10.12 Use of the data in Figure 10.11 to determine the phase inversion composition. Here, the experimental data are compared to calculated values based on the parallel and series Takayanagi models. Phase inversion for a continuous acrylic phase occurs at about 30% acrylic.

moduli. A phase inversion is seen to occur at about 30% acrylic component. In the middle of this phase inversion, both components are continuous in space, forming a mosaic in three dimensions.

10.8. EXTENT OF MISCIBILITY

10.8.1. Segregation Degree

Both the chemical reactions and the phase separation in IPNs proceed under nonequilibrium conditions. As suggested above, however, after some degree of chemical conversion and crosslinking, phase separation is impeded and the system fixes at a somewhat nonequilibrium morphology. Lipatov (40,41) developed a semi-empirical but useful quantification of the *segregation degree*, α, as a function of tan δ vs. temperature plots of the dynamic mechanical behavior of the material. The following discussion refers to Figure 10.13 (40).

The quantities l_1 and l_2 are the temperature shift of the lower and upper glass transitions, respectively. The quantities h_1 and h_2 represent the intensities of the transitions above the plateau region between the transitions, for the lower and upper glass transitions, respectively, and the superscript zero refers to the pure

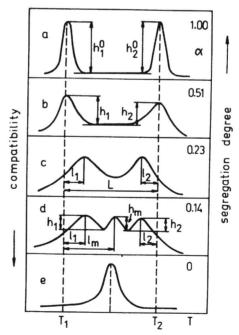

Figure 10.13 Polymer blends and IPNs exhibit various extents of miscibility and phase segregation. Here, the separation and heights of the loss modulus peaks are used to quantify the extent of phase separation.

Table 10.2   Dependence of segregation degree on reaction temperature for IPN-based on poly(butyl methacrylate) (PBMA) and a copolymer of styrene–divinylbenzene

Component Ratio (PBMA:PS)	333 °K	353 °K	363 °K
95:5	0.46	0.51	0.65
85:15	0.30	0.32	0.43
60:40	0.15	0.20	0.40

components. The quantity L represents the initial temperature difference between the peaks. The subscript m refers to the temperature and intensity of any intermediate mixed interphase peak. The general equation can be written

$$\alpha = [h_2 + h_1 - (l_1 h_1 + l_2 h_2 + l_m h_m)/L]/(h_1^0 + h_2^0) \qquad (10.5)$$

There are a few general comments. First, the quantity $h_m l_m$ is finite only for part d of Figure 10.13. Equation (10.5) does not apply to part e of Figure 10.13 because there is only one peak. The quantities $l_1 h_1$ and $l_2 h_2$ are measures of the area under the peaks. When $l_1 + l_2 = L$, a microheterogeneous morphology emerges. In the case of part b of Figure 10.13, equation (10.5) simplifies to

$$\alpha = (h_1 + h_2)/(h_1^0 + h_2^0) \qquad (10.6)$$

These relations outline a somewhat more sophisticated model than that proposed earlier by Curtius et al. (42).

Lipatov synthesized some semi-II IPNs based on poly(butyl methacrylate) and crosslinked polystyrene at different temperatures of polymerization. The resulting segregation degrees are summarized in Table 10.2 (41). The comparatively low degrees of segregation in most IPNs, 0.3–0.6, indicate that a large fraction of the material remains unseparated, that is, the IPNs are significantly more miscible by this measure than the corresponding polymer blends. Indeed, several workers in the field have observed that the glass transitions in IPNs are moved inward more than in the corresponding blends (17,43).

10.8.2. Thermodynamic Evidence

A quite different approach was taken recently. Instead of starting with immiscible polymer blends, and trying to create more miscible IPNs, the idea was to start with miscible blends and see if the corresponding IPNs were also miscible. The results are summarized in Table 10.3 (5).

System 1 of Table 10.3 provides an example. Polystyrene-H-*inter-net*-polystyrene-D were studied by SANS (44). This is a semi-II homo IPN of polystyrene, the second polymer deuterated. The H stands for natural hydrogen

Table 10.3 IPN miscibility

No.	System	Method of Measurement[a]	Blend Miscibility	IPN Miscibility	Comments	Ref.
1.	Linear PSH-*cross*-PSD semi-II	SANS	Miscible to 8×10^5 g/mol	Miscible to $3 \times 10^5 - 4 \times 10^4$ or $4 \times 10^4 - 1 \times 10^4$ g/mol	Semi-II IPN somewhat less miscible	b,c
2.	Phenolic-EVA copolymer	FTIR (EVA carbonyl)	Single phase	Crosslinking phenolic leads to phase separation	IPN less miscible	d
3.	*Cross*-PSD-linear PVME semi-II	SANS	Single phase	Moderately crosslinked PSD and above phase separate	IPN less miscible	e
4.	PSD-PVME crosslinked and grafted by γ-radiation	SANS	Single phase	Spinodal temperature increases from 140 to 430°C with 125 Mrad	IPN-Graft more miscible	f
5.	PPO-PS blends, SIPNs and IPNs	DSC, DMS, and TEM	Single phase	Single phase	All compositions miscible	g
6.	PVME-PS sequential IPNs	DSC, DMS	Single phase	Midrange compositions phase separate	IPNs less miscible	h

[a] SANS, small-angle neutron scattering; FTIR, Fourier transform IR spectroscopy; DSC, differential scanning calorimetry; DMS, dynamic mechanical spectroscopy; TEM, transmission electron microscopy.
[b] F. S. Bates and G. D. Wignall, *Phys. Rev. Lett.*, **57**, 1429 (1986).
[c] R. M. Briber and B. J. Bauer, *Macromolecules*, **24**, 1899 (1991).
[d] M. M. Coleman, C. J. Serman, and P. C. Painter, *Macromolecules*, **20**, 226 (1987).
[e] B. J. Bauer, R. M. Briber, and C. C. Han, *Macromolecules*, **22**, 940 (1989).
[f] R. M. Briber and B. J. Bauer, *Macromolecules*, **21**, 3296 (1988).
[g] H. L. Frisch, D. Klempner, H. K. Yoon, and K. C. Frisch, *Macromolecules*, **13**, 1016 (1980).
[h] J. J. Fay, C. J. Murphy, D. A. Thomas, and L. H. Sperling, in *Sound and Vibration Damping with Polymers*, R. D. Corsaro and L. H. Sperling, eds., ACS Symposium Series No. 424, American Chemical Society, Washington, DC, 1990.

(the ordinary stuff), while D stands for deuterated material. Because deuterium scatters differently than hydrogen, it serves as a probe material. There were two reasons given for the greater immiscibility of the deuterated-protonated polystyrene IPN:

1. A positive thermodynamic, binary interaction parameter, χ, which leads to phase separation in the blends at molecular weights greater than about 8×10^5 g/mol in the blend
2. An unfavorable contribution to the free energy from the elastic stretching of the first formed portion of the network

Thus, the problem actually involves ordinary rubber elasticity, the second reason. All networks formed under conditions where the first formed polymer is soluble in the remaining monomer will be subject to some elastic extension. The problem is the same as that encountered long ago in studying dilute solutions of a polymer in a thermodynamically good solvent: The chains are extended, preferring polymer–solvent contacts rather than polymer–polymer contacts. After polymerization, however, the extended chains of network II create a compressive type of stress on polymer I.

The overall conclusion from the data given in Table 10.3 is that IPNs may be less miscible than the corresponding blends. However, the two kinds of findings, IPNs being more and less miscible than the corresponding blends, is not necessarily contradictory. The presence of crosslinks may indeed cause smaller domains with larger interphases than the corresponding blends. Thus, more mixing occurs than in the blend. However, if the blend is thermodynamically miscible, rubber elasticity forces contribute a small positive value to the free energy of mixing, and hence cause some degree of demixing.

10.9. BEHAVIOR OF LATEX IPNS

10.9.1. Swelled Polymerization vs. Starvation Polymerization

Latex IPNs can be made in two general ways. Both start with a crosslinked seed latex. However, if the seed is then swelled with monomer mix II, a cellular morphology is likely to develop. If, however, starved polymerization is carried out, meaning that the monomer mix II is added more slowly than the rate of polymerization, then a core–shell morphology may develop. Of course, if the amount of monomer mix II added is greater than that which can swell into the crosslinked seed latex of polymer I, then both a cellular morphology and a core–shell morphology may develop. The *cellular morphology* terminology was developed after studying the salami slices of latex particles such as acrylonitrile–butadiene–styrene (ABS); see Figure 8.5. (Noting that the rubber in ABS latexes is normally crosslinked, these particles may be considered as latex semi-IPNs.) However, since these materials are thought to phase separate via spinodal

Figure 10.14 Morphology of latex IPN materials is modeled. After water evaporation, some degree of phase continuity of domains in individual phases is achieved.

decomposition kinetics, it may be that future work will establish that the domains are cylindrical in nature, rather than spherical.

Several additional factors contribute to the morphology. The crosslink density of the seed polymer determines both the swellability of the seed and the domain size of polymer II. The hydrophilicity of monomer II and polymer II relative to that of polymer I determines which component will tend to be on the surface facing the aqueous phase. There can even be a hydrophilic/hydrophobic inversion, with the seed latex polymer tending to be at the surface.

In a somewhat related case, there is a so-called inverted emulsion polymerization, with the *aqueous* phase dispersed, and the oil phase continuous. Used primarily with water-soluble monomers, it produces an inverse-phase morphology.

Silverstein et al. (45), Silverstein and Narkis (46), and Nemirovski and Narkis (47) investigated the morphology and rheological behavior of latex IPNs and related materials based on polyacrylates and polystyrene. After polymerizing and drying, the latex particles do not interdiffuse to form a monolithic solid because of the crosslinks. However, being moderately soft, they flow together as modeled in Figure 10.14 (46). The solid black spheres represent polystyrene domains. The rheological behavior of these materials resembles the flow behavior that might be expected for a room filled with underinflated beach balls.

Thus, the unique mechanical and rheological behavior of the latex IPNs are intimately related to their structure; the presence of reinforcing polystyrene intraparticle microdomains increases the modulus significantly over that expected for the elastomer alone, and the polystyrene interparticle ties yield a significant ultimate tensile strength, of the order of 3 MPa with 200%

elongation for 35% polystyrene content. Capillary rheometry indicates the existence of a stick, slip, roll flow mechanism in which the stable flow units involve the crosslinked latex particles. The viscosity is about 10 kPa-s at a shear rate of $10\,s^{-1}$, at 150°C. Injection molding shows that the latex IPN can be processed as a thermoplastic. Thus, the morphology of the elastomeric latex IPNs results in a dual thermoset–thermoplastic behavior, that is, another type of thermoplastic IPN.

Hourston and his co-workers have been very active in the field of latex IPNs (48–50). Latex IPNs made by Hourston and co-workers include polyurethane–acrylic, poly(vinyl isobutyl ether)–poly(methyl acrylate), polyurethane–modified polyester, and a large number of materials entirely acrylic in nature. For many of these compositions, they found that the glass transitions were shifted inward, suggesting significant molecular mixing. This was supported by a Small and Hoy analysis, see Section 2.4.6.4, of the solubility parameters. These workers also pointed out that since the magnitude of tan δ in the region between the two glass transitions was high, an overlap of the relaxations times distributions of the component materials was likely, that is, there was significant mixing of the two components.

10.9.2. Three Component Latex IPNs

There have been several studies of three-stage latex IPNs as well. Liucheng et al. (51) synthesized poly(n-butyl acrylate) crosslinked with ethylene glycol dimethacrylate as the seed latex. The second stage added styrene and divinyl benzene. The third stage was linear poly(methyl methacrylate). Starved polymerization conditions resulted in more regular shaped latex particles than batch addition of monomer.

Another three-stage latex IPN was polymerized by Hu et al. (52,53). Figure 10.15 (52) shows the loss and storage shear modulus for a three-stage latex IPN prepared as follows: A seed latex of poly(butadiene-$stat$-styrene), 90/10, was prepared as the core, polymerized to about 45% gel, via branching reactions. After evaporation of the remaining monomer, a mixture of ethylhexyl methacrylate, styrene and tetraethylene glycol dimethacrylate was added, and swollen in for 24 h at room temperature. The composition of the latter was intended to provide a glass transition of about +10°C at 1 Hz. The third component was a styrene–acrylonitrile (SAN) composition, 72/28 styrene/acrylonitrile. This last was added in a semicontinuous, starved polymerization manner to produce a shell of harder material.

Figure 10.15 shows three glass transitions, one for each component. The polybutadiene-based component was intended to provide ordinary toughening action against impact blows. The higher glass transition elastomer was supposed to act to absorb energy in slower deformations, such as room temperature fatigue. The highest glass transition material was selected because of its compatibility with polycarbonate, since these latex particles were intended to be used to toughen polycarbonate. Figure 10.15 should be compared with the

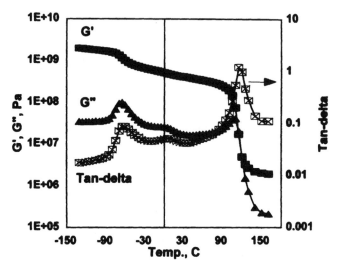

Figure 10.15 Three-component latex IPNs: Poly(butadiene-*stat*-styrene), poly(ethylhexylmetha-crylate-*stat*-styrene), and SAN, 0.5/0.5/1.0.

information in Figure 10.16 (52), prepared from a blend of two kinds of latex particles, with a core of poly(butadiene-*stat*-styrene), 90/10 as before, with SAN shell synthesized under starved polymerization conditions, and a poly(ethyl-hexyl methacrylate-*stat*-styrene), also prepared as before, but serving as the core of the second kind of particle, with SAN again serving as the shell, also synthesized under starved polymerization conditions. The blend of the two latex particles had an identical overall composition as the material shown in

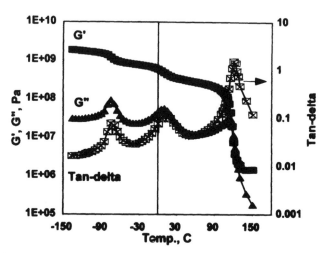

Figure 10.16 Blend of two latex particles, matching the overall composition of that in Figure 10.15. In this case, the three glass transitions are more distinctive, suggesting greater phase separation in a latex blend, rather than a three-component single latex system.

Figure 10.15. Figure 10.16 shows a better developed middle glass transition, and a lower temperature poly(butadiene-*stat*-styrene) glass transition. Both of these facts suggest that the mixture of the two latex particles might have less mixing at the molecular level. However, the morphology is different as well, as will be explored below in the damping section.

10.10. OTHER KINDS OF IPNS

The combination of two polymers in network form invites the imagination to invent yet new types. Several interesting materials already synthesized include:

1. In situ *sequential IPNs*. Meyer and Widmaier and co-workers (54–56) mixed two sets of vinyl monomers, crosslinkers, and initiators together, but the two monomers were polymerized sequentially. Both monomers can be polymerized independently via free radical reactions if the two monomers have quite different reactivities toward free radicals. This situation arises with vinyl or acrylic double bonds and allylic double bonds. The allylic double bonds are about 100 times less reactive than acrylic or methacrylic bonds. Furthermore, these workers used two different initiators, each activated at a different temperature. Then, the less activated monomer system can be reacted at a higher temperature. For example, in a system based on methyl methacrylate and diallyl carbonate of bisphenol A (DACBA), first, crosslinked PMMA is formed at a moderate temperature. Then, by just increasing the temperature after completion of the first polymerization, the formation of the allylic network follows. At 60°C, the first reaction was activated by azobisisobutyronitrile (AIBN), and at 95°C, the higher temperature reaction was activated by *t*-butyl peroxy isononanate (TBPIN). Theoretically, these materials can be distinguished from their true sequential counterparts because the polymer network I chains will be more relaxed in the in situ sequential case.

2. *Gradient IPNs*. The gradient IPNs are most often formed when polymer network I is formed; then monomer II plus crosslinker, and so forth, are swollen in from one side of a film or structure and rapidly polymerized before diffusion equilibrium takes place. Also, the crosslink level can be gradient in the material. These materials are obviously differing in properties from location to location. This field was recently reviewed by Lipatov and Karabanova (57). The most interesting proposed application is the formation of gradient composition optical lenses known as Gradans; see Section 10.12.3. Gradans are optical gradient elements used in optoelectronics, integrated optics, and medical engineering. Often, these materials are in the form of cylindrical rods with radial profiles of composition that ensure parabolic decreases in refractive index from the axis to the surface.

3. *Organic–inorganic IPNs*. Hybrid organic–inorganic IPNs are one of the newest creations in the field of IPNs. The preparation of composite materials from two components with vastly different physical properties has been a

recurrent concept in materials science, as few pure materials fulfill the strength, design, and cost requirements of modern applications. Jackson et al. (58) investigated materials made from SiO_2 prepared by sol–gel chemistry and poly(2-hydroxyethyl acrylate). In some cases, tetrakis(2-(acryloxy)ethoxy)silane was also added to promote phase mixing between the organic and inorganic phases. Small-angle X-ray scattering (SAXS) studies showed that the sizes of the SiO_2 domains were of the order of 100 Å or less.

The sol–gel method of making SiO_2 produces a finely divided morphology under mild synthetic conditions. If the solvent is a polymerizable monomer, then two independent reactions take place. Composites with up to 75 wt% glass content can be made this way, with reduced shrinkage during polymerization.

A special kind of organic–inorganic IPN involves the polyphosphazenes. The polyphosphazenes are characterized by containing both a nitrogen and a phosphorous atom in the backbone of each mer, having the general structure $\{N=P(-OR, -OR')\}_n$, where the two phosphorous ether moieties are frequently different and may be aliphatic or aromatic. Most recently, Allcock et al. (59) synthesized a series of poly(organophosphazenes) with various acrylics with unsaturated side groups for UV photocrosslinking and IPN formation. The morphologies of the IPNs derived by transmission electron microscopy (TEM) revealed domains ranging from 0.05 to 0.25 μm in many cases. Other inorganic polymer IPNs include the poly(dimethyl siloxane)-containing materials, discussed elsewhere in this chapter.

4. *Ionic IPNs.* In early work, IPNs and IPN-like materials were prepared by Pozniak and Trochimczuk (60,61), and Erbil and Baysal (62) who swelled polyethylene with styrene and divinyl benzene, followed by polymerization. At room temperature, the styrene only swells the polyethylene, dissolving in the amorphous portions. The Pozniak and Trochimczuk team carried out a chlormethylation with chloromethyl methyl ether, follow by subsequent amination with diethyl amine to produce an amine-bearing IPN for weak base anion exchange membranes. The Erbil and Baysal team sulfonated the polystyrene to make cation exchange membranes. The latter were used to study the electrolysis of borax solutions to produce boric acid and sodium hydroxide.

5. *Semicrystalline Commodity Polymers.* The most important semicrystalline commodity polymers are polyethylene, polypropylene, poly(ethylene terephthalate) (PET), and the nylons. At room temperature, these materials will not dissolve in monomers, only swelling as discussed immediately above. IPNs based on polyethylene and polystyrene were first investigated by Czarczynska and Trochimczuk (63). More recently, Borsig and co-workers (64–68) actually dissolved the polyethylene in hot styrene, butyl methacrylate, or comonomer mixes at 90–105°C. The monomers were polymerized and simultaneously crosslinked at temperatures above the melting temperature of the polyethylene. After the polymerization was complete and the temperature lowered, the polyethylene recrystallized, the extent of crystallinity was not

affected significantly by the presence of the second polymer. One might imagine that a slight swelling of polyethylene with styrenic or methacrylic monomer mixes would be useful in polymer recycling.

10.11. RENEWABLE RESOURCE IPNS

Ever since the invention of nylon and the development of synthetic polymers in the 1930s and 1940s, renewable resource polymers, sometimes called natural product polymers, have taken "a back seat" in research and development. And it is true that the new and exciting materials were the synthetic polymers. Of course, before that time, there was much research on cellulose and other polysaccharides, leather, silk, and other forms of proteins, triglycerides, and the like.

In the 1990s, however, the combination of pollution problems and the prospect of eventual exhaustion of petrochemical sources is forcing many to rethink the use of polymers for the future. There is good evidence that most, if not all, of the renewable resource polymers or polymerizable natural monomers can be made into as good materials as the synthetics, but first people must apply modern polymeric principles and research methods to these materials with the same intensity as the synthetic materials have had. After all, it must be remembered that the plastic toys of the 1930s lasted about one day. Today, most children outgrow them, still usable. The difference has been research and development.

In the following sections, several IPN materials based on renewable resource materials will be described. Several of those based on castor oil are commercial, which will be described in Section 10.11.1.

10.11.1. Castor-Oil-Based IPNs

Since castor oil is used widely in such materials as polyurethane foams, it is already an important commercial material. Castor oil IPNs have attracted much attention, especially in those countries where castor oil is grown commercially, but also in the United States. In India, Suthar and co-workers (69,70) synthesized a host of castor oil–urethane network-based IPN materials with various acrylics. The morphology of castor oil–polystyrene SINs was described in Figure 2.5. These utilized the pendent hydroxyl groups that occur naturally on each of the triglyceride side chains.

Among the more recent studies are those of Nayak et al. (71–74). They reacted toluene diisocyanate (TDI) with castor oil to form a viscous liquid prepolymer, then mixed the product with various acrylic monomers, using ethylene glycol dimethacrylate (EGDM) as the crosslinker. The main focus of the Nayak and co-workers studies was related to the temperatures and kinetics of high-temperature degradation. Three different weight loss ranges were found between about 275 and 620°C. These results are reminiscent of those obtained by Kim et al. (75) on PU/PMMA SINs. In both of these studies, there is some enhancement of the weight retention by the IPN. The most likely mechanism involves the acrylic polymer degrading first via an unzipping process. The improvement in thermal degradation stability arises in terms of the more easily degradable polymer absorbing the free radicals, thus stabilizing and protecting the less easily degradable polymer. Thus, the acrylics act as free-radical scavengers for the PU. Since this mechanism requires a close juxtaposition of the two polymers, it may be enhanced by IPN or SIN formation.

Tan (76) carried out studies of castor oil–urethane/acrylic SINs, using MMA, acrylonitrile (AN), and styrene monomer mixes. These were found to be hydrolysis and abrasion resistant, especially to jetting sand. These materials were used as a coating applied to the Ge Zhou Ba hydroelectric power station in China. Xie et al. (77) studied the relative rates of formation of castor oil urethanes and poly(methyl methacrylate). The rate of formation of the PU network was higher than that of the acrylic. The time of gelation decreased with increasing PU content. A broad glass transition temperature range was observed for many of these materials. Similar materials were reported to be commercial for the coating of iron, providing rust resistance (78).

In the United States, Barrett et al. (79–83) examined the properties of semi-IPNs prepared from castor oil or vernonia oil (has epoxy groups instead of hydroxyl groups) polyurethanes and poly(ethylene terephthalate). Typical time–temperature reaction conditions are shown in Figure 10.17 (84). One of

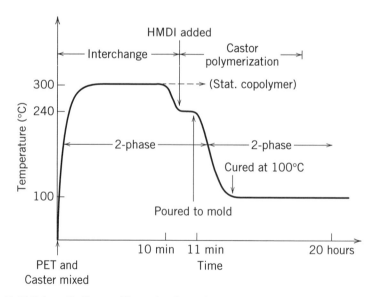

Figure 10.17 Schematic diagram illustrating the melt processing of poly(ethylene terephthalate) and castor oil with HMDI. Both components contain ester groups capable of bond interchange if held at elevated temperatures. After short times, grafting produces a compatibilized semi-IPN. After longer times, especially at elevated temperatures, a one-phased statistical copolymer will result.

the more interesting effects was the bond interchange that took place between the PET and the castor oil, probably via ester exchange reactions, which tended to make graft copolymers, thus lowering the interfacial tension.

10.11.2. Natural-Rubber-Based Latex IPNs

Natural rubber latexes have long been commercial, being the first source of rubber. While still in the latex form, the latexes can be treated just like synthetically made latexes and used as the core material of further compositions such as IPNs. Some early materials, still in commercial use, are called *Hevea Plus* (85). A particularly interesting material is Hevea Plus MG, prepared by polymerizing methyl methacrylate in the rubber latex. This makes an impact-resistant composition somewhat resembling ABS.

Hourston and Romaine (86) synthesized and characterized latex IPNs and semi-IPNs based on natural rubber latex and PMMA. More recently, natural-rubber-based latex IPNs have been considered for toughening plastics. For example, Schneider and co-workers (87,88) used linear and precrosslinked natural rubber latex particles that were made into IPNs with a shell of cross-linked poly(methyl methacrylate). Styrene was also used, both separately and as an intermediate layer between the core and the shell. This latter allowed for adjustments to the modulus to be made. While pure polystyrene had a notched Izod impact strength of about $1.5\,kJ/m^2$, the IPN containing material had an impact resistance up to $32.2\,kJ/m^2$. Similarly, pure PMMA had an impact strength of $1.1\,kJ/m^2$, while the toughened compositions had values of $4-10\,kJ/m^2$, with linear natural rubber (semi-IPN) doing better than the full IPN.

10.11.3. Cellulose Derivatives

Since the beginning of civilization until the present, cellulose and cellulose derivatives have been the largest tonnage single polymer sold in the world. Cellulose appears as the major constituent of wood, and cotton is substantially 100% cellulose. Actually, wood itself is a semi-IPN because the other major component, lignin, is a crosslinked polymer, forming a matrix for the cellulose fibers. However, just considering pulped cellulose, the major application is paper and paper-related products. Relatively large tonnages are also sold as rayon, "acetate" (i.e., cellulose acetate), cellophane, sodium carboxy methyl cellulose (viscosity thickener for foods, particularly ice cream), and many others. With three hydroxyl groups per mer, cellulose would be rather reactive, if it were not for its relatively high crystallinity and high melting temperature.

However, cellulose can be made into IPNs and related materials with highly interesting possibilities. Kamath et al. (89) prepared IPNs and related AB-crosslinked copolymers using compositions such as allyl cellulose cinnamonate and crosslinked polystyrenes, acrylics, or vinyl acetates. An improved thermal stability was noted. Semi-IPNs in hydrogel state using cellulose esters and polyacrylamide has also been reported (90,91).

10.12. COMMERCIAL APPLICATIONS

Most IPN compositions are crosslinked when fully synthesized. Thus, they are good for applications that demand resistance to creep and flow. Most IPNs have rather very small phase domain morphologies, and extensive mixing between the two components. This suggests uses that may capitalize on applications demanding such behavior. Further, it is relatively easy to make IPNs with dual-phase continuity, so that applications needing such a property, such as transport behavior, are well served. Many successful IPN compositions are composed of an elastomer and a plastic, making for mechanical behavior in the leathery range, suggesting energy absorption, electrical wire insulation, and many others.

It is a truism in science and engineering that people developing new applications are more willing to try new materials than people who already have a material successfully performing their needs. Since it is true that biomedical applications of polymers are relatively new, perhaps it should be no surprise that several IPN compositions have already been employed in this field. Other new areas interested in IPNs include the electronics industry and smart materials.

Up to 1979, there were about 125 scientific papers and about 75 patents related to IPNs, total. At that time, there were already several applications noted. Today, approximately that number of IPN papers and patents are produced per year, with a large and growing number of known commercial applications, and even more products developing in the wings.

Actually, IPNs are one of the oldest multicomponent polymer materials to be used commercially. The first known such material was invented by Aylsworth in 1914 (92). The Aylsworth material was composed of phenol-formaldehyde, a kind of Bakelite material, and natural rubber plus sulfur. It was used to toughen Edison's phonograph records from 1914 to about 1928 (93).

10.12.1. Applications of Thermoplastic IPNs

The thermoplastic IPNs are actually hybrids between the blends and the IPNs, thermoset at room temperature, but able to flow at elevated temperatures; see Section 10.7.2. One of the more important commercial thermoplastic IPN compositions involves polypropylene and ethylene–propylene–diene-monomer (EPDM) rubber. In this class of polyolefins, the EPDM is blended with the PP under conditions where the EPDM undergoes crosslinking through the diene moiety. The crosslinking reaction is stopped when the viscosities and volume fractions equate to nearly unity, according to equation (2.16). This is the point of phase inversion, and the reaction is thus stopped at the point of dual-phase continuity. The result is a leathery material of great energy-absorbing capacity, now widely used in automotive bumpers. The first PP-EPDM thermoplastic IPN was patented in 1974 by Fisher of Uniroyal (94). Since then, there have been many newer patents describing different approaches to the same problem,

Table 10.4   EPDM-PP thermoplastic IPN patents

Major Feature	Company	US Patent No.
Dynamically partly cured		
EPDM precured, then blended	Uniroyal	3,758,643
EPDM blended, then cured	Uniroyal	3,806,558
EPDM of high molecular weight	Uniroyal	3,835,201
Dynamically fully cured		
EPDM blended, then cured	Monsanto	4,130,535
EPDM semicrystalline		
Long ethylene segment length (with polyester)[a]	Goodrich	4,046,820
Long ethylene segment length (with PP)[a]	Goodrich	4,036,912
70–80% ethylene EPDM[a]	Uniroyal	4,031,169
Dual-phase continuity		
Blend of EPDM and PP[a]	Exxon	4,132,698

[a] No crosslinking required.

namely, how to achieve and control dual-phase continuity; see Table 10.4 (95). Some trade names are shown in Table 10.5.

Another thermoplastic IPN material was prepared by Davison and Gergen (34,35); see Figure 10.10. The combination of SEBS and polyamides, polyester, or polycarbonates results in materials that behave in the leathery range between about $-50°C$ and about $220°C$, making them extremely useful for under-the-hood wire insulation.

10.12.2. Current Applications

10.12.2.1. Biomedical Materials   Table 10.5 (95) summarizes a number of commercial IPN materials. The biomedical materials include compositions of silicone and polyurethane, useful as steam-sterilizable medical tubing (96,97). Semi-IPN membranes composed of silicone rubber and polytetrafluoroethylene, for example, Silon-TSR® (temporary skin replacement) created by Dillon (98–100), are useful for burn healing. Silon-TSR® features a high moisture–vapor transmission rate, while remaining impervious to liquid water. Being transparent, it also allows the physician to observe the healing progress directly. Related materials, Silon-SES® (silicon elastomer sheeting) and Silon-STS® (silicone thermoplastic splints), are useful for the treatment of scar tissue (101).

Roemer and Tateosian (102,103) of Dentsply developed prosthetic teeth made from acrylic sequential-IPN compositions. In one example, suspension-sized particles of crosslinked PMMA were mixed with linear PMMA, MMA monomer, crosslinker, and initiator. After polymerization, according to patent and advertising literature, the false teeth have two superior properties: Because a suspension polymerization route was used, the teeth grind to a fine powder,

Table 10.5 Selected commercial IPN materials

Manufacturer	Trade Name	Composition	Application
Shell Chemical Co.	Kraton IPN®	SEBS–polyester	Automotive parts
LNP Plastics	Rimplast®	Silicone rubber–nylon or PU	Gears or medical
ICI Americas Inc.	ITP®	PU–polyester–styrene	Sheet molding compounds
DSM N.V.	Kelburon®	PP–EP or rubber–PE	Automotive parts
Shell Research B.V.	—	Rubber–PP	Tough plastic
Uniroyal (Reichhold Chemical Co.)	TPR®	EPDM–PP	Auto bumper parts
Sun Marketing & Refining Co.	—	PE–PP	Low-temperature plastics
Rohm & Haas	—	Anionic–cationic	Ion-exchange resins
Monsanto	Santoprene®	EPDM–PP	Tires, hoses, belts, and gaskets
Du Pont	Somel®	EPDM–PP	Outdoor weathering
BF Goodrich	Telcar®	EPDM–PP or PE	Tubing, liners, and wire and cable insulation
Allied Chemical	ET polymer®	Butyl rubber–PE	Sheet molding compounds
Hercules	Profax®	EPDM–PP	Ultrahigh impact resistance
Exxon	Vistalon®	EPDM–PP	Paintable automotive parts
Freeman Chemical	Acpol®	Acrylic–urethane–polystyrene	Sheet molding compounds
Dentsply International	Trubyte,® Bioform®	Acrylic-based	Artificial teeth
Hitachi Chemical	—	Vinyl–phenolics	Damping compounds
Bio Med Sciences	Silon–TSR®	PDMS–PTFE	Burn dressing
Bio Med Sciences	Silon–SES®	PDMS–PTFE	Scar abatement

useful in the dentist's office when fitting to the opposing set of teeth. Macroscopic continuous crosslinking may cause charing under the dentist's burr. Second, because they are densely crosslinked, they do not swell significantly in such highly edible plasticizers as salad oils, margarines, or similar materials.

Contact lenses are another area of interest. A hydrophilic/hydrophobic SIN composition of hydroxy ethyl methacrylate and a polysiloxane were used in early work by Falcetta et al. (104). Kuzma and Odorisio (104a) patented SIN

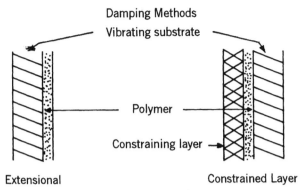

Figure 10.18 Extensional and constrained layer damping of sound and vibration with polymers. While the three-layer constrained layer system is more efficient, the extensional system is easier to install, especially in the correction of existing structures.

compositions of crosslinked collagen and water-swellable hydroxy ethyl metha-crylate polymers (HEMA) or related materials. These show improved oxygen permeability while maintaining mechanical strength. More recently, Chirila et al. (105) used melanin and HEMA. Melanins are biopolymers that strongly absorb ultraviolet and some visible light, useful to protect the retina from ultraviolet radiation damage. See also Section 10.12.3.

Gradient IPNs have been proposed as materials for controlled release of drugs (106,107). First, crosslinked suspension-sized particles are made, containing the drug to be delivered and one component monomer of a condensation polymer. The particles are then rolled in the second component monomer of the condensation polymer, which is made to react rapidly, forming a gradient IPN, with a shell rich in the condensation polymer. This permits a controlled permeability for the drug inside.

10.12.2.2. Sound and Vibration Damping   Sound and vibration damping with polymers makes use of the onset of molecular motion associated with the glass transition. Both loss moduli and tan δ are useful, albeit in different applications. IPNs are particularly useful for damping because the extent of mixing and phase continuity can be controlled to produce improved damping over broad but controlled temperature and frequency ranges (108,109).

There are two major modes of damping sound and vibration with polymers, extensional damping and constrained layer damping; see Figure 10.18 (110). An extensional damping system consists of two layers: the substrate to be damped and the polymeric damping layer. This works best if the polymer is at the stiff end of the glass transition temperature range. The constrained layer system is a three-layer affair, with the polymer in the middle, like the jelly in a sandwich. Either the constraining layer is part of the system, involving two half-thicknesses of sheet metal (or similar), and gluing them together with the damping layer, or the constraining layer can be applied afterward. In any case, the

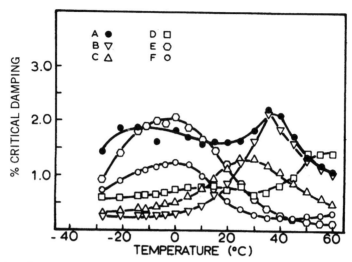

Figure 10.19 Damping characteristics of an IPN, composition A, vs. commercial homopolymers and statistical copolymers used for damping. Note the broad range of damping obtained in the IPN system, making it useful for outdoor or machinery applications.

composition should be near the soft end of the glass transition temperature range. Extensional damping is best characterized via E'' (or G''), while constrained layer damping is best characterized by $\tan \delta$.

Sound and vibration damping has been an important application for IPNs because they have broad but controllable temperature ranges of damping; see Figure 10.19 (111). In this last, composition A is an acrylic/methacrylic latex IPN with an epoxy constraining layer, while the remaining materials are selected from the commercial damping compositions of that day. Composition B was based on poly(vinyl acetate). While very effective at room temperature for the damping of automotive doors and the like, it fails at either cold winter temperatures or hot summer days. Hitachi Chemical has a new high-temperature sound and vibration damping material, based on vinyl-phenolic compositions (112). Such materials are useful for motors and the like. As will be explored below, currently there is much interest in IPNs as damping materials.

10.12.3. Proposed and Nascent Applications

Table 10.6 delineates more than 30 proposed or nascent IPN and SIN applications. Under smart materials, there are several highly novel ideas. *Smart materials* are supposed to do something special to respond to changes in their environment. One such smart material that has been around for a long time is natural rubber. At low extensions, it is soft and highly extensible. When it is highly stretched and about to break, it suddenly crystallizes, becoming much stronger. Several IPNs have now been shown to fit the category of smart materials.

Table 10.6 Proposed or nascent IPN and SIN applications

Composition	Application	Ref.
A. Nonlinear Optical Materials		
Polyurethane/polyacrylate	Good temporal stability	a
Epoxy/phenoxy	Optical wave guide	b
Epoxy/sol–gel processible phenoxysilicon polymer	Frequency doubling	c
B. Circuit Board Materials		
Acrylic/cellulosic	Photoresists for printed circuit boards	d
Epoxy/aromatic polyester	Printed wiring boards	e
C. Toughening Plastics		
SBR/acrylate latex IPNs	Lowers brittle ductile temperature in PC	f
Natural rubber/acrylics or styrenics	Toughen thermoplastics	g
D. Biomedical Area		
Acrylic/PEO–PPO	Controlled transdermal drug delivery	h
Poly(acrylic acid)/SAN	Blood compatibility and controlled permeability	i
PEG/dextran	Multistimuli-responsive drug delivery	j
Poly(ε-caprolactone)/PEG	Bioerodible drug release	k
PEG/polytriacrylate	Properties independent of solvent environment	l
PU/malimide–PU	Hydrophilic/hydrophobic blood compatibility	m
E. Sound and Vibration Damping		
Polystyrene/polyacrylate latex IPNs	Broad temperature range, fillers introduced	n
PU/Bu methacrylic-oligocarbonate gradient IPNs	Damping	o
PS/acrylic latex IPNs	Broad temperature damping range	p
PEA/PMMA	Controlled temperature range	q
Acrylic/methacrylic	Water-soluble damping coating	r
Wood/acrylics	Sandwich structures for damping	s
F. Smart Materials		
Poly(vinyl alcohol)/ poly(acrylic acid)	Permeability depends on pH	t
Poly(methacrylic acid)/polypyrrole	Molecular recogition of imprinted chemical	u
Poly(chloromethyl styrene)/ quaternized polyamine	Humidity sensor	v
Ion-bearing acrylic/EGDM	Humidity sensor	w

Table 10.6 Proposed or nascent IPN and SIN applications (continued)

Composition	Application	Ref.
PPO/styrene–HEMA	Solar heating with adjustable LCST, 30–150°C	x
G. Pervaporation Membranes		
Poly(acrylic acid)/SAN	Separation of ethanol and water	y
Poly(vinyl alcohol)/polyacrylamide	Separation of ethanol and water	z
H. General		
Polypyrrole/EPDM	Electrostatic shields/microwave absorbers	aa
Polystyrene/poly(methyl methacrylate) gradient IPNs	Gradient-refractive index optical lenses	bb
Acrylate/unsaturated polyesters	Photochemical stereolithography	cc
PDMS/Poly(acrylic acid)	High-permeability soft contact lenses	dd
Urethane/acrylic	High-impact strength coatings	ee
Castor oil–urethane/acrylic	Coating rusted iron	ff
Poly(vinyl pyrrolidone)/HEMA	Gradient-refractive index soft contact lenses	gg

[a] H. Q. Xie, X. D. Huang, and J. S. Guo, *Polym. Adv. Technol.*, **7**, 309 (1996).

[b] S. Marturunkakul, J. I. Chen, L. Li, X. L. Jiang, R. J. Jeng, S. K. Sengupta, J. Kumar, and S. K. Tripathy, *Polymers for Second-Order Nonlinear Optics*, ACS Washington, DC, 1995.

[c] S. Tripathy, J. Kumar, S. Marturunkakul, J. I. Chen, and L. Li, *SPE ANTEC '95*, 1611 (Boston) Vol. 2 (1995).

[d] M. Tate and D. F. Varnell, U.S. Pat. 4,855,212 (1989).

[e] A. Takahashi, T. Horiuchi, M. Nomoto, K. Nanaumi, and K. Yamamoto, *Polym. Adv. Technol.*, Special IPN Issue, **7**(4) (1996).

[f] V. Tannattanakul, E. Baer, A. Hiltner, R. Hu, M. S. El-Aasser, and L. H. Sperling, *J. Appl. Polym. Sci.*, **62**, 2005 (1996).

[g] M. Schneider, T. Pith, and M. Lambla, *Polym. Adv. Technol.*, **6**, 326 (1995).

[h] S. C. Cho, S. O. Choi, Y. H. Bae, and S. W. Kim, *Drug Delivery Syst.*, **10**(6), 437 (1995).

[i] Y. D. Kim, B. K. Lee, E. J. Jeon, Y. C. Shin, and S. C. Kim, *Macromol. Symp.*, **98**, 665 (1995).

[j] M. Kurisawa, M. Terano, and N. Yui, *Macromol. Rapid Commun.*, **16**(9), 663 (1995).

[k] S. O. Kim, J. H. Ha, Y. J. Jung, and C. S. Cho, *Arch. Pharmacal Res.*, **18**(1), 18 (1995).

[l] P. D. Drumheller and J. A. Hubbell, *J. Biomed. Mater. Res.*, **29**(2), 207 (1995).

[m] K. H. Hsieh, C. Y. Chen, D. C. Liao, and W. Y. Chiu, *Polym. Adv. Technol.*, Special IPN issue, **7**(4), 1996.

[n] J. Wang, R. Lui, W. Li, Y. Ki, and X. Tang, *Polym. Intern.*, **39**(2), 101 (1996).

[o] A. A. Brovko, L. M. Sergeeva, L. M. Karabanova, and L. A. Gorbach, *Ukr. Khim. Zh.* (Russian edition), **61**(1–2), 58 (1995).

[p] L. Shucai, P. Weijiang, and L. Xiuping, *Int. J. Polym. Mater.*, **29**(1–2), 37 (1995).

[q] S. Kim and J. H. An, *J. Appl. Polym. Sci.*, **58**(3), 491 (1995).

[r] Z. Xie, J. Yang, and Z. Li, *Gongneng Gaofenzt Xuebao*, **8**(2), 185 (1995).

[s] H. J. Lee and H. Mizumachi, *Mokuzai Gakkaishi*, **41**(1), 9 (1995).

[t] L. F. Gudeman and N. A. Peppas, *J. Membr. Sci.*, **107**(3) 239 (1995).

[u] D. Kriz, L. I. Andersson, M. Khayyami, B. Danielsson, P. O. Larsson, and K. Mosbach, *Biometrics*, **2**, 81 (1995).

[v] Y. Sadaoka, M. Matsuguchi, S. Masanobu, and H. Sakai, *Sens. Actuators*, **B25**(1–3), 689 (1995).

[w] Y. Sakai, M. Matuguchi, and Y. Sadaoka, *J. Electrochem. Soc.*, **140**, 432 (1993).

[x] W. Eck, H. R. Wilson, and H. J. Cantow, *Adv. Mater.*, **7**, 800 (1995).

[y] Y. D. Kim, B. K. Lee, E. J. Jeon, Y. C. Shin, and S. C. Kim, *Macromol. Symp.*, **98**, 665 (1995).

[z] L. Liang and E. Ruckenstein, *J. Memb. Sci.*, **106**(1–2), 167 (1995).

[aa] R. A. Zoppi and M. A. DePaoli, *Polim. Cienc. Tecnol.*, **5**(3), 19 (1995).

[bb] T. L. Bukhbinder and V. I. Kosjakov, *Vysokomolekuljarny Soedineniya, Ser. B*, **28**, 625 (1996).

[cc] A. Torres-Filho and D. C. Neckers, *Polym. Mater. Sci. Eng. (Prepr.)*, **72**, 532 (1995).

[dd] C. Robert, C. Bunel, and J. P. Vairon, Eur. Pat. Appl. 643083 (1995).

[ee] H. X. Xiao and K. C. Frisch, *J. Coat. Technol.*, **61**(3), 51 (1989).

[ff] H. Q. Xie, J. S. Guo, and G. G. Wang, *Eur. Polym. J.*, **29**, 1547 (1993).

[gg] I. Calderara, D. Baude, D. Jayeux, and D. J. Lougnot, *Polym. Mat. Sci. Eng. (Prepr.)*, **75**, 244 (1996).

10.12.3.1. Control of Solar Heating Temperatures Among these smart materials is the solar heating material with adjustable LCST (113). The IPN of poly(propylene oxide) and poly(styrene-*stat*-2-hydroxyethyl methacrylate) has been shown to exhibit an adjustable LCST in the range of 30–150°C, depending on the exact composition. When used with solar heating systems, it can help prevent overheating by clouding up and reflecting greater fractions of light (backward light scattering) when heated too hot. It is used in conjunction with transparent insulating materials, which can be attached to the dark-colored, light-absorbing outside walls of houses, permitting the conversion of the incident sunlight to heat. This provides a partial substitute for conventional, carbon-based heating energy, such as coal- or petroleum-based combustion. The crosslinking limits the size of the domains that form on phase separation, ensuring that phase separation and phase mixing are rapid and reversible.

10.12.3.2. Pervaporation Membranes The pervaporation process is highly useful in areas where conventional distillation techniques are difficult to apply, such as the fractionation of azeotropic or isomeric mixtures. Thus, alcohol and water can easily be separated via this method. Figure 10.20 (114) illustrates such a process, where a feed of 10% alcohol can be fed in liquid form over a PU/P(MMA-AA) IPN membrane. The water diffuses through faster and is removed in the vapor state. After three stages, 94% ethanol can be obtained.

Another relatively new area relates to nonlinear optics. This field is basically concerned with the interaction of optical frequency electromagnetic fields with materials, resulting in the alteration of the phase, frequency, or other propagation characteristics of the incident light. One of the more interesting embodiments of second-order nonlinear optics is second-harmonic generation, involving the doubling of the frequency of the incoming light. Such materials are being considered for the replacement of wires in electronic devices, since light travels faster, and light of different frequencies may be less prone to cross-talk than electrons flowing in closely fit wires (15).

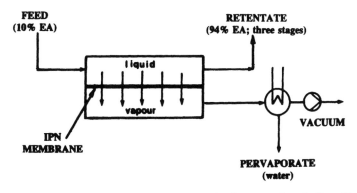

Figure 10.20 Illustration of the pervaporation process for the separation of ethanol from water.

10.12.3.3. Soft Contact Lens Gradient IPNs   Gradient refractive index IPNs provide another exciting new area. There are two ways of controlling and focusing light paths: the use of curved surface lenses is extremely well-known. The use of variable internal refractive index, used here, is less well known. A general way of producing such materials involves the swelling in of monomer II mix into polymer network I, followed by a space-controlling polymerization. In the bulk state, gradient refractive index lenses and fibers have been explored; see Section 10.10. Now, Calderara et al. (116) are developing a new gradient refractive index soft lens based on poly(vinyl pyrrolidone) and poly(hyroxyethyl methacrylate). These materials are synthesized using pin-point laser light beams and an amplitude mask to produce selective polymerization rates of monomer II mix. Thus, the variation in absorbance causes selective polymerization. The basic requirement involves the selection of two hydrophilic polymers with different refractive indices. These developments introduce a new dimension in the control of light paths.

10.12.3.4. IPN Coatings   In another development, Xie et al. (117) developed a castor-oil–urethane/acrylic or vinyl IPN for coating rusted iron. A redox reaction is employed for the vinyl or acrylic portion, with the rust providing part of the catalyst system. The monomer mix soaks through and on polymerization provides both significant adhesion to the iron as well as protective properties.

Many other kinds of IPNs and semi-IPNs are proposed for coatings or are now in service. For example, polyurethane–acrylic copolymers and related materials make impact-resistant coatings via a solution synthesis (118). Interpenetrating elastomer networks can also be made into a host of materials, including coatings (119).

Skinner et al. (120) made SIN coatings, some grafted, of vinyl and urethane compositions. Foscante et al. (121) patented a semi-IPN epoxy–siloxane composition (epoxy crosslinked) intended as a coating material. Sebastiano (122) invented a transparent polyurethane/polyacrylate SIN coating for safety glass. The monomer–polymer or prepolymer solutions are to be sprayed on the inside glass of automobiles. Thicknesses of $100–190\,\mu m$ yield good optical properties and tear resistance.

An interesting common feature of most of these materials is the incorporation of polyurethane technology. Among polymers, polyurethanes are easy to make and crosslink and provide a range of elastomeric to leathery compositions in IPN form.

10.13. OUTLOOK FOR IPNS

Interpenetrating polymer networks and related materials are one of the fastest growing fields in multicomponent polymer materials. Although in some ways they are still smaller in tonnage than polymer blends, blocks, and grafts, their potential applications cover areas not easily covered by the other areas, opening up new possibilities for the future.

REFERENCES

1. L. H. Sperling, *Interpenetrating Polymer Networks and Related Materials*, Plenum, New York, 1981.

2. D. Klempner, L. H. Sperling, and L. A. Utracki, Eds., *Interpenetrating Polymer Networks*, Adv. Chem. Ser. No. 239, ACS Books, Washington, DC, 1994.

3. D. Klempner and K. C. Frisch, Eds., *Advances in Interpenetrating Polymer Networks*, Technomic, Lancaster, PA, Vol. I, 1989; Vol. II, 1990; Vol. III, 1994; Vol. IV, 1994.

4. L. H. Sperling and V. Mishra, *Polym. Adv. Technol.*, **7**, 197 (1996).

5. L. H. Sperling, in *Interpenetrating Polymer Networks*, D. Klempner, L. H. Sperling, and L. A. Utracki, Eds., ACS Books, Washington, DC, 1994.

6. A. M. Fernandez, Ph.D. Dissertation, Lehigh University, 1984.

7. J. H. An and L. H. Sperling, in *Cross-Linked Polymers: Chemistry, Properties, and Applications*, R. A. Dickie, S. S. Labana, and R. S. Bauer, Eds., ACS Symposium Series No. 367, American Chemical Society, Washington, DC, 1988.

8. A. M. Fernandez, J. M. Widmaier, and L. H. Sperling, *Polymer*, **25**, 1718 (1984).

9. A. A. Donatelli, L. H. Sperling, and D. A. Thomas, *Macromolecules*, **9**, 671 (1976).

10. I. Ostromislensky, U.S. Pat. 1,613,673 (1927).

11. D. Sophiea, D. Klempner, V. Senjijarevic, B. Suthar, and K. C. Frisch, in *Interpenetrating Polymer Networks*, D. Klempner, L. H. Sperling, and L. A. Utracki, Eds., ACS Books, Washington, DC, 1994.

12. B. S. Kim, T. Chiba, and T. Inoue, *Polymer*, **34**, 2809 (1993).

13. H. H. Winter and F. Chambon, *J. Rheol.*, **30**, 367 (1986).

14. F. Chambon and H. H. Winter, *Polym. Bull.*, **13**, 499 (1985).

15. V. Mishra and L. H. Sperling, *The Polymeric Materials Encyclopedia: Synthesis, Properties, and Applications*, CRC Press, Boca Raton, 1996.

16. S C. Kim, D. Klempner, K. C. Frisch, W. Radigan, and H. L. Frisch, *Macromolecules*, **9**, 258 (1976).

17. S. C. Kim, D. Klempner, K. C. Frisch, and H. L. Frisch, *Macromolecules*, **9**, 263 (1976).

18. T. Hur, J. A. Manson, R. W. Hertzberg, and L. H. Sperling, *J. Appl. Polym. Sci.*, **39**, 1933 (1990).

19. T. Hur, J. A. Manson, R. W. Hertzberg, and L. H. Sperling, in *Multiphase Polymers: Blends and Ionomers*, L. A. Utracki and R. A Weiss, Eds., ACS Symp. Ser. No. 395, 1989.

20. I. Hermant and G. C. Meyer, *Polymer*, **24**, 1419 (1983).

21. A. Morin, H. Djomo, and G. C. Meyer, *Polym. Eng. Sci.*, **23**, 394 (1983).

22. P. Zhou and H. L. Frisch, *J. Polym. Sci, Part A; Polym. Chem.*, **31**, 3479 (1993).

23. H. L. Frisch and P. Zhou, *J. Polym. Sci., Part A; Polym. Chem.*, **31**, 1967 (1993).

24. V. Mishra, F. E. Du Prez, , E. Gosen, E. J. Goethals, and L. H. Sperling, *J. Appl. Polym. Sci.*, **58**, 331 (1995).

25. V. Mishra and L. H. Sperling, *Polymer*, **36**, 3593 (1995).

26. L. H. Sperling and V. Mishra, in *Proceedings of the 25th Anniversary Symposium of the Polymer Institute*, K. C. Frisch and E. W. Eldred, Eds., Technomic, Lancaster, PA, 1994.

27. F. E. Du Prez, P. Tan, and E. J. Goethals, *Polym. Adv. Technol.*, **7**, 257 (1996).

28. A. A. Donatelli, L. H. Sperling, and D. A. Thomas, *J Appl. Polym. Sci.*, **21**, 1189 (1977).

29. J. K. Yeo, L. H. Sperling, and D. A. Thomas, *Polymer*, **24**, 307 (1983).

30. U. Bianchi, E. Pedemonte, and A. Turturro, *Polymer*, **11**, 268 (1970).

31. J. M. Widmaier and L. H. Sperling, *Macromolecules*, **15**, 625 (1982).

32. H. Gankema, M. A. Hempenius, M. Moller, G. Johansson, and V. Percec, *Macromol. Symp.*, **102**, 381 (1996).

33. W. P. Gergen, *Kautsch. Gummi, Kunstst.*, **37**, 284 (1984).

34a. S. Davison and W. P. Gergen, U.S. Pat. 4,041,103 (1977).

34. W. P. Gergen and S. Davison, U. S. Pat. 4,101,605 (1978).

35. D. L. Siegfried, D. A. Thomas, and L. H. Sperling, *J. Appl. Polym. Sci.*, **26**, 177 (1981).

36. B. Ohlsson, H. Hassander, and B. Tornell, *Polym. Eng. Sci.*, **36**, 501 (1996).

37. V. Mishra, F. E. Du Prez, E. J. Goethals, and L. H. Sperling, *J. Appl. Polym. Sci.*, **58**, 347 (1995).

38. Y. Wei, H. Yuan, and Z. Pan, *Gaofenzi Xuebao*, **5**, 606 (1995)

39. M. Matsuo, T. K. Kwei, D. Klempner, and H. L. Frisch, *Polym. Eng. Sci.*, **10**, 327 (1970).

40. Y. S. Lipatov, *Pure Appl. Chem.*, **57**, 1691 (1985).

41. Y. S. Lipatov, in *Interpenetrating Polymer Networks*, D. Klempner, L. H. Sperling, and L. A. Utracki, Eds., ACS Books, Washington, DC, 1994.

42. A. J. Curtius, M. J. Covitch, D. A. Thomas, and L. H. Sperling, *Polym. Eng. Sci.*, **12**, 101 (1972).

43. G. M. Yenwo, L. H. Sperling, J. Pulido, J. A. Manson, and A. Conde, *Polym. Eng. Sci.*, **17**, 251 (1977).

44. R. M. Briber and B. J. Bauer, *Macromolecules*, **24**, 1899 (1991).

45. M. S. Silverstein, Y. Talmon, and M. Narkis, *Polymer*, **30**, 416 (1989).

46. M. S. Silverstein and M. Narkis, *J. Appl. Polym. Sci.*, **33**, 2529 (1987).

47. N. Nemirovski and M. Narkis, in *Interpenetrating Polymer Networks*, D. Klempner, L. H. Sperling, and L. A. Utracki, Eds., ACS Books, Washington, DC, 1994.

48. D. J. Hourston and J. A. McCluskey, *Polymer*, **20**, 1573 (1979).

49. D. J. Hourston and R. Satgurunthan, *J. Appl. Polym. Sci.*, **29**, 2969 (1984).

50. A. K. Holdsworth and D. J. Hourston, in *Interpenetrating Polymer Networks*, D. Klempner, L. H. Sperling, and L. A. Utracki, Eds., ACS Books, Washington, DC, 1994.

51. Z. Liucheng, L. Xiucuo, and L. Tianchang, in *Interpenetrating Polymer Networks*, D. Klempner, L. H. Sperling, and L. A. Utracki, Eds., ACS Books, Washington, DC, 1994.

52. R. Hu, V. L. Dimonie, M. S. El-Aasser, R. A. Pearson, A. Hiltner, S. G. Mylonakis, and L. H. Sperling, accepted, *J. Polym. Sci., Polym. Chem. Ed.*, (1997).

53. R. Hu, V L. Dimonie, M. S. El-Aasser, R. A. Pearson, L. H. Sperling, A. Hiltner, and S. G. Mylonakis, *J. Appl. Polym. Sci.*, **58**, 375 (1995).

54. V. Nevissas, J. M. Widmaier, and G. C. Meyer, *J. Appl. Polym. Sci.*, **36**, 1467 (1988).

55. X. He, J. M. Widmaier, and G. C. Meyer, *Polym. Internt.*, **32**, 295 (1993).

56. C. Rouf, S. Derrough, J. J. Andre, J. M. Widmaier, and G. C. Meyer, in *Interpenetrating Polymer Networks*, D. Klempner, L. H. Sperling, and L. A. Utracki, Eds., ACS Books, Washington, DC, 1994.

57. Yu. S. Lipatov and L. V. Karabanova, *J. Mater. Sci.*, **30**, 2475 (1995).

58. C. L. Jackson, B. J. Bauer, A. I. Nakatani, and J. D. Barnes, *Chem. Mater.*, **8**, 727 (1996).

59. H. R. Allcock, K. B. Visscher, and Y. B. Kim, *Macromolecules*, **29**(8), 2721 (1996).

60. G. Pozniak and W. Trochimczuk, *Angew. Makromol. Chem.*, **92**, 155 (1980).

61. G. Pozniak and W. Trochimczuk, *Angew. Makromol. Chem.*, **104**, 1 (1982).

62. H. Y. Erbil and B. M. Baysal, *Angew. Makromol. Chem.*, **165**, 97 (1989).

63. H. Czarczynska and W. Trochimczuk, *J. Polym. Sci., Polym. Symp.*, **47**, 111 (1974).

64. U. Schyulze, G. Pompe, E. Meyer, A. Janke, J. Pionteck, A. Fiedlerova, and E. Borsig, *Polymer*, **36**, 3393 (1995).

65. E. Borsig, A. Fiedlerova, K. G. Hausler, R. M. Sambatra, and G. H. Michler, *Polymer*, **34**, 4787 (1993).

66. J. Tino, E. Borsig, Z. Hlouskova, and A. Fiedlerova, *Polym. Internat.*, **35**, 389 (1994).

67. R. Greco, A. Fiedloerov, U. Schulze, and E. Borsig, *J. Macromol. Sci., A, Pure Appl. Chem.*, **A32**, 1957 (1995).

68. M. Lazar, L. Hrckova, U. Schulze, J. Piontech, and E. Borsig, *J. Macromol. Sci., A, Pure Appl. Chem.*, **A33**, 261 (1996).

69. P. Patel, T. Shah, and B. Suthar, *J. Appl. Polym. Sci.*, **40**, 1037 (1990).

70. P. Patel and B. Suthar, *Polym. Plast. Technol. Eng.*, **28**(1), 1 (1989).

71. P. L. Nayak, S. Lenka, S. K. Panda, and T. Patniak, *J. Appl. Polym. Sci.*, **47**, 1089 (1993).

72. T. Pattniak and P. L. Nayak, *Macromol. Reports*, **A31** (Suppls. 3 and 4), 447 (1994).

73. T. Pattniak, P. L. Nayak, S. Lenka, S. Mohanty, and K. K. Rao, *Thermochim. Acta*, **240**, 235 (1994).

74. D. Parida, P. Nayak, D. K. Mishra, S. Lenka, P. L. Nayak, S. Mohanty, and K. K. Rao, *J. Appl. Polym. Sci.*, **56**, 1731 (1995).

75. S. C. Kim, D. Klempner, K. C. Frisch, and H. L. Frisch, *J. Appl. Polym. Sci.*, **21**, 1289 (1977).

76. P. Tan, in *Interpenetrating Polymer Networks*, D. Klempner, L. H. Sperling, and L. A. Utracki, Eds., ACS Books, Washington, DC, 1994.

77. H. Q. Xie, C. X. Zhang, and J. S. Guo, in *Interpenetrating Polymer Networks*, D. Klempner, L. H. Sperling, and L. A. Utracki, Eds., ACS Books, Washington, DC, 1994.

78. H. Xie, private communication, September, 1993.

79. L. W. Barrett, O. L. Shaffer, and L. H. Sperling, *J. Appl. Polym. Sci.*, **48**, 953 (1993).

80. L. W. Barrett, G. S. Ferguson, and L. H. Sperling, *J. Polym. Sci., Part A, Polym. Chem.*, **31**, 1287 (1993).

81. L. W. Barrett, L. H. Sperling, J. Gilmer, and S. G. Mylonakis, *J. Appl. Polym. Sci.*, **48**, 1035 (1993).

82. L. W. Barrett and L. H. Sperling, *Polym. Eng. Sci.*, **33**, 913 (1993).

83. L. W. Barrett, L. H. Sperling, and C. J. Murphy, *JAOCS*, **70**, 523 (1993).

84. L. W. Barrett, L. H. Sperling, J. W. Gilmer, and S. G. Mylonakis, in *Interpenetrating Polymer Networks*, D. Klempner, L. H. Sperling, and L. A. Utracki, Eds., ACS Books, Washington, DC, 1994.

85. Natural Rubber Producers Research Association, Technical Information Sheet No. 9, Revised (1977).

86. D. J. Hourston and J. Romaine, *J. Appl. Polym. Sci.*, **39**, 1587 (1990).

87. M. Schneider, T. Pith, and M. Lambla, *Polym. Adv. Technol.*, **6**, 326 (1995).

88. M. Schneider, Ph.D. Dissertation, University Louis Pasteur, Strasbourg, France (1995).

89. M. Kamath, J. Kincaid, and B. Mandal, *J. Appl. Polym. Sci.*, **59**, 45 (1996).

90. P. H. Corkhill, J. H. Fitton, and B. J. Tighe, *Polymer*, **31**, 1526 (1990).

91. P. H. Corkhill, J. H. Fitton, and B. J. Tighe, *J. Biomater. Sci. Polym. Edn.*, **4**, 615 (1993).

92. J. W. Aylsworth, U.S. Pat. 1,111,284 (1914).

93. L. H. Sperling, *Polym. News*, **12**, 332 (1987).

94. W. K. Fischer, U.S. Pat. 3,806,558.

95. L. H. Sperling, in *Multicomponent Polymer Materials*, ACS Advances in Chem. Ser. No. 211, D. R. Paul and L. H. Sperling, Eds., American Chemical Society, Washington, DC, 1986.

96. B. Arkles, U.S. Pat. 4,500,688 (1985).

97. B. Arkles, *ChemTech*, 542, Sept., (1983).

98. M. E. Dillon, U.S. Pat. 4,832,009 (1989).

99. J. E. Dillon and W. J. Okunski, *Wounds*, **4**, 203 (1992).

100. T. Kupper, in *The Morning Call*, Allentown, PA, June 4, 1996, p. A10.

101. M. E. Dillon, PCT Int. App. WO 95 22,997.

102. F. D. Roemer and L. H. Tateosian, U.S. Pat. 4,396,377 (1984).

103. F. D. Roemer and L. H. Tateosian, U.S. Pat. 4,396,476 (1983).

104. J. J. Falcetta, G. D. Friends, and G. C. C. Niu, Ger. Offen. Pat. 2,518,904 (1975).

104a. P. Kuzma and G. Odorisio, U.S. Pat. 4,388,428 (1983).

105. T. V. Chirila, R. L. Cooper, I. J. Constable, and R. Horne, *J. Appl. Polym. Sci.*, **44**, 593 (1992).

106. K. F. Mueller and S. J. Heiber, *J. App. Polym. Sci.*, **27**, 4043 (1982).

107. K. F. Mueller and S. J. Heiber, U.S. Pat. 4,423,099 (1983).

108. D. J. Hourston and F. Schafer, *Polym. Adv. Technol.*, **7**, 273 (1996).

109. R. Hu, V. L. Dimonie, M. S. El-Aasser, R. A. Pearson, A. Hiltner, S. G. Mylonakis, and L. H. Sperling, accepted, *J. Polym. Sci., B, Polym. Phys. Ed.*, (1997).

110. L. H. Sperling, in *Sound and Vibration Damping with Polymers*, R. D. Corsaro and L. H. Sperling, Eds., ACS Symposium Series No. 424, American Chemical Society, Washington, DC, 1990.

111. L. H. Sperling, T. W. Chiu, R. G. Gramlich, and D. A. Thomas, *J. Paint Technol.*, **46**, 4 (1974).

112. K. Yamamoto and A. Takahashi, in *Sound and Vibration Damping with Polymers*, R. D. Corsaro and L. H. Sperling, Eds., ACS Symposium Series 424, American Chemical Society, Washington, DC, 1990.

113. W. Eck, H. R. Wilson, and H. J. Cantow, *Adv. Mater.*, **7**, 800 (1995).

114. Y. K. Lee, I. S. Sohn, E. J. Jeon, and S. C. Kim, *Polym. J.*, **23**, 427 (1991).

115. D. Williams, in *Electronic and Photonic Applications of Polymers*, M. J. Bowden and S. R. Turner, Eds., Adv. Chem. Ser., No. 218, American Chemical Society, Washington, DC, 1988.

116. I. Calderara, D. Baude, D. Joyeux, and D. J. Lougnot, *Polym. Mater. Sci. Eng. (Prepr.)*, **75**, 244 (1996).

117. H. Q. Xie, J. S. Guo, and G. G. Wang, *Eur. Polym. J.*, **29**, 1547 (1993).

118. H. X. Xiao and K. C. Frisch, *J. Coat. Technol.*, **61**(3), 51 (1989).

119. H. L. Frisch, K. C. Frisch, and D. Klempner, U.S. Pat. 4,302,533 (1981).

120. E. Skinner, M. Emeott, and A. Jevne, U.S. Pat. 4,247,578 (1981).

121. R. E. Foscante, A. P. Gysegem, P. J. Martinich, and G. H. Law, U.S. Pat. 4,250,074 (1981).

122. F. Sebastiano, Intl. Pat. 0035130 (1981).

MONOGRAPHS, EDITED WORKS, AND GENERAL REVIEWS

Kim, S. C. and L. H. Sperling, *IPNs Around the World: Science and Engineering*, Wiley, Chichester, 1997.

Klempner, D. and K. C. Frisch, Eds., *Advances in Interpenetrating Polymer Networks*, Technomic, Lancaster, PA, Vol. I, 1989; Vol. II, 1990; Vol. III, 1994, Vol. IV, 1994.

Klempner, D., L. H. Sperling, and L. A. Utracki, Eds., *Interpenetrating Polymer Networks*, ACS Books, Washington, DC, 1994.

Lipatov, Yu. S. and L. M. Sergeeva, *Interpenetrating Polymeric Networks*, Naukova Dumka, Kiev, 1979.

Mark, J. E., C. Y. C. Lee, and P. A. Bianconi, Eds., *Hybrid Organic-Inorganic Composites*, ACS Books, Washington, DC, 1995.

Sperling, L. H., *Interpenetrating Polymer Networks and Related Materials*, Plenum, New York, 1981.

Sperling, L. H. and S. C. Kim, Eds., *Polym. Adv. Tech.*, **7**(4), 1996, special issue devoted entirely to IPNs.

STUDY PROBLEMS

1. A HIPS polymerization is carried out at 70°C, using 10% polybutadiene and 90% styrene monomer. At 95% conversion of the styrene to polystyrene, the remaining monomer is partitioned between the polystyrene-rich phase and the polybutadiene-rich phase.
   a. What is the concentration (in moles per liter preferred) of the remaining styrene monomer in each of the two phases?
   b. What is the locus of polymerization in each of the two phases?

2. Briefly, how does ABS toughen polycarbonate?

3. How would you use maleic anhydride to toughen nylon-6,6 with polypropylene? Show the chemical reactions.

4. Why can't block copolymers have ordinary composition–temperature phase diagrams?

5. Surprisingly, the number of blocks making up the domains is a relatively small, countable number.
   a. What is the approximate number of separate blocks in one domain in Figure 9.3a?
   b. Assuming an A-B-A type of triblock copolymer with amorphous blocks and spherical domains for the A blocks, can you derive an equation expressing the relationship between block molecular weight(s) and the average number of chains in each domain?

6. What are some of the differences in the morphology and organization of glassy block vs. crystalline block copolymers? How does that affect the molecular weights of the blocks as used commercially?

7. Read a study published in the last 12 months on the science or engineering of block copolymers. How does the study improve on understanding or present new theories about block copolymers, not in this book? Provide the full reference.

8. Compare the expected morphologies of 80/20 polystyrene/polybutadiene HIPS, SBS triblock copolymers, and PB/PS sequential IPNs. What would you expect the stress–strain curves and impact resistances to be?

9. What specific pieces of information can be learned about block copolymers using (a) transmission electron microscopy, (b) SAXS, (c) SANS, and (d) wide-angle x-ray diffraction?

10. Show the steps necessary to make a diblock copolymer starting with methyl methacrylate and butadiene, each block to have a narrow molecular weight distribution, and a molecular weight for each block of 10,000 g/mol.

11. Assuming equal degrees of polymerization for the two blocks in problem 10 above, what is the critical total molecular weight for phase separation?

12. Polymer blends and block copolymers differ significantly in application areas. Can you name and briefly discuss one application for each that would be difficult for the other multipolymer combination to fulfill?

13. Invent a new multicomponent polymer material. What will its physical characteristics be? What will it be good for?

14. Noting Figures 10.7 and 10.8, what is the locus of polymerization and phase separation of a 50/50 "U"/MMA mix (each with crosslinker), if the "U" is polymerized first and the MMA is polymerized second? Draw arrows to indicate the path.

15. Again noting Figures 10.7 and 10.8, can you devise a composition and polymerization route that will have both polymers gel before phase separation? Draw arrows to indicate the polymerization route.

16. Gradient IPNs are useful for many things. Briefly describe three applications of these materials.

17. If a 50/50 PB/PS sequential IPN is made with 1% of divinyl benzene in each polymer, what is the diameter of the PS domains?

11

OVERVIEW AND FUTURE

11.1. INTEGRATED PICTURE OF INTERFACES

The preceding chapters have developed the concepts of multicomponent polymer materials, starting with basics, going through polymer interfaces, and on to thermoplastic elastomers, rubber-toughened plastics, and IPNs. Although at first these various subfields may seem divergent, they have much in common. One topic in common for each of these fields is their interface: No polymer blend or composite can be made without creating an interface!

Most people consider that surfaces and interfaces are what is *on top*. However, surfaces and interfaces in multicomponent polymer materials play key roles in determining the characteristics of the final system and have taken on an importance far beyond what was originally thought.

An examination of the literature shows that modern polymer interface science dates only from about 1989. In that year, several key studies of both theoretical and experimental nature were published. Instruments to measure surface properties down to the atomic scale were available. One no longer had to imagine surface and interface characteristics, they could be measured. The roles of nonrandom chain conformation, entanglements between different kinds of chains, interfacial bonding, the role of compatibilizers, and much more are being clarified.

Figure 11.1 summarizes the state of the art of multicomponent polymer materials from the point of view of the role of surfaces and interfaces. Again, the three major types of surfaces and interfaces are the free surfaces and the polymer blend and composite interfaces. While the dilute solution–colloid interfaces are not *free* in the ordinary sense, the fluid phase has a low viscosity and allows rapid diffusion, similar in some ways to the free air surface, and is classified as such for the present purposes. Since the actual extent of compatibilization in rubber-toughened plastics varies, the tie lines indicate only one possible route. In the case of IPNs, it is the crosslinking that provides the compatibilization, although some grafting is occasionally employed.

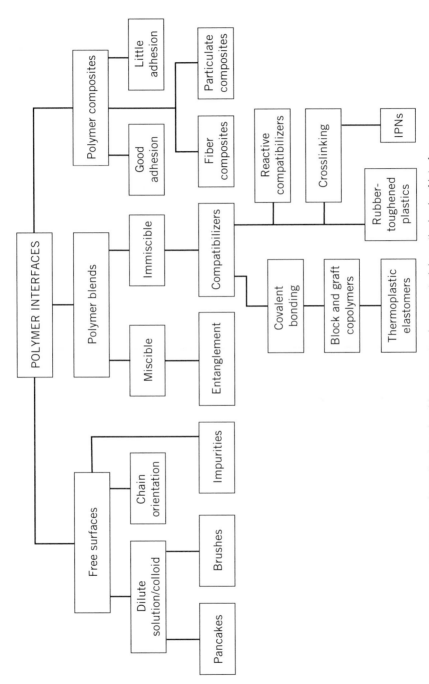

Figure 11.1. Classification of multicomponent polymer materials on the basis of interfaces.

383

By classifying all of the major known kinds of materials along side each other, their roles, similarities and differences become clearer. Many ideas still remain to be clarified. For example, do pancakes and brushes have their equivalents in polymer composites? Is there an optimum level of adhesion between the polymer and the filler in composites? Similarly, is there an optimum level of covalent or other bonding between two phases, constituting compatibilization?

11.2. SOME INTERRELATIONSHIPS AMONG THE BULK PROPERTIES

Major areas of interest in this book have been the description of how glass transitions, melting behavior, rheological phenomena, and mechanical behavior relate to one another to create interesting and useful engineering materials. In the case of polymer blends and composites, all of these features are complicated by component mixing, small phase domain size, and attractive or repulsive forces between the phases. Again, there are extensive interrelationships among these features for seemingly widely divergent systems. Thus, the use of the Takayangi relationships, equations relating to phase continuity and inversion, and so forth all are common for very many systems.

Nearly all of the composites employ a *harder* phase and a *softer* phase. This may be glass fiber and a plastic, for example, or carbon black and rubber. Very many polymer blends, grafts, blocks, and IPNs also utilize the same hard–soft concept, although the magnitude of the hard phase modulus is usually 10–1000 times softer than the corresponding composite systems. Having hard and soft phases, whether inorganic-glassy or glassy-rubbery, allows for the development of synergistic behavior, such as fiber composites, rubber-toughened plastics, or thermoplastic elastomers.

11.3. JUST AROUND THE CORNER

Where will the future lead? Of course, it is not clear. However, borrowing phraseology from the newer computer software terminology, the better polymer blends of tomorrow seem to be headed toward *seamless* morphology between phases. This is the equivalent of having rather thick, gradient composition interphases. The material composition must vary continuously from one phase to the other. Thus, the maximum use of entanglement to bond the phases together will be achieved. Such materials now being brought to reality in a number of polymer blends, blocks, grafts, and IPNs, but still appear wanting in most composite systems. A possible exception is the organic–inorganic polymer hybrids (Section 7.8).

It has been said that the probability of inventing or discovering a new simple homopolymer structure that would enter the *commodity polymer market* as a novel material is small (but not zero!). The solution adopted by industry has been through two general areas:

1. Statistical copolymerization, plasticization, and additives, generally beyond the scope of this book.
2. Through multicomponent polymer materials, the exact subject of this book. Almost literally, people can now make potato chip bags that cannot be breached, and actually do make composites lighter and stronger than steel.

The art of blending and composite formation is old, stretching back to the biblical days of bricks of clay strengthened by the addition of straw (*Exodus* **5**:7–19). But, the science began with the twentieth century. With the twenty-first century just around the corner, it is clear that the field is just beginning.

INDEX